Classroom Companion: Economics

The Classroom Companion series in Economics includes undergraduate and graduate textbooks alike. It welcomes fundamental textbooks aimed at introducing students to the core concepts, empirical methods, theories and tools of the field, as well as advanced textbooks written for students at the Master and PhD level seeking a deeper understanding of economic theory, mathematical tools and quantitative methods.

Florian Bartholomae · Marcus Wiens

Game Theory and Applications

A Guide for Students and Researchers

Springer

Florian Bartholomae
Munich Business School
München, Bayern, Germany

Marcus Wiens
TU Bergakademie Freiberg
Freiberg, Germany

ISSN 2662-2882 ISSN 2662-2890 (electronic)
Classroom Companion: Economics
ISBN 978-3-658-44607-9 ISBN 978-3-658-44608-6 (eBook)
https://doi.org/10.1007/978-3-658-44608-6

Translation from the German language edition: "Spieltheorie" by Florian Bartholomae and Marcus Wiens, © Springer Fachmedien Wiesbaden GmbH, ein Teil von Springer Nature 2020. Published by Springer Fachmedien Wiesbaden. All Rights Reserved.

This book is a translation of the original German edition "Spieltheorie," 2nd edition, by Florian Bartholomae, published by Springer Fachmedien Wiesbaden GmbH in 2020. The translation was done with the help of an artificial intelligence machine translation tool. A subsequent human revision was done primarily in terms of content, so that the book will read stylistically differently from a conventional translation. Springer Nature works continuously to further the development of tools for the production of books and on the related technologies to support the authors.

This Springer imprint is published by the registered company Springer Fachmedien Wiesbaden GmbH, part of Springer Nature.
The registered company address is: Abraham-Lincoln-Str. 46, 65189 Wiesbaden, Germany

Paper in this product is recyclable.

Preface: Motivation and Structure

Welcome to the profound realm of game theory—a discipline that unfolds through meticulous mathematical analysis and strategic decision-making. However, if you harbor preconceptions of a dry and formulaic journey, allow us to dispel that notion. This book is your scholarly guide to unraveling the complexities of game theory with a sense of gravity, intellectual rigor, and also accessibility.

In a world where strategic considerations permeate every facet of societal interaction, game theory emerges as an indispensable and very valuable tool. Envision delving into the intricacies of financial and social policies, deciphering the ramifications of market entries, navigating negotiations, and exploring the nuanced psychology underlying strategic decisions. This book serves as your gateway to understanding the omnipresence and significance of game theory in our daily lives.

From our vast experience of teaching numerous courses at a wide variety of institutions with diverse audiences, we are aware of the many potential challenges for newcomers which has enabled us to compile a comprehensive resource. We, thus, offer a systematic approach to traverse the complexities of game theory, making it accessible even to those without an extensive background in mathematical theory.

This book is not merely a theoretical compendium; it is a scholarly companion designed to facilitate a deep understanding of this exciting discipline. We have developed a structured "training program" that systematically breaks down complicated concepts, ensuring a differentiated understanding for readers with different mathematical abilities. To ensure this, the last chapter presents the most important mathematical basics so that even inexperienced readers can handle the mathematical concepts.

As you navigate the "Theory" and "Problems" sections, you will find a judicious blend of theoretical foundations and practical applications. The incorporation of comprehension questions and real-world problems serves to engage you actively in the learning process. Whether you seek to establish a solid foundation or aspire to refine your advanced understanding, this book accommodates a spectrum of academic needs.

Beyond passive consumption, the book encourages active participation, offering insights into setting up one's own games. This feature renders the book not merely a static repository of knowledge but an interactive toolkit, particularly valuable for those formulating their academic theses or seeking to independently delve into the intricacies of game theory.

Gratitude is extended to Prof. Dr. Karl Morasch, whose profound insights shaped the core of this book, and to Kirsten Johannemann, M. Sc. for her discerning advice. Special acknowledgment is also extended to the students at the University of the Bundeswehr Munich, whose constructive feedback propelled us to refine and elevate our teaching methodologies.

This book is based on a textbook originally published in German, which has been completely revised and adapted to the needs of an international audience. In addition, the content has been expanded and errors have been removed compared to the original editions.

Embark on this scholarly journey with an open mind. Whether you are a student of business and economics, a prospective lawyer, an engineer, or anyone with a keen interest in the strategic underpinnings of societal interactions, this book serves as your gateway to the intellectually enriching universe of game theory. Embrace the challenges, relish the progress, and allow the profound nature of strategic thinking to captivate you. Let the scholarly exploration begin.

Munich, Germany
Freiberg, Germany
January 2024

Florian Bartholomae
Marcus Wiens

Contents

Part I

Theory

Decision Theory

1

Since game theory analyzes the effects and interdependencies of decisions made by various agents, it is a special part of *decision theory* as the broader discipline. For this reason, it should be clarified in advance what is understood by decisions and how they are made by individuals in different situations. In short, a decision is a consciously made choice between different alternatives or actions.

1.1 Basics of Decision-Making

In a decision-making situation, an agent, the decision-maker, must choose between at least two options. To make a (rational) decision in such a (strategic) situation, the decision-makers must be able to assess the individual possible decision outcomes. To do this, they need to know all possible outcomes as well as how desirable they are (for them). Scientifically formulated, it must be possible to bring all decision results into a complete and transitive preference order. Let us assume there are n possible outcomes o, denoted by $\{o_1, o_2, \ldots, o_n\}$. In a *complete* order of results, the decision-maker knows all n possible outcomes and can weigh them against each other, meaning they can assess whether

- o_i is better than o_j, $o_i \succ o_j$,
- o_i is worse than o_j, $o_i \prec o_j$, or
- o_i is as good as o_j, $o_i \approx o_j$.

An order is *transitive* if the ranking is consistent and thus free of contradictions. For example, if outcome A is preferred to outcome B, $A \succ B$, and outcome B is preferred to outcome C, $B \succ C$, then in a transitive order, outcome A must also be preferred to outcome C, $A \succ C$.

F. Bartholomae and M. Wiens, *Game Theory and Applications*, Classroom Companion: Economics, https://doi.org/10.1007/978-3-658-44608-6_1

To fully represent a decision problem, we need knowledge about

- all n available actions $\{a_1, a_2, \ldots, a_n\}$,
- all m possible states of the world $\{w_1, w_2, \ldots, w_m\}$, and
- all outcomes resulting from the actions and states$\{o_{11}, o_{12}, \ldots, o_{21}, o_{22}, \ldots, o_{nm}\}$.

To understand how a specific decision situation can be represented by this, let us consider the following example: On a vacation day, we can decide whether to go to the outdoor pool in the afternoon (action a_1) or to the cinema (action a_2). Since the two leisure places are far apart, only one action is possible, and the decision made (i.e., by purchasing a corresponding ticket) cannot be revised afterwards. There are two *states of the world* possible: Either the weather will be warm and sunny (state w_1) or it will be cold and rainy (state w_2). Thus, there are four possible outcomes: Visit to the outdoor pool in warm weather (outcome o_1), visit to the outdoor pool in cold weather (outcome o_2), visit to the cinema in warm weather (outcome o_3), and visit to the cinema in cold weather (outcome o_4). To make a rational decision we must now evaluate these four outcomes. Let us assume that according to our preference order, we favor the four possible results in the following order: $o_1 \succ o_4 \succ o_3 \succ o_2$. To better represent the order, it is useful to specify the *utility* of the outcomes. With the help of utility, preferences can be described as numbers: If one outcome is preferred over another, it provides a higher utility and is assigned a higher numerical value. Thus, from the (transitive and complete) preference order, a *utility function* $u(\cdot)$ can be derived, which assigns a number to each possible outcome. For example, we can assign the following utility values to the outcomes[1].

Figure 1.1 represents the situation from Table 1.1 using a *decision tree*. This tree provides additional information and is read as follows: At the first point or *node*, nature draws the state of the world. "Nature" is not a decision-maker in the actual sense, but represents chance ("Lady Luck"), which blindly (and thus not strategically) draws a state from all possible states. Subsequently, the decision-maker—in this case we—decides between the two actions. The dashed line between the two decision nodes indicates that we do not know in which state we are. Formally speaking, both decision nodes are in the same information set, i.e., they are indistinguishable from each other with respect to the realized state.

So how should we decide? Obviously, the decision depends on the realized state of the weather: The visit to the outdoor pool (a_1) only makes sense in warm weather (w_1), while in cold weather (w_2) going to the cinema (a_2) is the best option. This means, if we know the state, we can make the optimal decision. In this specific situation, the case seems simple: We look out of the window, check the

[1] The assigned utility value is exclusively relevant in terms of the ranking of possible bundles, i.e., this is an *ordinal concept*. As long as the ranking (preferred outcome has a higher numerical value) remains unchanged, we can also choose any other numerical values without changing the decision of the individual. Try to replicate the results in the following sections also with $u(o_1) = 400$, $u(o_2) = -200$, $u(o_3) = 0$, and $u(o_4) = 200$!

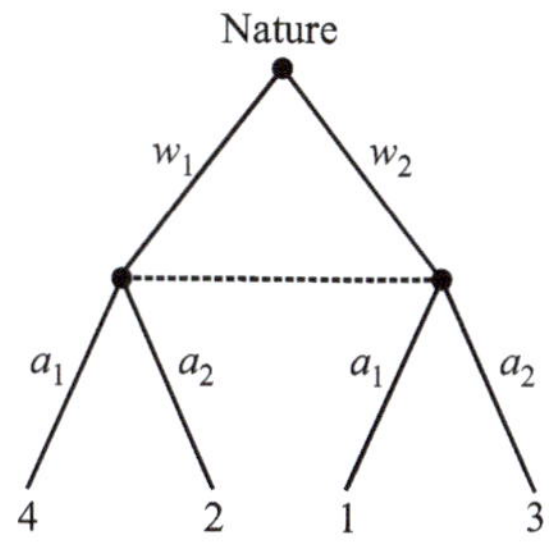

Fig. 1.1 Decision tree

Table 1.1 Utility matrix

	Warm & sunny (w_1)	Cold & rainy (w_2)
Outdoor pool (a_1)	4 [$= u(o_1)$]	1 [$= u(o_2)$]
Cinema (a_2)	2 [$= u(o_3)$]	3 [$= u(o_4)$]

weather, and make our choice. But what if we must make the decision before the state is observable, e.g., because the tickets must be purchased the day before? In this case, there are two possibilities: Either we have no idea how the weather will turn out, i.e., which state will be realized, or we know the (objective or subjective) probabilities of occurrence. If there is no knowledge about the probability of the states occurring, this is referred to as decision under *uncertainty*, if the probabilities are known, this is referred to as decision under *risk*. Depending on our level of knowledge, we must resort to different decision-making methods.

1.2 Decision Under Uncertainty

Let us first consider the case that we do not know the probability of the weather occurring, that is, the state of the world is uncertain. Since the decision must be made before the realization of the state, we must consider on what sound basis a decision can be made. There are some decision rules for solving such problems. We want to take a brief look at some of them:

Maximin Rule. The maximin rule (also known as *Wald's rule*) advises: Choose that action a_i that delivers the highest payoff/utility in the worst case:

$$\max_{a_i}\left(\min_{w_j} u\left(o_{ij}\right)\right), \tag{1.1}$$

where o_{ij} denotes the outcome that action a_i delivers in state w_j. This formula is to be read from the inside out: First search across all states w_j for the worst/lowest outcome ("min") for each action, and then choose the largest value, "max", from these minima.

In the present example, the lowest utility for action a_1 is 1 ($= u(o_2)$) and for action a_2 it is 2 ($= u(o_3)$). Therefore, we would decide for the cinema visit (a_2), as we get here the least bad payoff.

This rule represents a very pessimistic criterion, as it always assumes the occurrence of the most unfavorable state and therefore results in strong attempts to limit the damage.

Maximax Rule. The maximax rule advises: Choose that action a_i that delivers the highest payoff/utility in the best case:

$$\max_{a_i}\left(\max_{w_j} u\left(o_{ij}\right)\right). \tag{1.2}$$

Again, this formula is to be read from the inside out: First search across all states w_j for the best/highest outcome ("max") for each action, and then choose the largest value, "max", from these maxima.

In the present example, the highest utility of action a_1 is 4 ($= u(o_1)$) and of action a_2 is 3 ($= u(o_4)$). We will therefore opt for the visit to the outdoor pool (a_1), as in this case the highest utility is possible.

In contrast to the maximin rule, the maximax rule represents a very optimistic decision criterion.

Hurwicz Rule. The Hurwicz rule proposes a compromise between the maximin and maximax rules by considering the worst and best outcome of each decision:

$$\max_{a_i}\left[\lambda \cdot \max_{w_j} u\left(o_{ij}\right) + (1-\lambda) \cdot \min_{w_j} u\left(o_{ij}\right)\right] \tag{1.3}$$

Lambda λ, referred to as the optimism parameter, plays a major role and can be chosen by the decision-maker according to their own preferences between 0 and 1. For $\lambda = 1$, we get the optimistic suggestion of the maximax rule while for $\lambda = 0$ we get the pessimistic result of the maximin rule.

In the present example, we evaluate action a_1 with $\lambda \cdot 4 + (1-\lambda) \cdot 1$ and action a_2 with $\lambda \cdot 3 + (1-\lambda) \cdot 2$. For $\lambda > 0.5$, a_1 is rated higher than a_2 and we therefore opt for the visit to the outdoor pool. Accordingly, for $\lambda < 0.5$, we prefer going to the cinema while at $\lambda = 0.5$ we are indifferent between the two leisure activities, i.e., both alternatives are equally attractive.

The result of the decision therefore depends significantly on the optimism parameter and can hence be very sensitive with respect to the decision-maker's mood.

Laplace Rule. The Laplace rule assumes that all states occur with equal probability and advises: Choose that action a_i that has the highest expected value:

$$\max_{a_i}\left(\sum_{w_j=1}^{m} u\left(o_{ij}\right) \cdot \frac{1}{m}\right). \tag{1.4}$$

The expected value represents a weighted average value. The payoffs are weighted by their probability of occurrence and added together. As we assume that all states are equally probable, the probability of occurrence depends on the number of possible states. A simple example of this is the roll of a dice: there are six possibilities and therefore the probability of rolling a certain number, e.g., a three, is exactly one-sixth. In the formula, an addition $x_1 + x_2 + \ldots$ is generally represented by the sum sign $\sum$, i.e., $x_1 + x_2 + \cdots + x_m = \sum_{j=1}^{m} x_j$.

In our example, the expected value of the visit to the outdoor pool would be $E(a_1) = 4 \cdot \frac{1}{2} + 1 \cdot \frac{1}{2} = 2.5$ and the expected value of going to the cinema would be $E(a_2) = 2 \cdot \frac{1}{2} + 3 \cdot \frac{1}{2} = 2.5$. The same numerical value means that we are indifferent between both actions according to this rule.

1.3 Decision Under Risk: Expected Utility Theory[2]

A decision situation in which we know the probabilities of the states $\rho(w_j)$ is referred to as decision under risk. These probabilities can be subjective or objective. A subjective probability is based on personal experience and "gut instinct", whereas an objective probability can be derived from a (statistical) frequency distribution (for example, in dice rolling, on average every sixth roll yields a 6). The probabilities of all states must naturally add up to 1 (= 100%).

We represent a risky decision as *lottery* a_i (also called "prospect"), which is determined by

- the possible states of the world $\{w_1, w_2, \ldots, w_m\}$,
- their probabilities of occurrence $\{\rho(w_1), \rho(w_2), \ldots, \rho(w_m)\}$, as well as
- the realized (monetary) payoffs $\{y_{11}, y_{12}, \ldots, y_{21}, y_{22}, \ldots, y_{nm}\}$.

Please note that now the outcomes must be expressed in (monetary) units. In our example, we could generate such payoffs by asking ourselves what the single outcome is worth to us in each of the states—for example, how much we are willing to pay to be able or allowed to visit the outdoor pool in warm weather. For better comparability, we choose the amounts y_{ij} analogous to the utility values from Table 1.1, so that in Table 1.2 the monetary payoffs (expressed in euro values) in the individual states are represented.

In our example, let us now assume that the probabilities, ρ (rho), of the two states are equally high, that is, $\rho(w_1) = 0.5$ and $\rho(w_2) = 1 - \rho(w_1) = 0.5$. It is obvious, as with the Laplace rule, to consider the expected value and to decide

[2] This section deals with the expected utility theory and is rather technical. For the further understanding of the concepts discussed in this book, this knowledge is not necessarily required and can be skipped over. However, for a deeper understanding of the payoff structure in a game and its significance, it is recommended to read this section. The problems marked with * at the end of this chapter refer to this section.

Table 1.2 Payoff matrix

	Warm & sunny (w_1)	Cold & rainy (w_2)
Outdoor pool (a_1)	4 Euro ($= y_{11}$)	1 Euro ($= y_{12}$)
Cinema (a_2)	2 Euro ($= y_{21}$)	3 Euro ($= y_{22}$)

based on it. The *expected value* indicates the average payoff of a lottery with infinitely frequent repetition.

It is calculated by weighing all payoffs in the different states of the world with their probability of occurrence and then summing them up. Formally, the expected value of a lottery a_i is determined by

$$E(a_i) = \sum_{w_j} \rho(w_j) \cdot y_{ij}. \tag{1.5}$$

All possible payoffs of the lottery must be considered so that the probabilities sum up to 1 (=100%): $\sum_{w_j} \rho(w_j) = 1$. In the specific example, the expected value of the outdoor pool visit is $E(a_1) = 0.5 \cdot 4 + 0.5 \cdot 1 = 2.5$ and that of the cinema visit is $E(a_2) = 0.5 \cdot 2 + 0.5 \cdot 3 = 2.5$. Therefore, we would be indifferent, which means a clear decision is not possible.

However, if we only look at the expected value, we do not yet consider that both lotteries reflect different risk with respect to the dispersion or spread of the possible outcomes: While the visit to the outdoor pool has the highest (4 euros) and the lowest payoff (1 euro), less variation in payoffs is expected when visiting the cinema (2 euros and 3 euros). We can represent the risk of a random variable by its dispersion around the expected value, which is measured by variance or its square root, the standard deviation. In general, the standard deviation, σ (sigma), for an action a_i is calculated as

$$\sigma(a_i) = \sqrt{\sum_{w_j} \left(y_{ij} - E[a_i]\right)^2 \cdot \rho(w_j)}. \tag{1.6}$$

It considers the dispersion of each payoff, y_{ij}, from the total expected value, $E[a_i]$. As the dispersion can be positive and negative, the difference, $y_{ij} - E[a_i]$, is squared to obtain only positive results. Each squared difference is weighed by its probability and finally summed up. The expression within the root is the variance, while the standard deviation "unsquares" this value.

In the specific example, the standard deviation or risk of visiting the outdoor pool is

$$\sigma(a_1) = \sqrt{(4 - 2.5)^2 \cdot 0.5 + (1 - 2.5)^2 \cdot 0.5} = \sqrt{2.25} = 1.5$$

and for going to the cinema

$$\sigma(a_2) = \sqrt{(2 - 2.5)^2 \cdot 0.5 + (3 - 2.5)^2 \cdot 0.5} = \sqrt{0.25} = 0.5$$

Thus, the risk is higher when visiting the outdoor pool (a_1). A rational decision-maker will also consider the risk of a decision, which is why this should also be included in the evaluation of the lottery. It seems plausible that someone who avoids risk (perhaps because they are planning a family trip) would opt for the cinema visit (a_2) and someone who seeks risk would opt for the riskier visit to the outdoor pool.

Excursion: The St. Petersburg Paradox. A famous example to justify the expected utility theory is a gambling game that Daniel Bernoulli formulated in 1738: A fair coin is tossed, i.e., heads and tails have the same probability of 50% of occurring. The game continues until the first "head" appears. Until this event, the payoff is doubled for each "tail": If the coin shows "head" on the first toss, the payoff is 2 euros. If the coin shows "head" on the second toss, the payoff doubles to 4 euros, if "head" also appears on the third toss, the payoff doubles further to 8 euros etc. Thus, the payoff increases exponentially in the number of total tosses, until finally "head" appears. In general, the payoff, when "head" appears on the k-th toss, is 2^k euros. The question Bernoulli asked was: How much is a player willing to pay to participate in this lottery?

The obvious thing to do is to determine the upper bound for the *fair price*, i.e., you pay just enough so that you do not make a profit in the long run. In other words, we should be willing to pay the expected value of the lottery at most. This results in

$$E = \frac{1}{2} \cdot 2^1 + \frac{1}{2} \cdot \frac{1}{2} \cdot 2^2 + \frac{1}{2} \cdot \frac{1}{2} \cdot \frac{1}{2} \cdot 2^3 + \cdots =$$
$$= \frac{1}{2^1} \cdot 2^1 + \frac{1}{2^2} \cdot 2^2 + \frac{1}{2^3} \cdot 2^3 + \cdots = \sum_{k=1}^{\infty} \frac{1}{2^k} \cdot 2^k = \sum_{k=1}^{\infty} 1 = \infty.$$

Accordingly, we should be willing to pay up to an infinite amount for the lottery! However, this willingness to pay seems more than unrealistic. Surveys and experiments show that most subjects would be willing to pay a maximum of about 4 euros for it. Consequently, the players base their calculations on a value other than the nominal one. Bernoulli suspected that the marginal utility (= the utility that an additional euro would bring to the player) of the payoff amount decreases with increasing amount. He assumed that the players evaluate the payoff accordingly and proposed the utility function $u(x) = \ln x$. If we now calculate the expected value of the utilities from the monetary amounts, we obtain

$$E\left[u(y_i)\right] = \sum_{k=1}^{\infty} \frac{1}{2^k} \cdot \ln 2^k \cong \ln 4.$$

A payment of 4 euros therefore provides the same utility as the lottery and is consequently the maximum willingness to pay of the player. If such a utility function is assumed, the paradox that players should be willing to pay infinitively large amounts for this lottery, as suggested by the expected value, is resolved.$\triangleleft$

The concept of *(von Neumann-Morgenstern) expected utility* allows us to include the decision-maker's attitude towards risk. This utility concept was developed specifically to evaluate risky payoffs and lotteries in particular (see also the box "The St. Petersburg Paradox"). While in Sect. 1.2 a utility function was assumed that could order results in different states of the world (*ordinal concept*), we now need a *cardinal* utility concept, so that differences in utility levels can be measured and interpreted.

The *expected utility* of the lottery results from the sum of the lottery payouts, y_{ij}, in each state, w_j, which are evaluated with a utility function, $u(y_{ij})$, and weighed with their respective probabilities of occurrence, $\rho(w_j)$. The expected utility is formally calculated as

$$E[u(a_i)] = \sum_{w_j} [\rho(w_j) \cdot u(y_{ij})]. \tag{1.7}$$

For a utility function of the form $u(y_{ij}) = y_{ij}$, the expected value of the lottery corresponds to the expected utility. This specific utility function represents a *risk-neutral* individual, i.e., an individual who is indifferent to risk. A *risk-averse* individual, on the other hand, could have the utility function $u(y_{ij}) = \ln y_{ij}$. Their expected utility when visiting the outdoor pool would be $E[u(a_1)] = 0.5 \cdot \ln 4 + 0.5 \cdot \ln 1 \approx 0.69$ and when visiting the cinema $E[u(a_2)] = 0.5 \cdot \ln 3 + 0.5 \cdot \ln 2 \approx 0.90$. They would therefore opt for the action or lottery with the lower risk (the cinema visit). This tendency of the decision-maker to align the decision more towards lower risk is generated by the concave curvature of the logarithm function. If an individual is *risk-seeking*, their utility function could be $u(y_{ij}) = y_{ij}^2$. This individual would prefer the risky option "outdoor pool", $E[u(a_1)] = 0.5 \cdot 4^2 + 0.5 \cdot 1^2 = 8.5$, to the safer option "cinema", $E[u(a_2)] = 0.5 \cdot 3^2 + 0.5 \cdot 2^2 = 6.5$.

Hence, the risk attitude of an individual can show three forms:

- A *risk-averse* individual suffers a loss of utility due to risk.
- A *risk-neutral* individual is indifferent to risk.
- A *risk-seeking* individual experiences an increase in utility due to risk.

To generally determine the risk attitude implied by a certain utility function, there are three possibilities:

Comparison between expected utility and utility of expected value. The *utility of the expected value* is determined by the expected value of the lottery, $E(a_i)$—see (1.5), evaluated with the utility function, $u[E(a_i)]$. A risk-averse individual always prefers the *certain* (i.e., the probability of occurrence is 1) payoff of the expected value of the lottery to the uncertain payoff from the lottery. This is because the risk of the lottery causes a loss of utility. A risk-neutral individual is just indifferent between the expected value and the lottery, while a risk-seeking individual will prefer the lottery to the expected value, as it experiences additional utility through the risk. This is reflected in the curvature of the expected utility function (see also Fig. 1.2): The expected value results from a linear combination (i.e., the sum weighted with the probabilities of occurrence) of the payoffs: With a concave curvature of the expected utility function, the resulting value will always be lower than the expected value, which represents a risk-averse individual. Conversely, in

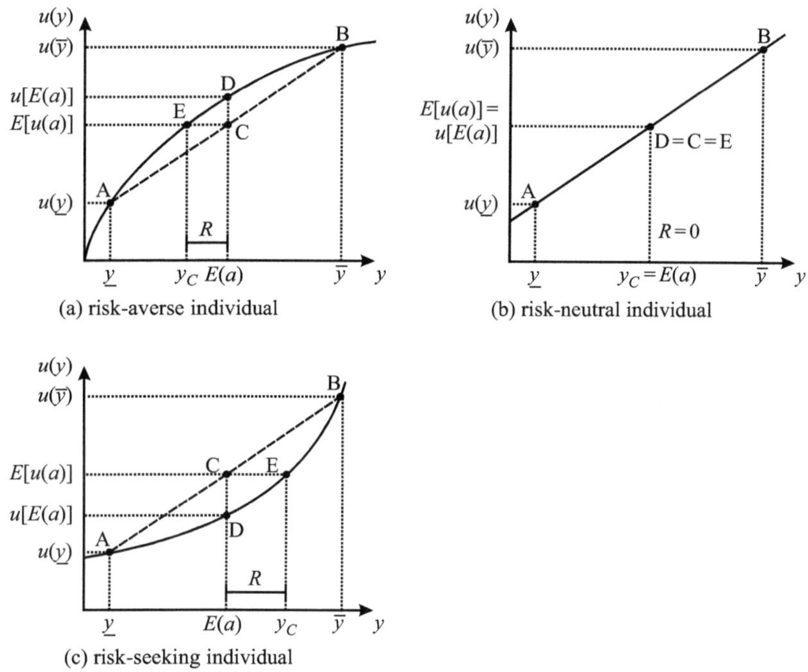

Fig. 1.2 Graphical relationship of the concepts based on the utility function (expected utility diagram)

the case of a convex course of the expected utility function, the expected utility is larger than the expected value, which reflects a risk-seeking attitude. It thus follows:

$$\begin{aligned} E[u(a_i)] &< u[E(a_i)] \quad \text{risk-averse} \\ E[u(a_i)] &= u[E(a_i)] \quad \text{risk-neutral} \\ E[u(a_i)] &> u[E(a_i)] \quad \text{risk-seeking} \end{aligned}$$

Comparison between expected value and certainty equivalent. The *certainty equivalent* of a lottery, y_C, is the certain payment that an individual must receive to be just indifferent between the lottery and the certain paymenty_C, i.e.,$u(y_C) = E[u(a_i)]$. In other words, it refers to the certain payoff that, from the individual's perspective, is equivalent to the lottery. A risk-averse individual is willing to forego the lottery for a lower, but certain payment compared to the expected value—as we know from the previous consideration, it would always prefer the expected value to the lottery. On the other hand, a risk-neutral individual will only accept the expected value of the lottery, while a risk-loving individual must be paid a higher certain amount for it to be willing to forego the risk of the lottery:

$$y_C < E(a_i) \quad \text{risk-averse}$$
$$y_C = E(a_i) \quad \text{risk-neutral}$$
$$y_C > E(a_i) \quad \text{risk-seeking}$$

Measurement of the risk premium. The *risk premium* of a lottery, R, is the amount an individual is willing to give up receiving the certain payoff of the expected value of the lottery instead of the lottery. In other words, it is calculated as the difference between the expected value of the lottery and the certainty equivalent, $R = E(a_i) - y_C$. A risk-averse individual will be willing to give up a portion of the expected value in favor of a certain payoff to avoid risk, i.e., it would be willing to pay a positive risk premium. A risk-neutral individual will not want to pay a risk premium, whereas a risk-seeking individual would demand a negative risk premium as compensation for the lost risk. A negative risk premium thus implies that a risk-seeking individual is willing to pay for the opportunity ("fun") to take risk.

$$R > 0 \quad \text{risk-averse}$$
$$R = 0 \quad \text{risk-neutral}$$
$$R < 0 \quad \text{risk-seeking}$$

To better illustrate the concepts and their interrelationships, let us consider a lottery a that pays out a small amount of $\underline{y}$ with probability ρ and a high payoff $\overline{y}$ with the counter-probability $1 - \rho$. Figure 1.2 depicts this lottery as well as the utility functions of a risk-averse, risk-neutral, and a risk-seeking individual in an expected utility diagram. In this figure, the horizontal axis shows the payoffs y while the vertical axis shows the respective utility values $u(y)$.

The expected value, as well as the expected utility, results from a linear combination of the payoffs or the utilities, respectively:

$$E(a) = \rho \cdot \underline{y} + (1 - \rho) \cdot \overline{y} \tag{1.8}$$

$$E[u(a)] = \rho \cdot u(\underline{y}) + (1 - \rho) \cdot u(\overline{y}). \tag{1.9}$$

Graphically, the linear combination corresponds to a straight line, which connects points A and B, whose coordinates $\{y, u(y)\}$ result from the payoff of the lottery and the associated utility. Point C, whose exact location is determined by the probabilities of the two states, is located on the resulting connection line: $\overline{AC}/\overline{AB} = 1 - \rho$ (or $\overline{CB}/\overline{AB} = \rho$), that is, the higher the probability ρ of the small payoff (point A), the closer point C is to point A (the shorter the distance $\overline{AC}$) and the further away is point C is to point B (the longer the distance $\overline{CB}$). If both states are equally probable, point C is exactly in the middle of the distance $\overline{AB}$, i.e., $\overline{AC} = \overline{CB}$.

From the coordinates of point C, both the expected value (to be read on the y-axis) and the expected utility of the lottery (to be read on the $u(y)$-axis) can be derived. If its $u(y)$-value (the expected utility of the lottery) is kept constant, starting from point C, point E can be found on the utility function. The y-value of point E indicates the certainty equivalent of the lottery. The difference between the expected value and the certainty equivalent finally results in the risk premium. If the y-value of point C is kept constant, point D can be found on the utility function. The $u(y)$-value of point D indicates the utility of the expected value.

To derive the individual concepts from the diagram, it is important to know how the values are measured in order to determine them at the right place: On the horizontal axis, the *monetary values* are determined (payoffs, expected value of the lottery, certainty equivalent, and risk premium), whereas on the vertical axis, the *utility values* are found (utility of the payoffs, expected utility of the lottery).

Relationship between utility function and indifference curve. The utility of a lottery results from both payoffs $u(y_1, y_2)$, i.e., it is a function that depends on two variables. Consequently, the representation requires three-dimensionality, which is somewhat difficult on two-dimensional paper. For this reason, we help ourselves with the graphical representation by fixing one of the dimensions. In Fig. 1.2, for example, we have fixed the second payoff and thus only considered the utility as a function of the change in one payoff. Another possibility is to fix the utility value. In this way, we find all payoff combinations that are equally good from the decision-maker's point of view.

Let us now take a closer look at Fig. 1.3. In the lower section, the diagram labeled "expected utility diagram", we see the representation of the utility function of a risk-averse individual, as we know it from Fig. 1.2. We have a fixed payoff in this case. If we consider the payoffs shown there as certain payoffs, i.e., they occur in the same amount in every possible state, we can transfer them directly to the utility graph and mark the corresponding points—the points marked in the figure are the same in all diagrams. The "utility mountain" indicates the utility values of all payoff combinations and is therefore three-dimensional: each of the two (two-dimensional) payout combinations is assigned a utility value. This means that the lottery can also be plotted on the mountain. It must have the same utility value as the certainty equivalent, i.e., be at the same height. We can depict the contour lines as *indifference curves*, as shown in the right section, the diagram labeled "state preference diagram". Indifference curves represent the geometric location of all payoff combinations that provide the same utility, and between which one is therefore indifferent. The lottery L lies on the same indifference curve as point C.

Let us summarize: The utility function is the side view of the utility mountain, while the indifference curve is the "top view" of the utility mountain.

To determine an indifference curve, all payoff combinations are sought that have the same utility. For example, consider the utility function $u(y) = \sqrt{y}$ and search for all payoff combinations that generate a fixed utility value of $\overline{u}$, i.e., $\overline{u} = \rho_1 \cdot \sqrt{y_1} + (1 - \rho_1) \cdot \sqrt{y_2}$. To draw the indifference curve, we solve this expression for y_2, $\sqrt{y_2} = \left(\overline{u} - \rho_1 \cdot \sqrt{y_1}\right)/(1 - \rho_1)$, and thus obtain the equation

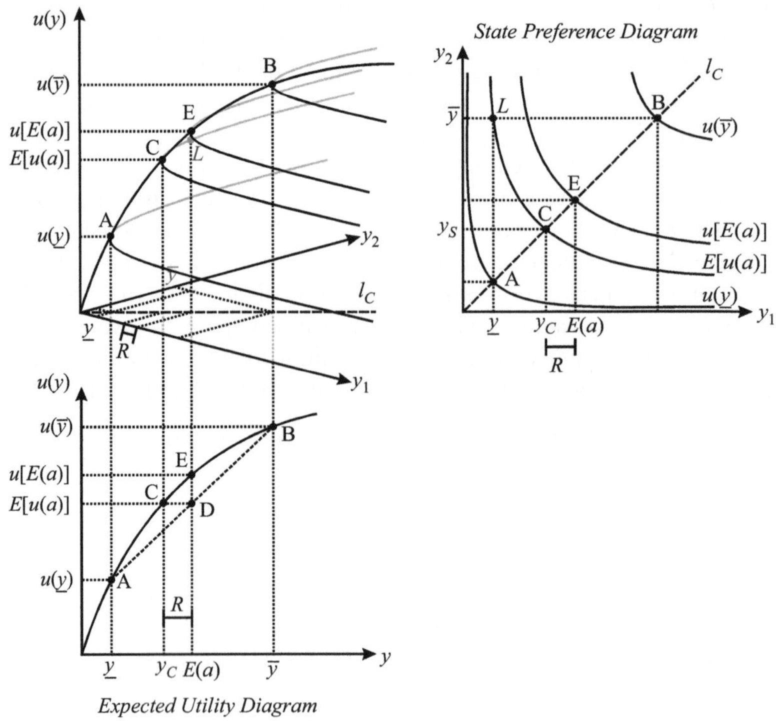

Fig. 1.3 Graphical relationship between utility function, utility mountain, and indifference curve

of the indifference curve,

$$y_2 = \left[\frac{1}{1-\rho_1} \cdot \overline{u} - \frac{\rho_1}{1-\rho_1} \cdot \sqrt{y_1}\right]^2.$$

The indifference curves are represented in the state preference diagram. In this diagram, all known concepts can be read off just like in the expected utility diagram. Every single (certain) payoff of the lottery lies on an indifference curve, but also the lottery itself with its risky payoffs is located on a particular indifference curve. The falling slope course of the indifference curve implies that the individual is only willing to give up part of the payoff in one state if it receives a higher payoff in the other state. The concave course (i.e., the flattening of the curve) means that if the individual already has a low payoff relative to the good state in the bad state and more and more is taken from them in the bad state, they must be compensated with an increasingly higher payoff in the good state. This increases the expected value of the lottery, but the increase in risk prevents the achievement of a higher utility (and thus a higher indifference curve).

The *certainty line* l_C drawn in the diagram represents all certain payoff combinations, i.e., all combinations that have the same value in both states. It therefore has a slope of 1 or 45°. The intersection between the indifference curve of the lottery and the certainty line, point C, gives us the certainty equivalent of the lottery which can be read on one of the axes. The expected value of the lottery can also be plotted on the certainty line (point E). As we can see, the expected value generates a higher utility than the lottery in both the utility mountain and the state preference diagram—it is either higher or lies on an indifference curve further out.

The difference between expected value and certainty equivalent provides the risk premium. However, this can only be read off the horizontal axis, as it is a monetary payoff. By contrast, the direct distance between point E and point C indicates the utility difference (i.e., the utility value of the risk premium)—only the axes allow the monetary value to be determined (it does not matter whether you focus on the y_1- or y_2-axis).

Arrow–Pratt Measure. The utility functions of the three risk types in Fig. 1.2 suggest that the curvature of the utility function is related to risk aversion. This is the basic idea of the Arrow–Pratt measure. The utility function of a risk-averse individual is *concave*,[3] that of a risk-neutral individual is *linear*, and that of a risk-seeking individual is *convex*. Hence, the risk attitude of an individual can be easily determined by the curvature or the second derivative of the utility function $d^2u(x)/dx^2$. Since the utility concept is cardinal, any positive linear transformation, $v(y) = a + b \cdot u(y)$, with $b > 0$, should not change the individual's decision between different lotteries. However, such a transformation changes the curvature of the utility function (by multiplication with b). Another issue that arises from this is that it is—in principle—not possible to compare the risk attitudes of different decision-makers. The *Arrow–Pratt measure of absolute risk aversion* is both robust against a positive linear transformation and—for this reason—also allows interpersonal comparisons of the individual risk attitudes. The measure is defined as the negative ratio between the second and the first derivative:

$$r_A(y) = -\frac{d^2u(y)/dy^2}{du(y)/dy} \tag{1.10}$$

The negative sign serves for standardization and leads to positive values for risk-averse individuals. We can easily verify that a positive linear transformation can indeed be neutralized by this:

$$r_A(y) = -\frac{d^2\left[a + b \cdot u(y)\right]/dy^2}{d\left[a + b \cdot u(y)\right]/dy} = -\frac{b \cdot d^2u(y)/dy^2}{b \cdot du(y)/dy} = -\frac{d^2u(y)/dy^2}{du(y)/dy}. \tag{1.11}$$

The risk attitude can be read from the sign of $r_A(y)$:

$$\begin{aligned} r_A(y) &> 0 \text{ risk-averse} \\ r_A(y) &= 0 \text{ risk-neutral} \\ r_A(y) &< 0 \text{ risk-seeking} \end{aligned}$$

[3] See also Sect. 12.3.

As mentioned, the measure allows for interpersonal comparisons. For example, a higher positive (negative) value of $r_A(y)$ means that the individual is more risk-averse (risk-seeking) compared to another individual with a correspondingly lower value.

The Arrow–Pratt measure also allows statements to be made about how an individual's attitude to risk can change depending on their (expected) income. If, for example, $r_A(y)$ increases with a rise in y, this is referred to as *increasing absolute risk aversion*, i.e., the richer the individual becomes, the more they avoid risk. A distinction can be made between three patterns:

$$\begin{aligned} dr_A(y)/dy &> 0 \text{ increasing absolute risk aversion (IARA)} \\ dr_A(y)/dy &= 0 \text{ constant absolute risk aversion (CARA)} \\ dr_A(y)/dy &< 0 \text{ decreasing absolute risk aversion (DARA)} \end{aligned}$$

Another measure is the *Arrow–Pratt measure of relative risk aversion*, which indicates utility elasticity. It is defined as

$$r_R(y) = -y \cdot \frac{d^2u(y)/dy^2}{du(y)/dy} = y \cdot r_A(y). \tag{1.12}$$

This measure changes depending on the expected income as well:

$$\begin{aligned} dr_R(y)/dy &> 0 \text{ increasing relative risk aversion (IRRA)} \\ dr_R(y)/dy &= 0 \text{ constant relative risk aversion (CRRA)} \\ dr_R(y)/dy &< 0 \text{ decreasing relative risk aversion (DRRA)} \end{aligned}$$

We can illustrate the two concepts with the following example: A decision-maker has a wealth of 1,000 euros and is willing to spend 20 euros on a lottery. If their wealth increases absolutely by 1,000 euros, while their willingness to pay only increases absolutely by 10 euros, there is increasing *absolute* risk aversion. If the decision-maker is willing to spend 20% of their wealth on a lottery, but this relative share falls to 10% of their new total wealth after an increase in their wealth, the decision-maker has an increasing *relative* risk aversion.

1.4 Decision with Opponent: Interdependent Decision Situation

We now want to change the decision-making situation by considering two decision-makers, whose decisions affect each other. This changes the situation significantly: the state of the world that a decision-maker faces is now consciously created by another decision-maker and not—as before—randomly determined by nature. Let us consider the situation where two decision-makers $i = A, B$ can choose between visiting an indoor swimming pool (a_{i1}) and going to the cinema

Table 1.3 Decision with opponent

Alice	a_{B1}	a_{B2}
a_{A1}	0	4
a_{A2}	2	0

Bob	a_{A1}	a_{A2}
a_{B1}	0	2
a_{B2}	4	0

(a_{i2}) but want to avoid each other (due to a "complicated" relationship). Accordingly, a joint leisure activity leads to a utility of 0 for both. Furthermore, we assume that decision-maker A, let us call her Alice, would rather be (alone) in the swimming pool than in the cinema and decision-maker B, let us call him Bob, prefers the cinema. The utilities of the two players are shown in Table 1.3.

The utilities of the two decision-makers are *interdependent*, that is, they influence each other: The decision of one determines how the utility of the other can turn out. For example, if Alice chooses a_{A2}, Bob's utility can either be 0 or 2. At the same time, however, Bob's decision has effects on Alice and thus determines whether Alice, when choosing a_{A2}, gets either 2 or 0. Note that since the outcome is risky, it is assumed that the utilities in this case are determined by the expected utility function discussed in Sect. 1.3 so that the (behavioral) risk faced by the decision-maker is already fully considered by the payoffs.

We can better illustrate the dependence of the decisions on each other if we combine both tables, as done in Table 1.4, into a matrix where the first number in the bracket describes the payoff of Alice and the second that of Bob.

This summary also changes the perspective: The decision of the other decision-maker is not perceived as a (random) state, but as a conscious decision, and the effects of the decisions on each other can be clearly seen. The form shown in Table 1.4 is referred to as a payoff matrix and represents a simple *game*.

This interdependence is also reflected in the utility function of the decision-makers: u_i (a_i, a_{-i}). In addition to the decision a_i from individual i their utility is also influenced by the decisions of the other decision-maker(s)—"$-i$" symbolically stands for all other individuals except i. How can a decision be made in such a situation? We now want to check whether our previous rules can serve to support decision-making.

Table 1.4 Decision with opponent: Payoff matrix

Alice, Bob	a_{B1}	a_{B2}
a_{A1}	(0, 0)	(4, 4)
a_{A2}	(2, 2)	(0, 0)

Let us first consider the maximin rule. Compared to its formulation in (1.1) we must modify it a little to consider the interdependence with the other decision-makers:

$$\max_{a_i}\left(\min_{a_{-i}} u_i(a_i, a_{-i})\right). \tag{1.13}$$

When applying this rule, decision-maker i assumes that all other individuals ("$-i$") decide in such a way that they want to minimize i's payoff, i.e., they decide in such a way that they cause i the greatest possible harm. Based on this premise, decision-maker i finally chooses the best possible action that has the least harmful outcome. In our example, however, this rule does not help Alice, as the worst payoff is 0 for both a_{A1} and a_{A2}. Bob is in the same situation.

The maximax rule from (1.2) can also be reformulated to serve as a decision-making aid here:

$$\max_{a_i}\left(\max_{a_{-i}} u_i(a_i, a_{-i})\right) \tag{1.14}$$

The very optimistic decision-maker i assumes now that all other individuals try to maximize i's payoff. Thus, Alice will opt for the outdoor pool (a_{A1}) according to this rule, as the utility here is 4, which is higher than the utility in the cinema (2). For Bob, the situation is exactly the opposite, i.e., the decision to go to the cinema (a_{B2}) has a higher utility (4) than a visit to the swimming pool (2). Both decision-makers are very happy with this outcome, as they can each pursue their preferred activity alone and each realizes a payoff of 4.

The Hurwicz rule from (1.3) is modified to

$$\max_{a_i}\left[\lambda \cdot \max_{a_{-i}} u_i(a_i, a_{-i}) + (1-\lambda) \cdot \min_{a_{-i}} u_i(a_i, a_{-i})\right]. \tag{1.15}$$

Without knowing the optimism parameter λ, we cannot make a reliable statement on the result. The more optimistic the two decision-makers are, the more likely they will be able to realize the advantageous outcome of the maximax rule.

Finally, the Laplace rule from (1.4) is modified to

$$\max_{a_i}\left(\sum_{a_i}^{n} u_i(a_i, a_{-i}) \cdot \frac{1}{n}\right). \tag{1.16}$$

If Alice chooses the swimming pool (a_{A1}) the expected payoff is $0 \cdot 1/2 + 4 \cdot 1/2 = 2$ and for a_{A2} accordingly $2 \cdot 1/2 + 0 \cdot 1/2 = 1$. The decision is therefore again in favor of the swimming pool. The situation is similar for Bob, who will again prefer going to the cinema (a_{B2} with 2) over going to the swimming pool (a_{B1} with 1).

In summary, it appears that these rules generally seem to be suitable as decision aids. However, these rules require the decision-maker to assume the deterministic, so to speak preprogrammed, behavior of the other decision-makers. According to the maximin rule, the very pessimistic, deterministic behavioral assumption regarding the other decision-makers is "All other decision-makers pursue the (only) goal of causing you maximum damage". Conversely, the maximax rule assumes the enormously optimistic (and perhaps naive) attitude that all others only aim to support the decision-maker to the maximum and wish them only the best. Applying Laplace's rule, the other decision-makers are depicted purely mechanically as they are assumed to make purely random decisions.

To complicate matters, we can ask the very legitimate question: What is the result if every decision-maker uses a different rule? Or even more fundamentally: On what basis do decision-makers choose their decision rule? Both questions show that the previous representation of the decision-making situation is inadequate, and we especially need to better represent the incentives and motivations of all decision-makers. This is exactly the task of game theory. In the next chapter, we will deal with the necessary assumptions and develop solutions for strategic decision problems.

1.5 Comprehension Questions

1. The preference order $o_1 \prec o_2$, $o_2 \approx o_4$, and $o_3 \succ o_4$ is transitive if...
   A $o_1 \succ o_3$
   B $o_2 \succ o_3$
   C $o_3 \succ o_1$
   D $o_4 \succ o_1$

2. Which of the statements are correct?
   A According to the maximin rule, the action that provides the worst payoff in the best case should be chosen
   B According to the maximax rule, the action that provides the highest payoff in the worst case should be chosen
   C According to the maximax rule, the action that provides the highest payoff in the best case should be chosen
   D Neither the maximin nor the maximax rule reflect the attitude of the decision-maker

3. The expected utility...
   A Considers the risk of a decision
   B Is always positive
   C Allows any linear transformation
   D Contains no information about the risk attitude of the decision-maker

4. The risk attitude of a decision-maker can be measured...
   A By comparing the expected value and the risk premium
   B By comparing the expected utility and the expected value

C By sole consideration of the risk premium
D By sole consideration of the certainty equivalent

1C,D; 2C [D: Yes! Optimist or pessimist]; 3A, [B: No! It can also be negative, depending on the specification of the utility function and the payoffs; C: No! Only positive transformations]; 4C.

1.6 Problems

1.6.1 Decision Under Uncertainty

Decide for one of the four actions based on the maximin, maximax, Laplace, and Hurwicz rule (with $\lambda = 0.7$)!

	w_1	w_2	w_3	w_4
a_1	1,100	700	1,200	300
a_2	600	1,000	1,500	1,700
a_3	1,000	1,000	1,000	1,000
a_4	400	700	1,500	1,800

Solution in Sect. 1.7.1.

1.6.2 Criticism of Maximin and Maximax Rule

An investor can choose between three investments A, B, and C, but does not know how likely the seven possible states are. How will the investor decide based on the maximin and the maximax rule? Do you consider the decision to be plausible?

	w_1	w_2	w_3	w_4	w_5	w_6	w_7
Investment A (a_1)	–10	−10	−10	−10	16	−10	−10
Investment B (a_2)	10	10	10	10	10	10	10
Investment C (a_3)	15	15	15	15	9	15	15

Solution in Sect. 1.7.2.

1.6.3 Lottery Comparison Using the Expected Utility*

There are three lotteries:

lottery a_1 64 euros with 55% and 144 euros with 45%
lottery a_2 100 euros with 50% and 100 euros with 50%
lottery a_3 256 euros with 35% and 16 euros with 65%

Determine the expected value and the risk (the standard deviation) of the three lotteries! Which lottery will be chosen by

i. individual A with utility function $u_A(y) = \sqrt{y}$,
ii. individual B with utility function $u_B(y) = 0.1 \cdot y$, and
iii. individual C with utility function $u_C(y) = 10 \cdot (0.1 \cdot y)^2$?

What is the risk attitude of the three individuals?

Solution in Sect. 1.7.3.

1.6.4 Graphical Representation of Expected Utility*

A decision-maker with utility function $u(y) = \sqrt{y}$ faces the lottery a, where the payoffs $y_1 = 4$ and $y_2 = 16$ occur with equal probability.

a. Calculate the expected payoff $E(a)$, the expected utility of the lottery $E[u(a)]$, the utility of the expected payoff $u[E(a)]$, as well as the certainty equivalent of the lottery y_C, and the risk premium R!
b. Draw the expected utility function and determine graphically the concepts calculated in a)!
c. Explain the idea of risk aversion using your graph from b)!
d. Represent the lottery a, the certainty equivalent y_C, the risk premium R, as well as the corresponding indifference curve of the expected utility function in a state preference diagram! Also, draw the certainty line and the indifference curve of a risk-neutral decision-maker that includes the lottery and explain the idea of risk aversion analogous to subtask c)!
e. What would change in the state preference diagram if the decision-maker's utility function is $u(y) = y^2$?
f. How is the position of the indifference curves affected if the probability for the payoff y_1 changes to $\rho_l = 1/3$ or $\rho_h = 2/3$? How can this be intuitively explained? Which of the three lotteries would the decision-maker prefer?

Solution in Sect. 1.7.4.

1.6.5 Arrow–Pratt Measure*

Consider the following utility functions:

(i) $u(y) = \sqrt{y}$,
(ii) $u(y) = 100 + 6 \cdot y$,
(iii) $u(y) = \ln y$,
(iv) $u(y) = \frac{y}{a+y}$,

(v) $u(y) = a \cdot y - b \cdot y^2$ with $a, b > 0$,
(vi) $u(y) = 1 - e^{-r \cdot y}$ with $r > 0$.

Determine the Arrow–Pratt measure of absolute and relative risk aversion! In which cases is there risk aversion and in which cases is there risk neutrality?

Solution in Sect. 1.7.5.

1.6.6 Interdependent Decision

The semester is over, and the exams are coming up. A (representative) student (decision-maker A) is considering how to prepare for the upcoming exam. He can either invest a lot of time in a good preparation (a_{A1}) or hardly prepare at all (a_{A2}). The professor (decision-maker B) is considering whether to set a difficult (a_{B1}) or an easy (a_{B2}) exam. Basically, the professor prefers an easy exam, as this means less correction effort for her, but also wants the students to study for the exam, while the student would rather spend less time on the preparation. The utilities in each situation are: $u_A(a_{A1}|a_{B1}) = 3$ (read: the utility of decision-maker A when choosing a_{A1} given a_{B1} was chosen by decision-maker B), $u_A(a_{A1}|a_{B2}) = 2$, $u_A(a_{A2}|a_{B1}) = 1$, and $u_A(a_{A2}|a_{B2}) = 4$ as well as $u_B(a_{B1}|a_{A1}) = 3$, $u_B(a_{B1}|a_{A2}) = 1$, $u_B(a_{B2}|a_{A1}) = 4$, and $u_B(a_{B2}|a_{A2}) = 2$.

Represent the situation in a payoff matrix! What result occurs when both players make their decision based on the maximin, the maximax, or the Laplace rule? What result arises when one decision-maker chooses according to the maximin rule and the other according to the maximax rule?

Solution in Sect. 1.7.6.

1.7 Solutions

1.7.1 Decision Under Uncertainty

Problem Text in Sect. 1.6.1

- Maximin rule: The minimum payoff in a_1 is 300, in a_2 is 600, in a_3 is 1,000, and in a_4 is 400. Thus, a_3 has the highest value and will be chosen.
- Maximax rule: The maximum payoff in a_1 is 1,200, in a_2 is 1,700, in a_3 is 1,000, and in a_4 is 1,800. Thus, a_4 has the highest value and will be chosen.
- Hurwicz rule: For $\lambda = 0.7$, the weighted payoff for a_1 is 930, for a_2 is 1,370, for a_3 is 1,000, and for a_4 is 1,380. Thus, a_4 has the highest value and will be chosen.
- Laplace rule: The expected value for a_1 is 825, for a_2 is 1,200, for a_3 is 1,000, and for a_4 is 1,100. Thus, a_2 has the highest value and will be chosen.

1.7.2 Criticism of Maximin and Maximax Rule

Problem Text in Sect. 1.6.2

- Maximin rule: The minimum payoff for A is -10, for B is 10, and for C is 9. As investment B (a_2), has the highest value, the investor will decide for it.
- Maximax rule: The maximum payoff for A is 16, for B is 10, and for C is 15. As investment A (a_1), has the highest value, the investor will decide for it. Note that it does not matter that it is a high-risk investment that in most cases leads to negative payoffs.

This example illustrates very well some of the fundamental problems with these two rules: It turns out that in no case is investment C chosen, which on average is better than the other two investments and would probably be the obvious decision for many. This is because in both rules the basis for the decision is essentially limited to state w_5. While in this state, when using the maximin rule, investment B is better than investment C (exceptionally), investment C is outperformed by investment A when applying the maximax rule. Unless state w_5 occurs with certainty, the two rules lead to distorted decisions. When applying these two rules, we must therefore always be aware that we are ultimately only deciding based on one extreme value (maximum or minimum) and thus ignoring everything else. The more possible states there are, the more serious this problem becomes.

1.7.3 Lottery Comparison Using Expected Utility*

Problem Text in Sect. 1.6.3

Calculation of the expected values of the three lotteries:

$$\text{lottery 1: } E(a_1) = 0.55 \cdot 64 + 0.45 \cdot 144 = 100$$
$$\text{lottery 2: } E(a_2) = 0.50 \cdot 100 + 0.50 \cdot 100 = 100$$
$$\text{lottery 3: } E(a_3) = 0.35 \cdot 256 + 0.65 \cdot 16 = 100$$

All three lotteries have the same expected value; thus, no distinct decision can be made.

Risk or standard deviation of the three lotteries:

$$\text{lottery 1: } \sigma(a_1) = \sqrt{0.55 \cdot (64 - 100)^2 + 0.45 \cdot (144 - 100)^2} = \sqrt{1584} \approx 39.80$$
$$\text{lottery 2: } \sigma(a_2) = \sqrt{0.50 \cdot (100 - 100)^2 + 0.50 \cdot (100 - 100)^2} = 0$$
$$\text{lottery 3: } \sigma(a_3) = \sqrt{0.35 \cdot (256 - 100)^2 + 0.65 \cdot (16 - 100)^2} = \sqrt{13104} \approx 114.47$$

Lottery 2's payoff is certain (risk is zero), while lottery 3 has the highest risk.

i. Expected utility of Individual A:

$$\begin{aligned}
&\text{lottery 1: } E[u_A(a_1)] = 0.55 \cdot \sqrt{64} + 0.45 \cdot \sqrt{144} = 9.8 \\
&\text{lottery 2: } E[u_A(a_2)] = 0.50 \cdot \sqrt{100} + 0.50 \cdot \sqrt{100} = 10 \\
&\text{lottery 3: } E[u_A(a_3)] = 0.35 \cdot \sqrt{256} + 0.65 \cdot \sqrt{16} = 8.2
\end{aligned}$$

The decision-maker will choose lottery 2 with a certain payoff. Therefore, this is a risk-averse individual.

ii. Expected utility of Individual B:

$$\begin{aligned}
&\text{lottery 1: } E[u_B(a_1)] = 0.55 \cdot (0.1 \cdot 64) + 0.45 \cdot (0.1 \cdot 144) = 10 \\
&\text{lottery 2: } E[u_B(a_2)] = 0.50 \cdot (0.1 \cdot 100) + 0.50 \cdot (0.1 \cdot 100) = 10 \\
&\text{lottery 3: } E[u_B(a_3)] = 0.35 \cdot (0.1 \cdot 256) + 0.65 \cdot (0.1 \cdot 16) = 10
\end{aligned}$$

The decision-maker is indifferent between the three lotteries. Therefore, this is a risk-neutral individual.

iii. Expected utility of Individual C:

$$\begin{aligned}
&\text{lottery 1: } E[u_C(a_1)] = 0.55 \cdot \left[10 \cdot (0.1 \cdot 64)^2\right] + 0.45 \cdot \left[10 \cdot (0.1 \cdot 144)^2\right] = 11.584 \\
&\text{lottery 2: } E[u_C(a_2)] = 0.50 \cdot \left[10 \cdot (0.1 \cdot 100)^2\right] + 0.50 \cdot \left[10 \cdot (0.1 \cdot 100)^2\right] = 10 \\
&\text{lottery 3: } E[u_C(a_3)] = 0.35 \cdot \left[10 \cdot (0.1 \cdot 256)^2\right] + 0.65 \cdot \left[10 \cdot (0.1 \cdot 16)^2\right] = 23.104
\end{aligned}$$

The decision-maker will choose the risky lottery 3. Therefore, this is a risk-seeking individual.

1.7.4 Graphical Representation of Expected Utility*

Problem Text in Sect. 1.6.4

Part a)
Since both states are equally likely, the probability for each of the two states is $\rho_1 = (1 - \rho_1) = \rho_2 = 1/2$. We calculate the expected value of the lottery as

$$E(a) = \rho_1 \cdot y_1 + \rho_2 \cdot y_2 = 0.5 \cdot 4 + 0.5 \cdot 16 = 10$$

and the expected utility according to

$$\begin{aligned}
E[u(a)] &= \rho_1 \cdot u(y_1) + \rho_2 \cdot u(y_2) = 0.5 \cdot \sqrt{4} + 0.5 \cdot \sqrt{16} = \\
&= 0.5 \cdot 2 + 0.5 \cdot 4 = 3.
\end{aligned}$$

From the utility of the expected value

$$u[E(a)] = \sqrt{10} \approx 3.16$$

and the certainty equivalent

$$\sqrt{y_C} = 3 \iff y_S = 3^2 = 9$$

we finally derive the risk premium

$$R = 10 - 9 = 1.$$

Part b)

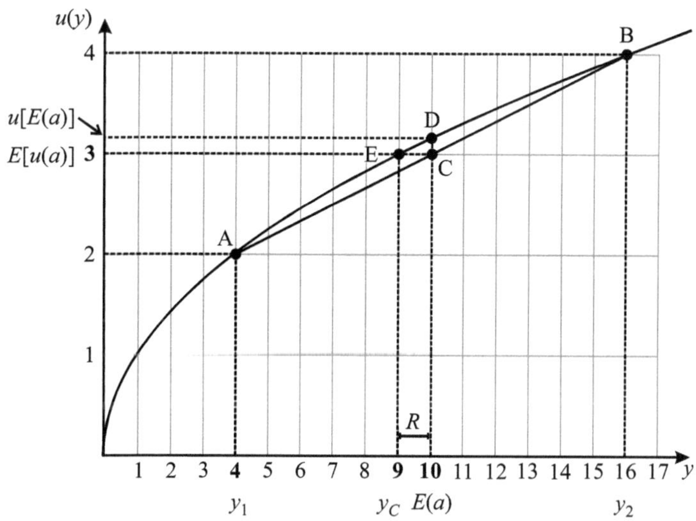

The payoffs in both states of the world result in points A and B, where the decision-maker has a utility of 2 and 4, respectively.

Part c)

In simple terms, a risk-averse individual has a loss of utility due to risk. In the graph, this is represented by the concave curve of the utility function. This concavity can be interpreted in two ways—in monetary terms and in utility units.

The monetary measurement is done on the horizontal axis. The risk attitude is shown by comparing the expected value and the certainty equivalent, or by sole consideration of the risk premium. Since the expected value is higher than the certainty equivalent, $E(a) = 10 > 9 = y_C$, the decision-maker would forego the lottery for a lower but certain payment compared to the expected value.

The measurement in utility units is done on the vertical axis. Here we see that the utility of the expected value is greater than the expected utility, $u[E(a)] \approx$

$3.16 > 3 = E[u(a)]$. The decision-maker would forego the lottery for the certain payment of the expected value, as they suffer a utility loss due to the risk of the lottery.

A risk-neutral utility function has a linear slope so that in monetary terms the expected value corresponds to the certainty equivalent, $E(a) = y_C$, and in utility units the utility of the expected value of the lottery equals the expected utility of the lottery, $u[E(a)] = E[u(a)]$. A risk-seeking decision-maker, on the other hand, gains utility from risk, so their utility function is convex. Thus, $E(a) < y_C$ and $u[E(a)] < E[u(a)]$.

Part d)

The state preference diagram represents the payoff combination in both states. This allows us to graphically represent the decision-maker's indifference curves. Here, we have two options: we can sketch the indifference curves through already known points, or we can formally determine them from the utility function.

Since the indifference curve we are looking for includes all payoff combinations that provide the same utility as the lottery, the lottery itself, point A (16, 4), lies on the indifference curve. Given that both states are equally likely, we can also conclude that the exactly reverse combination (4, 16) will also be on the indifference curve (point B). Finally, the certainty equivalent is also part of the indifference curve. A certain payoff is determined by the fact that it is the same in every state, i.e., this payment is described by point C with (9, 9). With these three points, we can approximately draw the indifference curve.

We can formally determine the indifference curve by calculating all combinations that provide the same utility as the lottery. In general, an indifference curve with a given utility $\overline{u}$ for two possible states is determined by

$$\overline{u} = \rho \cdot \sqrt{y_1} + (1 - \rho) \cdot \sqrt{y_2}.$$

To be able to draw the indifference curve, we solve the equation for y_1. This allows us to determine the payoff in the first state of the world (y_1) that provides the same utility as the lottery for a given payoff in the second state of the world (y_2):

$$\rho \cdot \sqrt{y_1} = \overline{u} - (1 - \rho) \cdot \sqrt{y_2}$$
$$\Rightarrow y_1 = \left[\frac{1}{\rho} \cdot \overline{u} - \frac{1 - \rho}{\rho} \cdot \sqrt{y_2}\right]^2.$$

The lottery has an expected utility of $E[u(a)] = 3$ and the respective probability of the states is $\rho = 1 - \rho = 0.5$, so the indifference curve with the utility $\overline{u} = E[u(a)] = 3$ that we are looking for is

$$y_2 = \left(6 - \sqrt{y_2}\right)^2.$$

The certainty line intersects the indifference curve at the certainty equivalent. There is also a special feature here: Since both states occur with equal probability, the indifference curve runs symmetrically around the certainty line.

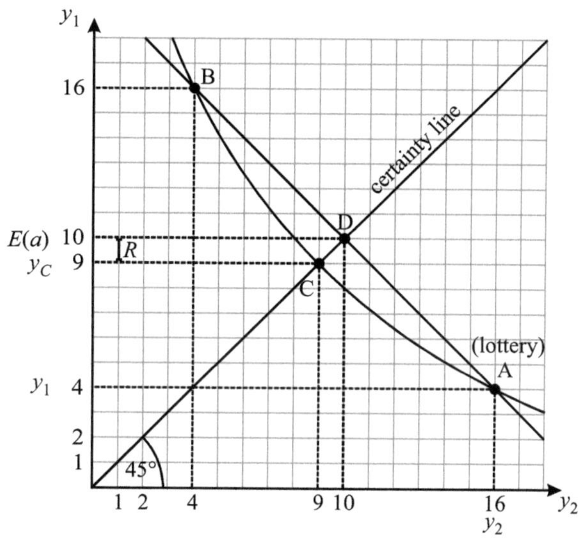

We can also determine the utility function of a risk-neutral individual in the two ways described. For the graphical determination we use the knowledge that both the lottery and the expected value of the lottery (10, 10) (point D) must lie on the indifference curve. As in the case of the utility function, the indifference curve of a risk-neutral individual is linear, so the indifference curve is already fully described by two points.

For the formal determination of the indifference curve, we need the utility function of a risk-neutral individual. In its most general form, this is $u(y) = a \cdot y + b$. Since the utility function is a cardinal concept and thus any positive linear transformation is permissible, we can also simplify it to $u(y) = y$. Along with the knowledge that the expected utility of the lottery for a risk-neutral individual corresponds to the expected value, the indifference curve can now be determined:

$$\begin{aligned} 0.5 \cdot y_1 + 0.5 \cdot y_2 &= E(a) = 10 \\ 0.5 \cdot y_1 &= 10 - 0.5 \cdot y_2 \\ y_1 &= 20 - y_2. \end{aligned}$$

When determining the risk premium, it should be noted that it is read off the axes (the distance $\overline{CD}$ does not indicate the risk premium). However, it does not matter whether the risk premium (as the difference between the expected value and the certainty equivalent of the lottery) is read off on the y_1- or on the y_2-axis.

The decision-maker's risk attitude can finally be determined by the curvature of the indifference curve. The convex (left-curved) course of the indifference curve indicates a risk-averse decision-maker. The expected value is therefore on a higher indifference curve than the lottery, which means that the individual suffers a loss of utility from the lottery. The more curved the indifference curve, the higher the decision-maker's risk aversion.

Part e)

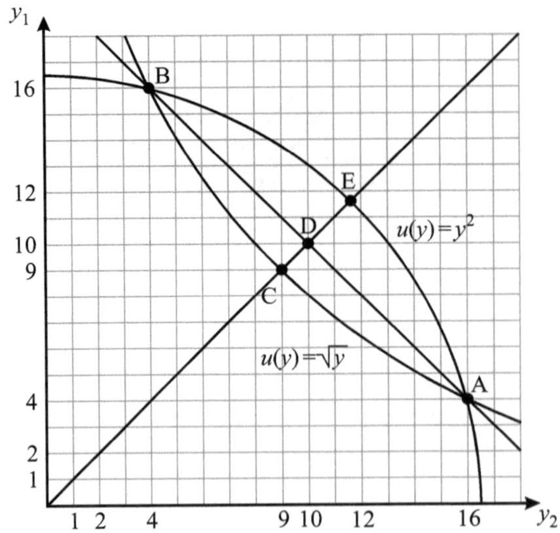

This utility function describes a risk-seeking decision-maker. As the expected utility of the lottery, we calculate $E[u(a)] = 0.5 \cdot 4^2 + 0.5 \cdot 16^2 = 136$. This results in a certainty equivalent of $y_C = \sqrt{136} \approx 11,7$ and thus point E on the certainty line. The equation of the indifference curve is given by $y_1 = \sqrt{272 - y_1^2}$. The indifference curve runs concave, as can be seen in the figure (similarly, the associated utility function would run convex), meaning that the expected value of the lottery lies on a lower indifference curve than the lottery, as the decision-maker derives a utility from risk.

Part f)

To determine the respective course of the indifference curves, it is recommended to set up their formal equation. Analogous to part d) for general ρ this is

$$y_1 = \left[\frac{1}{\rho} \cdot \left(\rho \cdot \sqrt{4} + (1-\rho) \cdot \sqrt{16}\right) - \frac{1-\rho}{\rho} \cdot \sqrt{y_2}\right]^2.$$

For $\rho = \rho_l = 1/3$ the indifference curve is

$$y_1(\rho_l) = \left[10 - 2 \cdot \sqrt{y_2}\right]^2$$

and for $\rho = \rho_h = 2/3$ it becomes

$$y_1(\rho_h) = \left[4 - 0.5 \cdot \sqrt{y_2}\right]^2.$$

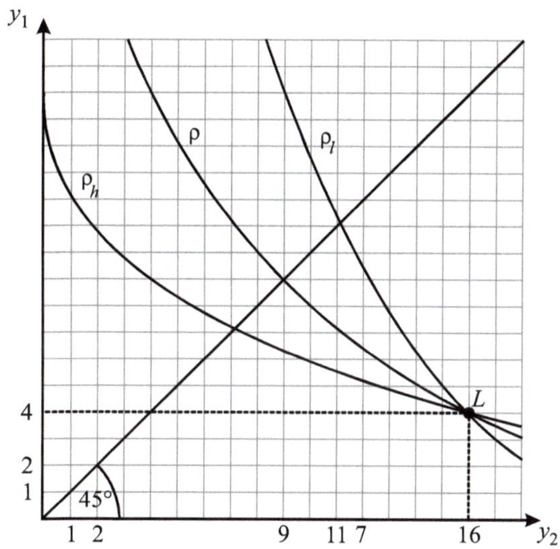

The new indifference curves now no longer run symmetrically to the certainty line. If state 1 becomes more probable (as in the case of ρ_h) the indifference curve turns inward, i.e., for a given payoff in state 1, a lower payoff in state 2 is already sufficient to keep the decision-maker indifferent (this state is less likely and therefore does not weigh as heavily). Conversely, the indifference curve turns outward with a lower probability of state 1: In the more probable state 2, given a payoff in state 1, we must offer more to keep the decision-maker indifferent.

The decision-maker will opt for the lottery with ρ_l. Graphically, this can be read from the fact that the certainty equivalent for this lottery with the same utility function is higher at 100/9 than for the others. Specifically, the expected utilities of the three lotteries are $E[u(a_l)] = 10/3$, $E[u(a)] = 9/3$, and $E[u(a_h)] = 8/3$.

1.7.5 Arrow–Pratt Measure*

Problem Text in Sect. 1.6.5

Utility function	$r_A(y)$	Degree	$r_R(y)$	Degree
(i) $\sqrt{y}$	$+\frac{1}{2 \cdot y} > 0$	DARA	$\frac{1}{2} > 0$	CRRA
(ii) $100 + 6 \cdot y$	0	Neutral	0	Neutral
(iii) $\ln y$	$\frac{1}{y} > 0$	DARA	$1 > 0$	CRRA

(continued)

(continued)

Utility function	$r_A(y)$	Degree	$r_R(y)$	Degree
(iv) $\frac{y}{a+y}$	$\frac{1}{a+y} > 0$	DARA	$\frac{y}{a+y} > 0$	IRRA
(v) $^*a \cdot y - b \cdot y^2$	$\frac{2 \cdot b}{a-2 \cdot b \cdot y}$		$\frac{2 \cdot b \cdot y}{a-2 \cdot b \cdot y}$	
(vi) $1 - e^{-r \cdot y}$	$r > 0$	CARA	$r \cdot y > 0$	IRRA

*In case (v), a distinction must be made: The expression is positive (and the individual is therefore risk-averse) if the denominator meets the condition $a - 2 \cdot b \cdot y > 0$ or $y < a/(2 \cdot b)$. For $y > a/(2 \cdot b)$, the individual is instead risk-seeking. Thus, the individual's risk attitude in this utility function depends on their income: At a specific income threshold, the formerly risk-averse individual suddenly becomes risk-seeking. Such a utility function is often used in CAPM models,[4] which is why we would like to point out the need for careful handling in their modeling at this point.

1.7.6 Interdependent Decision

Problem Text in Sect. 1.6.6

The payoff matrix is as follows:

A, B	a_{B1}	a_{B2}
a_{A1}	(3, 3)	(2, 4)
a_{A2}	(1, 1)	(4, 2)

The following tables show the minimum and maximum payoffs relevant for the decision rules:

Decision-maker A (Student)		
	Minimum	Maximum
a_{A1}	2	3
a_{A2}	1	4

Decision-maker B (Professor)		
	Minimum	Maximum
a_{B1}	1	3
a_{B2}	2	4

[4] CAPM = Capital Asset Pricing Model is a theoretically founded model of the capital market.

- Both use the maximin rule: The student chooses a_{A1} and the professor a_{B2}, the result is (2, 4).
- Both use the maximax rule: The student chooses a_{A2} and the professor a_{B2}, the result is (4, 2).
- Both use the Laplace rule: The student is indifferent (a_{A1}: 2.5 and a_{A2}: 2.5), the professor chooses a_{B2} ($3 > 2$). This results in either (2, 4) or (4, 2).
- The student chooses the maximin rule (i.e., a_{A1}), and the professor chooses the maximax rule(i.e., a_{B2}): The result is (2, 4).
- The student chooses the maximax rule (i.e., a_{A2}), and the professor chooses the maximin rule (i.e., a_{B2}): The result is (4, 2).

Literature Recommendation

Hansson, Sven Ove (2005) Decision theory: A brief introduction.

Basic Concepts 2

The preceding chapter dealt with decision problems of a single decision-maker who must make an optimal choice in a situation characterized by uncertainty (probabilities of the states of the world are unknown) or risk (probabilities are known). It was only in Sect. 1.4 that the decision situation was given a strategic character by adding another decision-maker, which brings us to the core issue of game theory: the analysis of interactive and strategic decision situations. Henceforth, there will always be at least two decision-makers or players interacting with each other. "Interaction" means that each player always influences the other player's game outcome to some extent. Therefore, each player needs to anticipate the actions of the other player when making their own decision and adjust to it as best as possible.

Before we explain the established solution methods in game theory, we first describe and clarify the essential parameters of a game. This chapter therefore explains important fundamental terms and assumptions that are of essential importance for understanding the games discussed in the rest of the book.

2.1 Game, Players, and Strategies

When people talk about "games" in everyday life, they often mean leisure activities like board games, computer games, or sports competitions. These generally understood games are also games in the sense of game theory. However, games are defined much more broadly in game theory. Here, a game is understood to be a precisely described strategic interaction of agents (players). Thus, any form of strategic interaction between two or more persons, groups, or institutions can

F. Bartholomae and M. Wiens, *Game Theory and Applications*, Classroom Companion: Economics, https://doi.org/10.1007/978-3-658-44608-6_2

be considered a game, whether it involves board games, military conflicts, business competition, or political elections. Specifically, a game $\Gamma(N, S, u)$, with Γ is Greek Gamma, is formally determined by

- the set $N = \{A, B, \ldots, n\}$ of players (n = number of players);
- the set S of possible strategies $S = \{s_A, s_B, \ldots, s_i, \ldots, s_n\}$ for each individual player (s_i = strategy of player i from their strategy set S_i);
- the utility or payoff vectors $u(s) = \{u_A(s), u_B(s), \ldots, u_n(s)\}$ which indicate the payoff each player receives depending on the strategies chosen by all players ($u_i(s)$ = utility of player i at strategy combination s); and
- the rules of the game, which specify, e.g., the sequence in which the players take their turn or the available information of each player at a specific point in time.

Who Are the "Players"? The *set of players* determines both the number of players and the information about who exactly becomes an active strategic party. Players are understood to be the decision-makers in the considered strategic situation. As already hinted above, we do not necessarily understand a player to be a single person. For example, if we want to theoretically examine the behavior of a tax-paying citizen towards the tax office, we can model the interaction appropriately as a game between a "representative" citizen and the tax authority. The first player, the taxpayer, thus represents the "totality of all tax-paying citizens" (in the sense of their interests) and the second player, the tax office, represents the government (or the interests of the government). Basically, a player can also represent a group or institution, if they behave like a single decision-maker. This assumes that there is (1) a clear preference order within the group (such as the interest in a fair taxation of all citizens) and (2) the decision-making processes within the group do not matter (so it is only the mere implementation of the tax laws that is relevant and not the political decision about their design). Thus, we are not interested in the actual decision-makers interacting in a game. After all, every game is a model that is only tailored to the strategically relevant features of an interaction and therefore must abstract from reality in a suitable way.

What Is a Strategy? The *set of possible strategies* includes the action options of the considered players. In analogy to decision theory (Chap. 1), we refer to a single strategic move of a player as an action. Accordingly, a_i is the action chosen by player i. Then, a strategy is the complete plan of all actions.

In the simplest case, a strategy consists of only a single action ($s_i = a_i$). Consider the following example: An energy company is faced with the choice of either investing in the expansion of conventional or renewable energies. It hopes that the government will also align its energy policy with the same focus (i.e., on conventional or renewable energy). Let us first assume that both players, the company and the government, are not able to observe what the other player chooses at the time of their decision. Thus, both players cannot align their decision with the decision

Table 2.1 Game matrix for the simultaneous game "Future Energy"

Company (A), Government (B)	Conventional (s_{B1})	Renewable (s_{B2})
Conventional (s_{A1})	(2, 1)	(0, 0)
Renewable (s_{A2})	(−1, 0)	(1, 2)

of the other due to lack of information. This situation is referred to as a *simultaneous game*, which is usually represented as a game matrix as shown in Table 2.1 and corresponds to the payoff matrix in Table 1.4.

Let us first focus on the players' strategies. The company's strategy set has two elements—the two strategies "expansion of conventional energy" and "expansion of renewable energy". To write down the strategies in a space-saving way, we assign numbers to the players and the strategies, in the same way, as we have already indexed the actions and states with numbers in the part on decision theory. The company is player A and the government is player B. Formally, the strategy set of the energy company can then be described as follows:

$$S_{A=\text{Company}} = \{s_{A1=\text{conventional energy}}, s_{A2=\text{renewable energy}}\}$$

Regardless of player A, the government as player B will decide on its energy policy, i.e., whether to promote the expansion of conventional or renewable energies (for example, through subsidies or public infrastructure investments). The government's strategy set would then be analogous:

$$S_{B=\text{Government}} = \{s_{B1=\text{conventional energy}}, s_{B2=\text{renewable energy}}\}$$

In a nutshell, we can specify the strategy sets of both players as follows: $S_A = \{s_{A1}, s_{A2}\}$ for the energy company and $S_B = \{s_{B1}, s_{B2}\}$ for the government.[1] Similarly, in this case, we could also write $S_A = \{a_{A1}, a_{A2}\}$ or $S_B = \{a_{B1}, a_{B2}\}$ since in the considered example the strategies of the players coincide with their actions.

The fact that a player's strategy set consists of only a single action is the simplest possible case, which is mainly important for simultaneous move games. In contrast, in *sequential games*, the players make their decisions one after the other. The crucial difference to a simultaneous game is that the player who moves second—the *second mover* or *follower*—can optimally adapt to the observed decision of the other player—the *first mover* or *leader*. Since the second mover is perfectly informed about the strategy choice of the first mover at the time of their decision, they can make their decision without strategic uncertainty. Sequential games are usually represented in *extensive form* or in the form of a *game tree*. Figure 2.1 shows the game tree for the game "Future Energy", which we now want to consider as a sequential game.

[1] Note: The first letter in the index of the strategy always denotes the player, the second number their strategy.

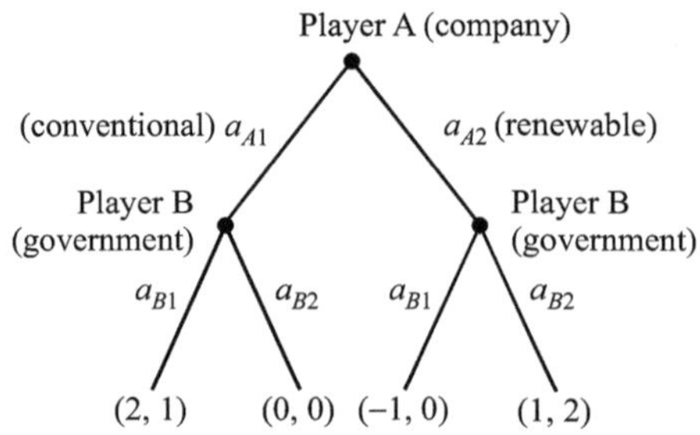

Fig. 2.1 Game tree for the sequential game "Future Energy"

As already briefly mentioned in the preceding chapter on decision theory, using Fig. 1.1, each decision point or decision node of a game tree indicates which player gets to move at this point in the game. At the starting point, the first decision node, the company makes a decision. Thereafter, the government decides at one of two different decision nodes, depending on how the company has decided. Since the company is the first mover in the sequential game and chooses the opening move, its strategy set is identical to that of the simultaneous game. We can therefore again refer to this set as $S_A = \{s_{A1}, s_{A2}\}$ or likewise $S_A = \{a_{A1}, a_{A2}\}$. For the government, however, the strategy set changes compared to the simultaneous game, as it, as the second mover, reacts to a game event that has already occurred, the so-called *history of the game*. The government has the choice between its two actions "promotion of conventional energy" and "promotion of renewable energy" at each decision node, but it will make the choice between both actions dependent on what has happened before. In technical terms of game theory, it bases its decision on the history of the game. Keep in mind that, by definition, its strategy must be a *complete plan* of all actions, meaning that each government strategy must contain a clear instruction for action for both the left and right decision nodes: players must know at any point in the game what they have to do. A strategy of the government could be

$$s_{B1} = \{a_{B1}|a_{A1},\ a_{B1}|a_{A2}\}$$

This strategy contains two *conditional actions* as instructions for action. The conditional action $a_{B1}|a_{A1}$ tells the government: "Promote conventional energy (a_{B1}), *if* (or: *under the condition that*) the company has previously invested in the expansion of conventional energy (a_{A1})". In the game tree, the government should choose the left branch (a_{B1}) if it is at the left node (a_{A1} was chosen by the company before).

In contrast, the conditional action $a_{B1}|a_{A2}$ tells the government what to do at the right node: "Promote conventional energy (a_{B1}), *if* the company has previously invested in the expansion of renewable energy (a_{A2})". Only the two instructions for action together form the strategy s_{B1}. If one of the two conditional actions

Table 2.2 Game matrix for the sequential game "Future Energy"

A, B	$\{a_{B1}\|a_{A1}, a_{B1}\|a_{A2}\}$	$\{a_{B1}\|a_{A1}, a_{B2}\|a_{A2}\}$	$\{a_{B2}\|a_{A1}, a_{B1}\|a_{A2}\}$	$\{a_{B2}\|a_{A1}, a_{B2}\|a_{A2}\}$
$s_{A1} = a_{A1}$	(2, 1)	(2, 1)	(0, 0)	(0, 0)
$s_{A2} = a_{A2}$	(−1, 0)	(1, 2)	(−1, 0)	(1, 2)

were missing, then the government would not know what to do at the respective decision node and its strategy would be incomplete.[2]

Overall, the government as a second mover in this sequential game has four strategies—this number results combinatorically from the fact that it can choose from two actions at two nodes, resulting in $2 \cdot 2 = 4$ combinations. Each of these strategies thus represents a combination of two conditional actions:

- $s_{B1} = \{a_{B1}|a_{BA1}, a_{B1}|a_{A2}\}$ Select "left" at the left node, select "left" at the right node
- $s_{B2} = \{a_{B1}|a_{A1}, a_{B2}|a_{A2}\}$ Select "left" at the left node, select "right" at the right node
- $s_{B3} = \{a_{B2}|a_{A1}, a_{B1}|a_{A2}\}$ Select "right" at the left node, select "left" at the right node
- $s_{B4} = \{a_{B2}|a_{A1}, a_{B2}|a_{A2}\}$ Select "right" at the left node, select "right" at the right node

The government's strategy set is completely described with these four strategies: $S_B = \{s_{B1}, s_{B2}, s_{B3}, s_{B4}\}$.

It is also possible to represent a sequential game in matrix form, as shown in Table 2.2 for the sequential game "Future Energy". Due to the four strategies of player B, the resulting matrix has more columns, i.e., it becomes larger in dimension. To transfer sequential games into matrix form, we need all combinations of conditional actions, which—as we have seen already this rather simple game—can be quite complex. For this reason, sequential games are presented in the much more intuitive game tree representation. Furthermore, it will also become apparent that the matrix form can also contain implausible solutions. Anyhow, we just wanted to point out that the form of presentation (game tree or matrix) is not necessarily tied to a specific type of game (sequential or simultaneous) but can be freely chosen. However, we generally recommend using the tree representation (extensive form) for sequential games and the matrix representation (strategic form) for simultaneous games.

A strategy, according to our definition, represents a *plan* for all possible action options, but it does not yet make a statement about whether it is an *optimal* choice. We can only identify optimal strategies of the players if we solve the game, i.e.,

[2] Imagine you had to program a robot to explore an alien planet and had to provide instructions in its program code for every imaginable (and unimaginable) eventuality—otherwise there would be an error message and the project would fail.

determine the equilibria of the game, which we will deal with in more detail later. However, before we can identify the optimal strategies from a set of strategies, we first need to understand what a *set of strategies* is and how to represent it.

The description of strategies is already quite complex in relatively simple games with few stages and few actions. In games like chess, the strategy space eventually becomes so complex that even with the support of the most modern supercomputers, it is not possible to determine the optimal (winning) opening move.

We will deal with the description and solution of multi-stage games in more detail in Chap. 4 (Sequential Games).

Discrete versus Continuous Strategies. The strategy space can be divided into *discrete* and *continuous* in terms of its scope. To put it simply, discrete strategies correspond to a countable list of actions that can be assigned to a quantitative or qualitative scale. Consider the game between a taxpayer and the tax office. If the description of the taxpayer's strategies is limited to the two options "pay taxes" or "evade taxes", then these would be discrete strategies on a quantitative scale, as the taxpayer chooses the measurable amount of money T, which they pay to the tax office, from a two-element list $T \in \{0, T^{full}\}$: Either they pay the full amount of taxes ($T = T^{full}$) or they pay nothing ($T = 0$). Since both are measurable quantities, the scale is quantitative. Examples of discrete strategies on a qualitative scale would be the tax office's decision between "control" and "no control", the investments in conventional or renewable energies by the energy company, or the decision from the previous chapter between the leisure activities "swimming pool" or "cinema". Thus, a qualitative scale consists of names and labels.

In the case of continuous strategies, however, there is no such list as a strategy space, but instead a continuous spectrum from which the player—at least theoretically—can choose between potentially infinitely many possibilities. For example, the taxpayer could choose the amount of tax they want to pay from the entire, continuous spectrum $T \in [0, T^{full}]$. In contrast to the curly brackets above, the square brackets indicate a continuous interval, which means that the taxpayer can choose 0, T^{full}, or any amount in between.[3] Most of the games discussed in this book involve discrete strategies (such as all matrix games). In cases where we prefer a continuous formulation of the strategy space, this is mainly for two reasons: First, we are interested in game-theoretical applications for precisely quantifiable strategies, such as the choice of quantities and prices by a company. Second, with continuous formulations, we can take advantage of differential calculus and derive optimal strategies. If you need a brief introduction to this method or a refresher, we recommend reading Chap. 12.

[3] In the strict mathematical sense, however, this is also a discrete strategy set, because amounts of money can ultimately only be represented in countable cent steps and are not infinitely divisible. However, this point is not relevant for game theoretical modeling.

Mixed Strategy. A mixed strategy assigns a probability to each action of the player (sum = 1): Thus, the taxpayer can consider paying their taxes with probability p or evading their taxes with the counter-probability $(1-p)$. They, thus, now only decide on the (continuous) probability $p \in [0; 1]$. Since there are infinitely many possibilities for how the probabilities can be assigned, the taxpayer now has a continuous strategy space despite originally discrete choices: If, for example, the tax amounts to $T^{full} = 1000$ euros and is only paid with a probability of 25%, the expected tax payment is 250 euros. Thus, with the probability choice, all continuous amounts between 0 and 1000 can also be generated as expected payment. This type of strategy is discussed in detail in Sect. 3.2. Since a pure strategy always exists when a certain action is played with probability 1 and all others with probability 0, pure strategies ultimately represent special cases of mixed strategies.

Tips for Identifying Relevant Strategies. If we want to translate some concrete, strategically interesting, question into a game, we must figure out which actions we consider relevant for the respective player, and which should therefore be included in the strategy set. In board games, such as chess or nine men's morris, this question does not arise, as the space of all actions may be very complex but is fixed. But what about, e.g., penalty shooting, a seemingly simple variant of sporting confrontation? In the simplest case, the shooter's decision could be limited to the two shooting directions "left" and "right" (referring to the left and right corners of the goal). For the goalkeeper, these would be the corresponding directions in which they can throw themselves to defend the shot. However, limiting the strategy space to "left" and "right" seems very restrictive, as a shooter can of course also shoot into the middle, choose between "top" and "bottom", and shoot with different strength and shooting technique. At this point, it should be emphasized once again that a game is always just a model. When formulating a game, the aim is therefore not to depict all facets of the decision-making situation as realistically as possible, but to capture the *central strategic aspect of the interaction* as economically as possible. This shifts the focus to the question: What is the central, strategic aspect of penalty shooting? Due to the direct conflict of interest between the shooter and the goalkeeper, the most important aspect initially lies in the tactical phenomenon of "mutual stalking". The implications of this stalking can be most easily and clearly represented by the simple choice between "left" and "right". We will discuss the solution of the penalty game in more detail in Sect. 3.2 and in this context also discusses in what respect the extension of the strategy set by the option "shoot in the middle" could change the central strategic aspect of the interaction. In this case, the explanatory power of the game can be increased from an empirical point of view, thus justifying the acceptance of this third strategy.

When transferring a real decision situation into game-theoretical language, we must therefore carefully consider in advance which action options are strategically relevant for the respective context. Our experience shows that students often make two types of mistakes: Either the strategy already anticipates the outcome of the game, or events are declared to be strategies that the player in question cannot influence at all or only with considerable restriction. An example of the first case would be to choose "prevent goal" and "enable goal" as the goalkeeper's strategies. Whether the goalkeeper ultimately manages to fend off the ball or not depends on the strategy choice *of both players*, goalkeeper and shooter. Whether the ball ends up in the goal is therefore dependent on the course of the game and is the *result* of the game and not the strategy. The second mistake is of similar importance and is particularly relevant to questions of decision theory, as discussed in Chap. 1. Here, the influence of coincidence is underestimated or ignored. Suppose the innovation strategy of a company is to be modeled and the two strategies "create innovation" and "create no innovation" are chosen. Certainly, the result of the investment in a product innovation will ultimately come down to one of these two possibilities (*ex post* view), but the company must decide on its investment at a time when it is by no means foreseeable whether the investment will lead to an innovation at all (*ex ante* view). This uncertainty component is a fundamental element of entrepreneurial risk. However, the company may be able to positively influence the *chances* of an innovation with the amount of the investment. In such a case, it is therefore important to distinguish between the influenceable and the non-influenceable component of the decision. The latter could be considered in the modeling of the game as follows: The entrepreneur determines their expenditure on research and development (R&D) as the variable they can control, which in turn influences the *probability* of an innovation being successful. The company can therefore only influence the chances of an innovation with its investment, the innovation itself is brought about by chance. What the player decides should not be confused with what chance or, in game-theoretical language, nature decides.

2.2 Payoffs as the Players' Target Variable

The third essential element of a game is determined by the *utility functions* or by the *payoffs of the players.* As described in Chap. 1, we assume for simplification that each player evaluates each possible outcome with a utility function, i.e., assigns utility values to the game outcomes. In accordance with standard practice in game theory, we assume the expected utility function discussed in Sect. 1.3 as the adequate type of utility function, since the outcome of a game is risky, and this risk must be considered by the players according to their risk attitude. We also synonymously refer to the utility values as the "payoffs" of the players and the entirety of all payoffs as the *payoff structure* of a game. The payoffs are, in a sense, the "engine" of a game, as they determine the goals of the players. Only these goals reveal a player's motivation to choose a particular strategy. Therefore, the payoff structure of a game can also be vividly understood as the *motivation structure* or *incentive structure.* We usually assume that each player only pays attention to their own utility value and is not interested in the utilities that the other players receive, meaning each player is (strictly) *selfishly motivated.*

When determining the payoffs, it should be noted that these represent a *plausible motivation* of the players for the respective modeled context. Let us consider again the simultaneous game "Future Energy" shown in Table 2.1, where we have already assumed specific values for payoffs. In the game, the symmetric cases, where the government and the company choose the same energy source, were assigned the highest payoffs (2, 1) and (1, 2). If you choose the payoffs this way, you assume that the main interest of both players (company and government) is that the direction of investment and promotion coincides. The government prefers slightly more than the company that renewable energy prevails, as the government can thus achieve its (self-set) climate goal. The company, on the other hand, is in the best situation when it can maintain its existing technology under subsidy conditions and does not have to switch. The worst case for the company (payoff–1) occurs accordingly when it must change the technology but does not receive any funding. This small example shows that when mapping player motivation, it is less about the absolute amount of the payoffs, but primarily about the *ranking* of the payoffs. So, if we want to fill a 2 × 2 matrix with payoffs, we must consider for all four possible game results how the respective player evaluates these outcomes *in comparison.* Which assumed motivation is the "correct" one ultimately depends on the context and is therefore primarily an empirical question.

The most important criterion for describing the incentive structure is the *degree of conflicting interests* of a game. Games that involve an extreme conflict of interest include war-like confrontations or face-off situations. In these cases, the players have completely opposed interests, so that in every game outcome, one of the players always suffers a loss. These conflict-prone games include the so-called constant-sum games, which will be examined in more detail below. In other games, compromise solutions are indeed possible, but they are difficult to enforce. An example of this is the so-called prisoner's dilemma, which we will analyze in more detail in Sect. 3.3. And finally, there are games where at best a very weak conflict of interest exists and where the players can reach a favorable solution for them with very little communication or minimal intuition. We will go into more detail in Chap. 3 on how to recognize the scope for communication and what the term "favorable solution" can mean.

Determining the payoffs of the players is straightforward in some games, while in others it is a particularly challenging task. Let us consider, for example, the students in a school class who are faced with the choice of preparing intensively or loosely for an exam. The exam result is always better for each student with intensive preparation than with loose preparation, and of course, each student will be interested in a good grade. However, intensive learning also involves an effort that requires at least a time commitment, and this time commitment can be perceived and evaluated very differently by students: Some students enjoy learning, have fun acquiring new learning material, and look forward to the exam in anticipation of a good result. For other students, however, learning is associated with a strong mental strain, whether through frustration, slow learning progress, or fear of failure. For some students, therefore, the temporal and psychological costs of learning are very high, for others very low and perhaps even negative (negative psychological costs are "psychological gains" that could be interpreted as anticipation and fun).

In addition, there are the "overambitious" students who not only want to achieve a good grade, but only feel truly satisfied when they have a better grade than the others. In the payoff of the overambitious student, a *social motivation* comes into play. Unlike a player with strictly selfish motivation, a player with social motivation evaluates their situation not only based on their own performance but also on the performance of the other players.

The social motivation of a player can have a negative or positive effect on the other players, depending on whether the social attitude of the player in question is more cooperative or more adversarial. This in turn depends crucially on the context of the game. In the case of the overambitious student, it is an adversarial, competitively oriented attitude towards the other students. Social motivations are particularly relevant in the context of *social norms*, such as in relation to helpfulness or fairness. Social norms tend to lead to cooperative behavior towards a cooperative player, but also to adversarial behavior towards uncooperative players. A player who is fair-minded would find it unsatisfactory if their fellow player is (significantly) worse off than they are and will therefore strive to compensate for this disadvantage (cooperative attitude). However, if they find that his generosity is being exploited, this creates an adversarial attitude, and they would behave more aggressively towards their fellow player than a player who is not subject to social motivation would. The sub-discipline of so-called *behavioral game theory* primarily investigates such questions of social motivation, which is strongly linked to social psychological research.

The case of the school class shows first that in a game both hard and soft factors can flow into the total payoff of a player: The time or the grade in the exam would be rather hard factors, as they are both measurable and interpersonally comparable. Psychological payoffs such as joy or fear, on the other hand, are of course very subjective and thus rather soft factors. Secondly, the example discussed illustrates that it can sometimes make sense to model several "types" of players, just as the school class also consists of different "student types". Instead of a *homogeneous* group of students, which can be represented by a single, representative student (as a kind of "student prototype"), one could model a *heterogeneous* group, which consists of weaker, rather overwhelmed students (we then speak of the weaker "type") and very high-performing, confident students (the high-performing "type"). When it comes to the question of how many and which types of players should be distinguished, we must again keep in mind that a game is just a model. So, one should not assume a possibly high number of player types just to make the game more colorful and thus seemingly more realistic. No model in the world can fully realistically capture a school class. The decision for heterogeneity should (and must) only be made when the specific difference between the modeled player types is particularly important for the question to be investigated. For example, if we want to investigate how a particular grading scheme (e.g., for an exam) affects the motivation of stronger and weaker students as well as the interaction between the two groups of students—do they tend to learn together, do they motivate each other, or does the success of one group tend to frustrate the other?—distinguishing explicitly between the types of students is useful. Regardless of whether the payoffs are determined more by hard or by soft factors, the resulting total payoff is always a subjective evaluation, i.e., it is fundamentally not possible to make

Table 2.3 Payoffs for the constant-sum game "Penalty Shootout"

Shooter (A), Goalkeeper (B)	Left	Right
Left	(0, 1)	(1, 0)
Right	(1, 0)	(0, 1)

Shooter (A), Goalkeeper (B)	Left	Right
Left	(−1, 1)	(1, −1)
Right	(1, −1)	(−1, 1)

statements about the comparison of payoffs of two players. Because even with hard, measurable factors, such as time loss or the achieved grade, we ultimately do not know how the respective student evaluates these factors. For example, if 6 h of studying leads to achieving 60% of the maximum score, this can be seen as an excellent result by one student, while another sees it as a bitter disappointment. For this reason, we may only ever compare a player's payoffs with their own payoffs (*intra*personal utility comparison), but not the payoffs of different players (*inter*personal utility comparison). The same applies to the utility function known from microeconomics for evaluating different bundles of goods.

Constant-Sum Games and Zero-Sum Games. In contrast to the example of student motivation just presented, the determination of payoffs in social games or sports competitions is much easier, as we can make the plausible behavioral hypothesis here that each player exclusively aims to win the competition. The left matrix in Table 2.3 exemplarily shows a possible payoff structure in a penalty shootout.

The penalty shooter is player A, who chooses between the vertically listed strategies: They can either shoot into the left or into the right corner of the goal. The goalkeeper as player B chooses between the horizontally listed strategies and can accordingly throw themself to the left or right to deflect the ball.[4] The motivation of the two players in this simple, clearly structured game is self-explanatory: The shooter wants to score a goal and the goalkeeper wants to prevent this. The player who wins the duel receives a utility or a payoff of 1; the loser receives a payoff of 0. Whenever both choose the same corner of the goal (both left or both right), the goalkeeper wins the duel, and the shooter loses it. If both choose different corners, the shooter wins the duel, and the goalkeeper is the loser. If we look at each individual cell of this matrix more closely, we recognize that the sum of the payoffs of both players always amounts to 1. This is because there are only two mutually exclusive outcomes for each player in this confrontation, "victory" and "defeat": Whenever one player loses, the other wins and vice versa. Since the

[4] The directions "left" and "right" refer to the perspective of a third, uninvolved observer, such as that of a cameraman or a sports journalist. Since the shooter and the goalkeeper face each other, they understand "left" and "right" from their respective individual perspectives, of course, in opposite ways.

sum of the payoffs in each cell is constant (in this example the constant sum is 1), such a game is also referred to as a *constant-sum game*. A special variant of a constant-sum game occurs when the payoff sum in each cell is 0. In this special case, we speak of a *zero-sum game*. The right matrix from Table 2.3 represents the penalty game as a zero-sum game. To turn the constant-sum game with a payoff sum of 1 into a zero-sum game, we simply reduced the payoff of the loser from 0 to -1. To describe the player motivations in face-off situations, zero-sum games are usually chosen, as a zero-sum game well describes the harsh "all or nothing" fate of the players: Either you win (positive payoff) or you lose (negative payoff), with the winner's gain exactly matching the loser's loss.

Ultimately, it does not matter whether the situation is modeled as a (more general) constant-sum game or as a zero-sum game: The absolute amount of the payoffs is almost never important for the solution of the game. What matters in the solution is rather the relation, i.e., the distance or the comparison, of certain payoffs. To solve and interpret a game, it is only important to know *that it is* a constant-sum game. Whether the sum of the cell payoffs is now exactly 0, 1, 42, or −372, however, is (almost always) irrelevant. The reason for this is obvious: You can turn *any* constant-sum game with any payoff sum into one with a different payoff sum by simply increasing or reducing each payoff from each player by a fixed value—after all, the utility function behind the payoffs allows any positive linear transformation (see Sect. 1.3). Since this change affects all payoffs equally, the *payoff structure* remains unchanged: As the relative attractiveness of the respective strategy for a player does not change, no player changes their strategy choice.

Preference Parameters. In addition to preferences for game outcomes, there are also several preference parameters that relate to other dimensions of the game (i.e., not the outcomes) but can certainly influence the decision of the players. An example of such a preference parameter would be the already known from Sect. 1.3 *Arrow–Pratt measure*, which captures the degree of risk aversion of a decision-maker through the curvature of the expected utility function. So, if we want to investigate the influence of the players' risk attitude on the outcome of a game, it makes sense to vary the risk aversion and to examine the relevant game for the assumption of risk-seeking, risk-neutral, and risk-averse players. However, if we are not interested in this issue, it is appropriate to ignore the influence of risk in the analysis in line with the principle of sparse modeling. This is most easily achieved by assuming that the payoffs occurring in the game have already been evaluated by an expected utility function, which then are already *risk-adjusted payoffs*. Since we are primarily interested in strategic interaction and less in risk implications in the following, we make this assumption for all subsequent chapters. If we did not make this assumption, we would have to explicitly state the utility function (and thus also the risk preference) of each player for each game and take it into account when solving the games, which makes the analysis unnecessarily complex and complicated.

Other examples of preference parameters would be the optimism–pessimism parameter of the Hurwicz decision rule (Sect. 1.2), or the *discount factor*. A discount factor indicates how a decision-maker evaluates and compares payoffs that occur at different times. Unlike the risk attitude, which is captured by the Arrow–Pratt measure, the discount factor is thus a measure of a decision-maker's time preference. The discount factor will play an important role in Chap. 5, as the temporal dimension, i.e., the duration or repetition of a game, becomes relevant there.

Figure 2.2 represents the various dimensions that can influence the determination of payoffs. Following the principle of spare modeling, only those payoff dimensions should be explicitly represented that are relevant to the game in question and that accordingly make a significant contribution to explaining the game outcome.

Symmetry. Many games exhibit a *symmetry* in the payoff structure, either because the game is the same from the point of view of both players (e.g., rock paper scissors) or because it simplifies the analysis considerably (e.g., competition between companies). The latter becomes obvious when we look at how we can take advantage of this. We distinguish between mirror symmetry and point symmetry. Figure 2.3 graphically illustrates the properties of mirror and point symmetry using the example of a 2 × 2 matrix with generally held payoffs.

Point symmetry implies that when the players change their strategies (i.e., both switch to their alternative strategy), they also change their positions in the game

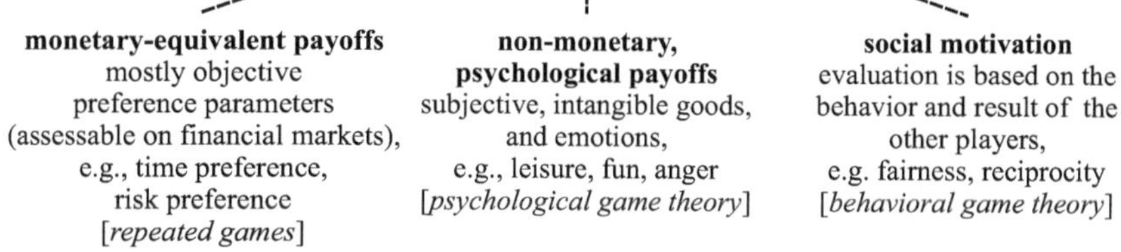

Fig. 2.2 Payoff dimensions

Fig. 2.3 Mirror and point symmetry of the payoff structure

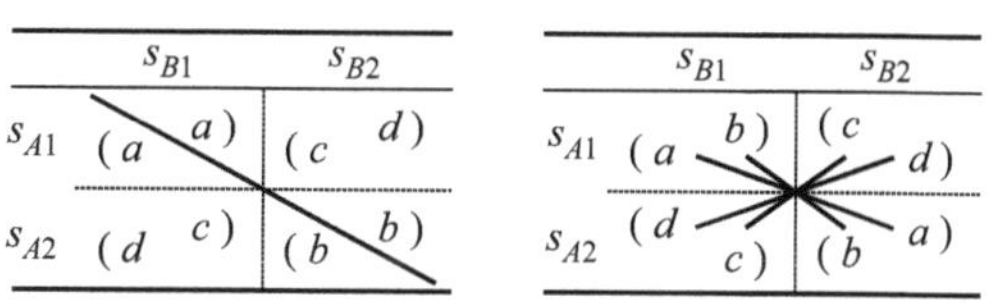

(both "swap" their payoffs, so to speak). Suppose both players choose their first strategy and thus achieve the payoff combination (a, b). If they both switched now, i.e., both chose their second strategy, this would lead to an exchange of payoffs, i.e., to the payoff combination (b, a). If the payoff structure is mirror symmetric, then both players have the same perspective on the game, i.e., they face the same payoffs when they swap roles (player A becomes player B and player B becomes player A). The property of mirror symmetry can greatly simplify the solution of a game, especially in multiplayer games, as it is sufficient to solve the game from the perspective of one (representative) player. The optimal strategy of this player is then automatically also the optimal strategy *of all* players, as the game looks the same from the perspective of each player. When we use the symmetry properties of a game in the following chapters, we always refer to mirror symmetry.

2.3 Game Structure and Rules

In addition to the players, the strategies, and the payoffs, the *rules of the game* also play an important role. The rules primarily include the *temporal structure* of a game, the *information structure*, as well as certain *behavioral and expectation assumptions*. The temporal structure of a game describes the number and order of moves as well as the number of game repetitions. Of fundamental importance is in this context, the already discussed distinction between a *simultaneous game*, in which the players make their strategy choice simultaneously, and a *sequential game*, in which one player makes their decision (first mover) and the other player (second mover) reacts to the previous decision.

The information structure determines whether the follower can observe the leader's moves and whether the follower knows the leader's strategies and payoffs. The observability of moves is referred to as *perfect* or *imperfect information*: If a player can observe the other player's strategy choice, the information is perfect, otherwise we call it imperfect information.

In the second category of information, which concerns strategy variants and payoffs, we distinguish between *complete* and *incomplete information*. Simplified, this is about the recognizability of the respective *player type* one is facing. A player type is characterized by their strategies and payoff structure. If player A is informed about the type of player B, it is complete information. If, on the other hand, A does not know the "type" of B, it is incomplete information. An example of the latter case would be a teacher who does not know whether a particular student is a high- or low-performing "type" of student, or a customer who wants to book an individual trip but cannot assess the quality of the travel provider.

If all information assumptions (both regarding observability and type distinction) apply equally to all players, then a game has *symmetric information*. If, on the other hand, the players each have a different level of information, there is *asymmetric information*.

There are important connections between the temporal structure and the information structure. A simultaneous game is a game with imperfect information and symmetric information status. The information is imperfect because none of the players can observe the moves of the opponent(s). Since this information restriction applies equally to all players, information is symmetric. The term "simultaneous game" is therefore somewhat misleading, as it is not about the players deciding simultaneously, but rather that each player *is not informed* about the other player's decision. Acting precisely at the same time, as in the well-known tactical game "rock paper scissors", is just one possible variant of how imperfect information can arise. The discussed "Penalty Shootout" can be considered as an example of a "nearly simultaneous" move order. In a real penalty shootout, the goalkeeper waits as long as possible in the hope of being able to assess the goal shooter's behavior. However, the final decision is made almost simultaneously, as the speed of the ball is too fast to wait for the shot.[5] Many strategic situations, where imperfect information is present and therefore represented as a simultaneous game, occur in fact sequentially. For example, in a television contest, where two candidates must complete a certain task one after the other, fairness can only be maintained if the candidate who goes second cannot see how the first candidate has acted. Everything possible is therefore done to prevent observation, such as blindfolding with headphones or a shielded waiting room. This contest interaction would therefore be—despite the de facto sequential game sequence—a simultaneous game, as it is ensured that the second candidate cannot observe what the first one is doing.

Since a simultaneous game only depends on the property of imperfect information, such a game can be represented both as a matrix and as a game tree. If the game tree representation is chosen, a dashed connection line between the two decision nodes of the second mover indicates that both decision points are in an *information set*. This approach was already used in Chap. 1 (see Fig. 1.1). There, coincidence was depicted in this way: The decision-maker had imperfect information about the "decision" or drawing of nature. In Fig. 2.4, "Future Energy" is used as an example to illustrate how a simultaneous game can be represented both as a matrix and as a game tree. Both representations are completely equivalent to each other. No matter which representation is chosen, it is always the same game. While we have already learned that the game tree representation is more advantageous for analyzing sequential games, the matrix form is better for simultaneous games or, respectively, games with imperfect information.

Company (A), Government (B)	s_{B1}	s_{B2}
s_{A1}	(2, 1)	(0, 0)
s_{A2}	(−1, 0)	(1, 2)

[5] The flight time of the ball in a penalty kick is approximately 0.32 s, which corresponds approximately to the time the goalkeeper would need just to determine in which direction the ball is flying. Therefore, the goalkeeper must decide on a corner at the same time as the shooter and hope that it is the same one.

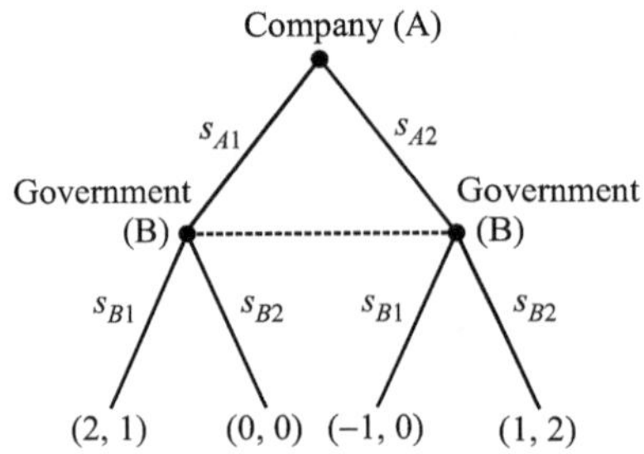

Fig. 2.4 Simultaneous game "Future Energy" in matrix and game tree representation

For sequential games, there is a corresponding relationship between game form and information structure as well.[6] Here, the second mover can observe the strategy choice of the first player, meaning he has perfect information about it. The first mover, however, does not yet know at the time of her decision, how the second player will react to her move (she can make assumptions, but this does not give her any reliable information about how the second mover will decide). The first mover thus has imperfect information. Since the information situation is different for both players, a sequential game inevitably has asymmetrical information.

Behavioral and Expectation Assumptions: Rationality and Common Knowledge. Our information assumptions so far determine for a game whether each player is informed about the relevant characteristics of the other player and about their previously chosen moves. To optimally assess the behavior of the other player and to be able to adjust accordingly, each player must also have an idea of how predictable the behavior of the other players is and to what extent the player can assume that these game-relevant details are generally known. Game theory makes two relatively pragmatic assumptions that are essential for solving a game.

The prerequisite about the players' knowledge or awareness is the *common-knowledge* assumption. Common knowledge requires that the players' level of knowledge is known to each other. It is not enough to determine what a player knows about the other, but they all must also know about each other that they know it. This assumption therefore goes one information level higher and requires symmetrical information about the information level itself.

The following example illustrates the problems that can arise when the common-knowledge assumption is not met: Two Roman troop units are in two opposite forests and want to attack the enemy together the next day. However, the attack will only be successful if they strike in a coordinated manner. As the enemy is in the middle, they have no way of communicating directly with each other, but they can send a messenger to the other unit to convey the message of exactly when the attack should take place. However, each unit will only begin preparations for the attack when it has confirmation that the messenger has reached the other unit

[6] In Sect. 2.1 we have already shown that a sequential game can be represented both as a game tree and as a matrix, with the game tree proving to be the more sensible variant.

and that the message has been received (and not intercepted by the enemy). Specifically, this means that troop A sends a messenger to troop B with the instructions. When this messenger arrives at B, it is sent back to A to say that the message has been successfully delivered. However, A will then send it to B to say that B knows that A is aware of the plans. B will then send the messenger back to A to say that they are informed that A knows that they (B) know that A is informed that they (B) have received the information.

The problem for the two troops is that they can only send messages to each other one after the other. This means that the confirmation requested by both can never be received by both at the same time, i.e., the information about the planned attack can never become common knowledge. One of the troops will always be unsure whether the other has received the last message. Since the other knows this, they will never be able to launch an attack. We can conclude from this that the Romans must have always been very lucky to launch successful attacks.

The aspect of predictability involves the assumption of *(strict) rationality* of the players; this assumption was already more or less implicitly assumed in Chap. 1. It implies that each player makes a consistent (i.e., contradiction-free) goal-oriented utility-maximizing decision. "Goal-oriented" means that the decision-makers focus exclusively on their payoffs (including preference parameters) and "consistent" assumes that it is possible for the decision-makers to put the possible game outcomes into a complete and transitive order (see Sect. 1.1). Rational decision-makers always try to maximize their payoff under the given circumstances (information status, available alternatives, opponent's decision). Of course, the assumption of (strict) rationality is an extreme assumption, making us view the behavior of each player as if they were a perfectly calculating computer. Critics of the rationality assumption therefore advocate for behavioral descriptions that consider human limits and weaknesses more strongly.

In the subfield of behavioral or experimental game theory, the behavior postulated in the model is tested using laboratory experiments. These experiments help to distinguish under which conditions the assumption of rationality can be maintained without problems and under which it becomes critical, as the theoretical and experimental game results diverge significantly. Figure 2.5 shows a game where it is quite likely that the theoretical prediction will differ from the actual observed game behavior in the experiment. In this sequential three-person game, which is a shortened version of the centipede game—the name refers to the appearance of the game tree—three players can sequentially choose between the strategies "continue" and "end". As soon as one of the players chooses "end", the game is over, and the corresponding payoffs are realized. If all three players choose "continue", the payoff vector $(1 + \varepsilon, 1 + \varepsilon, 1 + \varepsilon)$ is realized. The variable ε (epsilon), as usual in mathematics, stands for an extremely small value; if measured in terms of money, this would be the smallest possible value (about one cent).

If all players are *strictly rational* and rationality is *common knowledge*, then each player should choose "continue", as this would improve everyone's situation by the small amount ε compared to unilateral deviation—after all, it is 1 cent more! However, if we allow for the possibility that players do not behave strictly

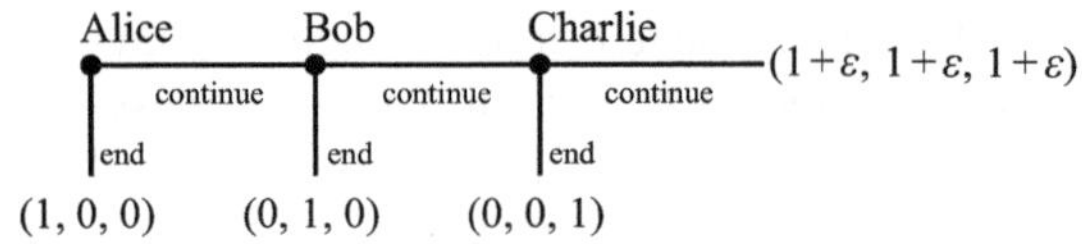

Fig. 2.5 Small centipede game with three players

rationally, then there are at least three reasons why it is very plausible in this game that players might also choose "end". Let us take player A(lice) as an example.

- The first possibility would be that ε is too small in Alice's perception to be decisive for her decision. Experimental economics refers to this as *salience*, i.e., the payoff is not conspicuous or prominent enough to influence the decision. If this were the case, then Alice would view the payoffs 1 and $1 + \varepsilon$ as approximately equal and would then have no reason to wait for the moves of the other players. This variant would be an example of *boundedly rational behavior* by Alice.
- The second possibility would be that Alice is strictly rational herself but doubts the strict rationality of the other players. Alice would then have *incomplete information* about the rationality of the other players. If one of the subsequent players had bounded rationality and would therefore choose "end", this would be disadvantageous for Alice, regardless of which of the subsequent players deviated. So, if she distrusts the rationality of the other players in this sense, it would be rational for her to choose "end" just to be safe.
- The third possibility would be that Alice is strictly rational herself and does not doubt that the others are strictly rational. But if strict rationality is *not common knowledge*, she cannot be sure whether the next player, Bob, (unjustifiably) doubts Carol's rationality and then chooses "end" just to be safe. In this case, too, Alice's rational decision would be to choose "end" just to be safe.

2.4 Equilibrium as a Solution to a Game

The three components of players, strategies, and payoffs in conjunction with the game rules and assumptions about information and rationality can uniquely describe any game. The correct and complete description of a game is the prerequisite for the actual goal of game theory: the solution of a game. Therefore, the focus in the following chapters is on the presentation of solution concepts for different types of games. The solution of a game is usually synonymous with the systematic search for *equilibria of a game*. We will deal intensively with this in the next chapters, but in this section, we would like to clarify what exactly is meant by an equilibrium in game theory, especially the *Nash equilibrium*.

If we assume that players follow certain rules of behavior, then an equilibrium is a combination of strategies for which these rules of behavior are compatible with each other. Normally, these rules of behavior are derived from an optimization approach, which, however, is not necessarily required. Let us consider as an example a couple that follows the following procedures in their leisure activities on weekends out of a certain daily routine: Alice joins her boyfriend Bob on Saturday evening if he goes to the cinema, otherwise she spends the evening alone at home. Alice then spends Sunday with Bob only if she did not spend Saturday evening with him (regardless of the activity); otherwise, she goes alone to the cinema or to football. Bob goes to the cinema or to football on Saturday evening and Sunday evening, preferring the activity on Sunday that he did not choose on Saturday. This game has two equilibria, i.e., two mutually compatible constellations of the rules of behavior: In the first equilibrium, Alice stays at home on Saturday, while Bob goes to football; on Sunday, Alice goes to the cinema with Bob. In the second equilibrium, Alice goes to the cinema with Bob on Saturday; on Sunday, Alice goes (alone) to the cinema again, while Bob goes to football. There are only these two constellations for which the plans for the evening are compatible with each other.

In this example, no optimization approach is necessary, as the evaluation of all possible constellations for the evening is already "built into" the rules of behavior (for example, desire for sociability in her, desire for variety in him). However, since we will always explicitly state the payoffs for all possible strategy combinations in the future, the determination of compatible strategy profiles is done by *optimization.*

It is obvious to maximize one's own payoff for a given strategy of the other player(s). Mutually compatible strategy profiles that follow this approach represent a Nash equilibrium. Formally expressed, in a Nash equilibrium, the payoff of each player i is maximal given the strategy of all other players $(-i)$[7]:

$$u_i\left(s_i^*, s_{-i}^*\right) \geq u_i\left(s_i, s_{-i}^*\right) \quad \forall s_i \in S_i, \forall i.$$

The equation is to be read as follows: The utility of player i utility, u_i, if i chooses the equilibrium strategy s_i^* and all other players also choose their equilibrium strategies s_{-i}^* as well, i.e., $u_i\left(s_i^*, s_{-i}^*\right)$ is at least as large as if i had chosen any other strategy s_i while all other players had chosen their equilibrium strategies s_{-i}^*.

The Nash equilibrium is the most firmly established solution concept in game theory. Let us consider the game "Future Energy" in Table 2.4, in which a company can choose between an investment in conventional or renewable energy sources and the government can decide between promoting these two energy sources. Since we want to consider the game here as a simultaneous game, it is again represented in matrix form. How can a Nash equilibrium be found in this game? Let us first consider the options of player A (the company), who can choose between the vertically plotted strategies s_{A1} and s_{A2}. Assuming that player B (the government) promotes conventional energy (s_{B1}), it is better for player A to also invest in conventional energy (s_{A1}): If we are in the first column of the matrix, the company receives a payoff of 2 (upper row) when choosing s_{A1} and only a payoff of -1

[7] The mathematical symbol "∀" means "for all".

Table 2.4 Nash equilibria in the simultaneous game "Future Energy"

Company (A), Government (B)	conventional (s_{B1})	renewable (s_{B2})
conventional (s_{A1})	(2, 1) ←	(0, 0) ↓
renewable (s_{A2})	↑ (−1, 0) →	(1, 2)

(lower row) when choosing s_{A2}. In relation to the first column, the investment in conventional energy sources is therefore better than the investment in renewable energy sources, which is expressed in the figure by the arrow pointing upwards. The optimal payoff of B for this constellation is underlined for identification. The same consideration of the company now also applies to the second column, i.e., in the case that the government promotes renewable energy (s_{B2}). Given that the government promotes renewable energy sources, it is now optimal for the company to also invest in renewable energies, as the payoff is larger here ($1 > 0$). The small downward arrow indicates this and the larger payoff, in this case 1, is again underlined. With these two steps, i.e., the column-wise search for the best option, we have now found the best response of the company for both possible decisions of the government. It is very important to recognize that we have not sought a generally optimal strategy for the company, but always only the best response of the company for a given decision of the government. We can also formally represent the optimal reaction of the company to the given strategy choice of the government as in expression (2.1):

$$r_A^* = \left\{ r_A^*(s_{B1}) = s_{A1}, r_A^*(s_{B2}) = s_{A2} \right\}, \tag{2.1}$$

where r_A^* is referred to as the reaction correspondence or best response correspondence of player A. The two equations are to be read as follows: The best reaction of the first player (r_A^*) to the first strategy of Player B (s_{B1}) is to choose the first strategy (s_{A1}). In contrast, Player A's best response to the second strategy of Player B (s_{B2}) is to choose the second strategy (s_{A2}).

Now we consider the government's decision problem in the same way. While we proceeded column by column for the company, the examination for the government is done row by row. If Player A (company) invests in conventional energy (top row), it is optimal for the government to promote conventional energy sources. This is made clear in the top row by the leftward pointing, dashed arrow, which again points to the higher payoff (in this case 1) ($1 > 0$). However, if the company invests in renewable energy sources (bottom row), then it is optimal for the government to also promote renewable energy sources (s_{B2}). The dashed arrow points to the right in this case and the higher payoff 2 is underlined. Thus, we have also found the best response for the government to a given strategy of the company. Expression (2.2) represents the government's reaction function (r_B^*) and is to be

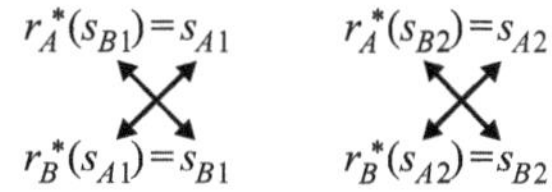

Fig. 2.6 Mutually consistent reaction correspondences in the Nash equilibrium

read analogously to expression (2.1):

$$r_B^* = \{r_B^*(s_{A1}) = s_{B1}, r_B^*(s_{A2}) = s_{B2}\}. \tag{2.2}$$

A Nash equilibrium exists where the two best response correspondences match. In Fig. 2.6, we have depicted the reaction correspondences of both players for better illustration.

Let us first consider the reaction correspondences shown on the left. The arrow cross indicates that the players' reaction correspondences are consistent with each other: Given that the government promotes conventional energy, s_{B1}, it is optimal for the company to also invest in conventional energy (top line), s_{A1}. The government's consideration follows directly: Given that the company invests in conventional energy, s_{A1}, it is best for the government to also promote conventional energy, s_{B1}. A Nash equilibrium thus describes a profile of mutually best responses. Once a Nash equilibrium is reached, none of the players has an incentive to deviate from it.

The strategy profile $(s_{A1}, s_{B1})^*$ therefore represents a Nash equilibrium of the game. We always mark a Nash equilibrium with a star symbol ($*$). We can also see from the reaction correspondence shown in Fig. 2.6 that there is another strategy profile $(s_{A2}, s_{B2})^*$ that is mutually optimal—both opt for renewable energy.[8] In the matrix from Table 2.4, the Nash equilibria are easy to find using both of our marking methods (arrows and underlining): The strategies describing the cell of the matrix represent a Nash equilibrium when the arrows *of both players* point into this cell. The arrow direction somewhat indicates the direction of deviation: Neither of the two players would deviate from this cell, provided the other player maintains the strategy. The Nash equilibrium can also be found via the underlining approach: Since we have always marked the optimal payoff given the other player's strategy by underlining, the strategies describing the cell of the matrix in which both payoffs are underlined must be a Nash equilibrium. Arrows and underlining mark the same fact, so it does not really matter which of the two markings we use as a tool.

Note and Tip If you are still unsure about the assignment of columns and rows to the players and perhaps do not know exactly whether you must select the rows for a given column or

[8] The two equilibria determined by reaction correspondences in this game are not the only Nash equilibria, but they are the only ones that can be directly read from the matrix. The latter are referred to as Nash equilibria in *pure strategies*. In contrast, the concept of mixed strategy is explained in Chapter 3. It is possible that a matrix game also has (at most) one Nash equilibrium in mixed strategies, which is also the case here.

vice versa, we recommend the arrow marking. At each arrow, you can directly read which payoffs were compared and whether it is a row-wise or column-wise comparison. Once you have gained more confidence, you should switch to the underlining method. This marking is more economical and especially for larger matrices significantly clearer.

The Nash equilibrium is a very convincing solution concept, as each player tries to maximize their payoff while simultaneously considering the incentives of the opponent as a constraint of their decision. It is therefore fundamentally better suited as a basis for decisions than the decision rules discussed in Sect. 1.4, such as the maximin or maximax rule, which assume deterministic behavior of the opponents.

2.5 Comprehension Questions

1. A game is among others determined by …
   A The number of players
   B The degree of risk aversion
   C The strategies of the players
   D Number and type of Nash equilibria
2. Information in a …
   A Simultaneous game is always symmetric and includes imperfect information
   B Simultaneous game is asymmetric and always includes incomplete information
   C Sequential game is always symmetric and includes perfect information for the second mover
   D Sequential game is always asymmetric and includes imperfect information for the first mover
3. Which value of Z, which is added to each payoff in a constant-sum game with the payoff sum X, turns it into a zero-sum game?
   A $Z = X$
   B $Z = -X$
   C $Z = -0.5 \cdot X$
   D $Z = 2 \cdot X$
4. The preference parameters of a game include, for example, …
   A The discount factor of the players
   B The number of strategies of a player
   C The rationality of the players
   D The risk aversion of the players
5. In a Nash equilibrium …
   A Mutually best responses are present
   B Both players achieve the highest payoffs of the game
   C All players always choose the same strategies
   D None of the players has an incentive to deviate

1a,c; 2a,d; 3c; 4a,d; 5a,d

2.5.1 Problems

2.5.2 What is a Game?

Suppose you want to analyze a real-life problem that you encounter in practice using game theory. For this purpose, you would like to translate the interaction that is of interest to you into a game. What elements and properties of a game do you first need to define to describe the game correctly?

Solution in Sect. 2.7.1.

2.5.3 Description of Specific Games

Indicate for the following games whether they are simultaneous or sequential games, what the sets of players and strategies are, and whether the strategy space is discrete or continuous!

a. Speed trap (game between a motorist and the traffic police at an accident-prone point in a city).
b. Tic-tac-toe (also known as "noughts and crosses" or "Xs and Os").
c. Rock paper scissors.

Solution in Sect. 2.7.2.

2.5.4 Representation of a Game

Alice and Bob play the game of rock paper scissors once. Represent the game in both strategic form (matrix) and extensive form (game tree)!

Solution in Sect. 2.7.3.

2.5.5 Multiplayer Game: Game Structure, Information Status, and Common Knowledge Assumption

Consider the following centipede game:

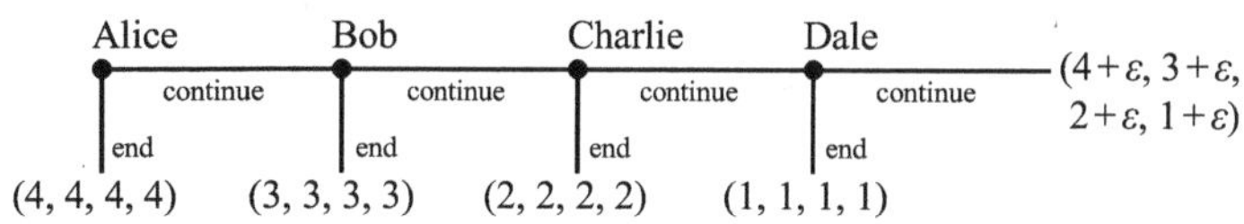

In this sequential game, four players choose one after the other between the strategies "continue" and "end". As soon as one of the players chooses "end", the game is over, and the corresponding payoffs are realized. Only if all players choose "continue", the payoff vector $(4+\varepsilon, 3+\varepsilon, 2+\varepsilon, 1+\varepsilon)$ is realized. The parameter ε (epsilon) stands for an extremely small payoff value. What game result do you expect if all four players behave strictly rationally and strictly selfishly, and this would also be known to all? Then consider whether Alice should decide to choose "end" if

1. Alice does not act strictly rationally.
2. Alice acts strictly rationally but must expect that (at least) one of the other players does not act strictly rationally.
3. Alice acts strictly rationally and expects all other players to also act rationally, but rationality is not common knowledge.
4. Alice does not act strictly selfishly but is strongly oriented towards fairness.
5. Alice acts strictly selfishly but must expect that (at least) one of the other players is strongly oriented towards fairness.
6. Alice acts strictly selfishly and expects all other players to also act strictly selfishly, but strict self-interest is not common knowledge.

Solution in Sect. 2.7.4.

2.5.6 Maximin Strategy and Nash Equilibrium

Consider the following simultaneous game:

Alice, Bob	s_{B1}	s_{B2}
s_{A1}	(3, 0)	(2, 1)
s_{A2}	(1, 2)	(0, 3)

Which two distinctive features does this game have regarding the payoff structure? Formally determine the equilibrium if both players decide according to the maximin rule! Also, determine the Nash equilibrium via reaction correspondences! Compare the results and explain them!

Solution in Sect. 2.7.5.

2.5.7 Payoffs and Equilibria in Simple Matrix Games (Pure Strategies)

Consider the simultaneous game "Future Energy" (Table 2.1) but adjust the payoff structure so that it reflects the following motivation of the company: The company would prefer to maintain its existing technology and ideally be supported by the government for this. Investing in renewable energy is the least attractive choice for the company, even if it receives government funding for it.

Represent the modified game in matrix form and solve it graphically using the arrow method, the underline method, and formally via the representation of best responses over the reaction correspondence!

Solution in Sect. 2.7.6.

2.6 Solutions

2.6.1 What is a Game?

Problem Text in Sect. 2.6.1

You should first ask yourself which agents or groups of agents are of strategic relevance (players), which strategies are available to them as options for action, and what interests they pursue (payoffs). Make sure you do not confuse strategies with possible outcomes (e.g., "make an offer of X" is a strategy, but "enforce interests" is not). Finally, it remains to clarify how the structure of the game is designed, which assumptions about information are plausible, and whether other framework conditions influence the actions of the players.

2.6.2 Description of Specific Games

Problem Text in Sect. 2.6.2

Part a)
Simultaneous game: Unless the driver is warned by oncoming vehicles, they have imperfect information about the police's action. Note: The actual process is sequential, as the police determine the location beforehand and position the camera. However, for the strategic interaction between the police and the driver, it is only decisive whether the driver is informed about this action (sequential game with perfect information of the driver) or not (simultaneous game with imperfect information on the side of the driver).

Two-person game: Representative driver versus traffic police as a representative of government authority.

Driver's strategy: Discrete choice between "complying with the speed limit" and "exceeding the speed limit". Note: The choice of speed is indeed a continuous strategy choice, but the strategically relevant question is not how fast exactly the driver is driving, but whether the permissible maximum speed is exceeded or not.

Police strategy: Discrete choice between "setting up a speed trap" and "not setting up a speed trap".

Part b)

In this game, two players place a mark (usually cross, X, and circle, O) on a square, 3 × 3 field game board in turn. The aim of the game is to be the first player to place three signs in a row, a column, or a diagonal.

Sequential game: The start of the game is drawn by lot, then the players place their markers one after the other for all to observe. The game has a maximum of nine rounds until all fields are occupied.

Two-person game: It is always played by two players.

Strategies: The game is (mirror) symmetrical, so both players have the same strategy space, which comprises a maximum of nine squares, but is period dependent: The strategy of a player i at time t is to place a marker in one of the fields still free $(= 10 - t)$ at this time t. In the first round $(t = 1)$, the starting player has $9\ (= 10 - 1)$ possibilities to mark, the other player 8 in the second round $(t = 2)$, and so on. These are discrete strategies.

Part c)

This game is a "duel with hand movements" between two players, in which the three objects, scissors (spread middle and index finger), rock (clenched fist), or paper (flat hand), are symbolized with the hand. The ranking or value of the objects is intransitive: Scissors cut paper (scissors win), paper wraps rock (paper wins), and rock blunts scissors (rock wins).

Simultaneous game: The hand movements occur simultaneously; both players therefore have imperfect information about the opponent's strategy choice at the time of their decision.

Two-person game: It is usually played in pairs.

Strategies: The game is (mirror) symmetrical, so both players have the same strategy space. They have the three discrete strategies: scissors, rock, and paper.

2.6.3 Representation of a Game

Problem Text in Sect 2.6.3

Rock paper scissors in strategic form (matrix):

Alice, Bob	Rock	Paper	Scissors
Rock	(0, 0)	(−1, 1)	(1, −1)
Paper	(1, −1)	(0, 0)	(−1, 1)
Scissors	(−1, 1)	(1, −1)	(0, 0)

Rock paper scissors in extensive form (game tree):

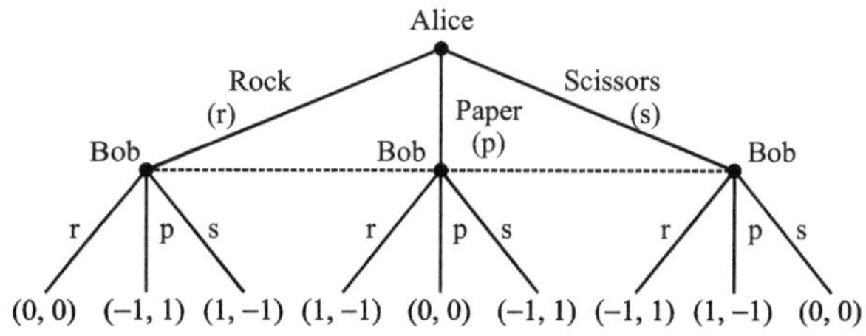

2.6.4 Game with Multiple Players: Game Structure, Information Status, and Common Knowledge Assumption

Problem Text in Sect. 2.6.4

If all players are strictly rational and rationality is common knowledge, each player should choose "continue", as this will make everyone better off by the small amount ε than by unilateral deviation.

1. Alice perceives the payoffs 4 and $4 + \varepsilon$ as approximately equal and thus has no reason to wait for the moves of the other players. This effect would be most likely to be expected for Alice, as ε in the payoff sum $4 + \varepsilon$ weighs proportionally less than in $3 + \varepsilon$, $2 + \varepsilon$, or $1 + \varepsilon$.
2. Alice is strictly rational but doubts the strict rationality of the other players. She would then have incomplete information about the rationality of the other players. If one of the subsequent players were boundedly rational and therefore chose "end", it would be the rational response for Alice to preempt and choose "end" for safety.
3. Alice is strictly rational and does not doubt that all others are also strictly rational. However, if strict rationality is not common knowledge, Alice cannot be sure whether, for example, Bob (unjustifiably) doubts the rationality of Charlie and chooses "end" for safety. The rational decision for Alice would then be to choose "end" for safety.
4. This variant has nothing to do with rationality but with the motivation, i.e., the utility or payoff, of the players. In this example, a fairness norm is assumed, where the players value that the resulting payoffs can be considered fair according to certain criteria. In the simplest case, fairness consists of a simple equal distribution, i.e., equal payoffs for all players. If Alice were a "fair" player in this sense, she would choose "end". This way, Alice could provide all other players with the high payoff of 4 and would only forego the very small amount ε.
5. Alice is strictly self-interested, but suspects that one of the other players could be a fair "type". In this case, she would also have incomplete information about at least one of the other players, but the incomplete information would relate to their motivation and not their rationality. Why would Alice have to fear that a

fair-minded Bob choose "end"? Bob could achieve that everyone gets the same payoff, with only Alice losing significantly. However, Bob might well accept Alice's loss: Players concerned with fairness also tend to punish other players if they do not behave fairly. If Alice chooses "continue", this would be an "unfair" move, as she misses the opportunity to give the others the high payoff of 4, just to secure the minimal additional amount ε.

6. Alice is informed about the self-interested motivation of the other players, but there is no common knowledge about this motivation. Self-interested Alice cannot rule out that self-interested Bob is uncertain about the motivation of Charlie (which Alice might know) and therefore chooses "end" for safety. To prevent this problem, strictly rational and self-interested Alice should choose "end" at the beginning.

2.6.5 Maximin Strategy and Nash Equilibrium

Problem Text in Sect. 2.6.5

In the payoff structure, the following two distinctive features are recognized: It is a constant-sum game, and the game is point symmetric.

When applying the maximin strategy, the result for Alice is: $\max\{\min\{3;\underline{2}\},\min\{1;\underline{0}\}\} = 2$. She will therefore choose s_{A1}: If she plays s_{A1}, the payoffs 3 and 2 are achieved; the worst that could happen to her would therefore be the payoff 2 (underlined). If she plays s_{A2} payoffs 1 and 0 result, with 0 being a pessimistic expectation (underlined). Alice should therefore play s_{A1}.

When applying the maximin strategy, the result for Bob is: $\max\{\min\{\underline{0};2\},\min\{\underline{1};3\}\} = 1$. He will therefore choose s_{B2}.

Note: You must determine the optimal strategy of Bob and cannot take advantage of the symmetry properties of the game because the game is point symmetric. Only if it were mirror symmetric, you can transfer the results of Alice to Bob.

The equilibrium in maximin strategies is thus in the strategy combination (s_{A1}, s_{B2}) with the payoff combination (2,1).

Alice's reaction correspondence is $r_A^* = \{r_A^*(s_{B1}) = s_{A1}, r_A^*(s_{B2}) = s_{A1}\}$. Explanation: The best response of Alice (r_A^*) to s_{B1} is s_{A1} ($3 > 1$). Also, with s_{B2}, s_{A1} remains her best response ($2 > 0$). Alice will therefore always choose s_{A1} (see Sect. 3.1: dominant strategy).

Accordingly, Bob's reaction correspondence is $r_B^* = \{r_B^*(s_{A1}) = s_{B2}, r_B^*(s_{A2}) = s_{B2}\}$. He also will always choose only one strategy: for both decisions of Alice, s_{B2} is his best response. Hence, this is again a dominant strategy.

The Nash equilibrium is therefore $(s_{A1}, s_{B2})^*$ with the payoff vector (2, 1). As we will see in Sect. 3.1 this is a Nash equilibrium in dominant strategies.

We thus see that the equilibrium in maximin strategies and the Nash equilibrium coincide. This is no surprise for this game because the following general rule applies: In a constant-sum game, an equilibrium in maximin strategies is always also a Nash equilibrium. This can be easily explained: If a player determines the maximin strategy, the highest payoff from all lowest payoffs is chosen. However, this player's lowest payoffs in a constant-sum game are the highest payoffs of the opponent. This means that the maximin strategy of a player in a constant-sum game always includes the optimal response of the opponent. An equilibrium in maximin strategies is thus a profile of mutually best responses, so by definition a Nash equilibrium.

2.6.6 Payoffs and Equilibria in Simple Matrix Games (Pure Strategies)

Problem Text in Sect. 2.6.6

Modified simultaneous game "Future Energy".

Company (A), Government (B)	conventional (s_{B1})	renewable (s_{B2})
conventional (s_{A1})	(2, 1)	(1, 0)
renewable (s_{A2})	(−1, 0)	(0, 2)

The changed motivation of the company can be most easily achieved by swapping the payoffs in the second column (swap the 0 for the 1).

By underlining and using the arrow method, the strategy combination $(s_{A1}, s_{B1})^*$ emerges as the unique Nash equilibrium of the game. Formally, the equilibrium can be determined as follows using the reaction correspondence:

The company's optimal strategy profile is: $r_A^* = \{r_A^*(s_{B1}) = s_{A1}, r_A^*(s_{B2}) = s_{A1}\}$. The company's best response strategy s_{B1} is to choose the first strategy (s_{A1}). The strategy s_{A1} is also the best response in the modified game to s_{B2}. Therefore, player A always chooses the first strategy, s_{B2}, regardless of the government (player B).

The government's optimal strategy profile is: $r_B^* = \{r_B^*(s_{A1}) = s_{B1}, r_B^*(s_{A2}) = s_{B2}\}$. Since the payoffs of the government were not affected by the modification of the game, the reaction correspondences remain unchanged: The best response to s_{A1} is the first strategy of player B and the best response to s_{A2} is the second strategy of player B.

If one compares the reaction correspondences of the players with each other, it turns out that now only the strategy combination $(s_{A1}, s_{B1})^*$ consists of mutually best responses and thus represents a Nash equilibrium.

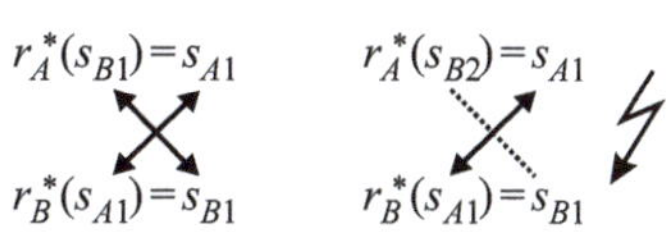

Literature Recommendations

Dixit A, Skeath S (2014) Games of strategy 4th ed. New York. (Chaps. 1–3).
Dutta PK (1999) Strategy and games, theory and practice. Cambridge.
Rieck C (2007) Game theory: An introduction. Eschborn.

Simultaneous Games

3

In Chap. 2, the central elements of a game were described, and the Nash equilibrium was already introduced as the most important solution concept of game theory. In this chapter, we turn to the systematic solution of games and first consider simultaneous games. Since simultaneous games are presented in strategic form, i.e., as game matrix, this chapter could also have been called "Matrix Games". However, the focus here is not on finding Nash equilibria in matrix games, but rather the systematic approach as well as the recognition and assessment of the underlying conflict structure of the games. The first part of this chapter deals with the identification of dominant or dominated strategies, which should always stand at the beginning of the equilibrium determination. Subsequently, we want to deal with the Nash equilibrium in more detail and discuss under what circumstances such an equilibrium exists. Especially in games with discrete strategies, this is not always guaranteed, which is why we will discuss the concept of mixed strategies and show how a Nash equilibrium is determined there. Finally, we will deal intensively with the conflict structure in matrix games, where Nash equilibria should be evaluated regarding efficiency and stability.

3.1 Dominance

When determining Nash equilibria in matrix games, it is appropriate to start with *dominance considerations*. Dominance considerations include the determination of dominant and dominated strategies, with a general distinction being made between strict and weak dominance.

Strictly dominant strategies. If a player i has a strictly dominant strategy $s_i{}'$, then a strictly higher payoff/utility is achieved with this strategy for any strategy of the

F. Bartholomae and M. Wiens, *Game Theory and Applications*, Classroom Companion: Economics, https://doi.org/10.1007/978-3-658-44608-6_3

Table 3.1 Equilibrium in dominant strategies: symmetric cases

Alice, Bob	Left	Right
Top	(4, 4)*	(2, 2)
Bottom	(2, 2)	(0, 0)

Alice, Bob	Left	Right
Top	(4, 4)*	(2, 2)
Below	(2, 2)	(2, 2)

opponent s_{-i}, $u_i(s_i\prime, s_{-i})$, compared to any other strategy $s_i^{''}$, $u_i\left(s_i^{''}, s_{-i}\right)$:

$$u_i(s_i\prime, s_{-i}) > u_i\left(s_i^{''}, s_{-i}\right) \ \forall s_{-i} \in S_{-i} \tag{3.1}$$

The choice of a strictly dominant strategy is therefore by definition always optimal. Hence, if such a strategy is available a rational player will certainly choose it. This means, if a player has a dominant strategy, they will not worry about the behavior of the other players—it does not matter whether a deterministic, rational, or completely arbitrary behavior of the opponents is assumed because the other player's strategy is simply irrelevant.

The presence of a strictly dominant strategy in a two-person game greatly simplifies the solution of the game, as there is no strategic uncertainty in such a game. In the simplest case, the second player also has a strictly dominant strategy, as shown in the left matrix in Table 3.1.

In this game, the strategy "Top" is strictly dominant for Alice, as she always receives a genuinely higher payoff with this strategy than when choosing her alternative strategy "Bottom" (the relevant payoff comparisons are 4 > 2 in the first column and 2 > 0 in the second column). If we again mark the higher payoffs by underlining, we recognize the strictly dominant strategy of Alice by the fact that in the first row *both payoffs* of Alice are underlined. If we use the arrow method, the strictly dominant strategy of Alice can be identified by both vertical arrows pointing "up".

Since the game is symmetric, Bob faces the same payoffs as Alice, which means that the first strategy "Left" is also a strictly dominant strategy for him. We can therefore assume that Bob will certainly choose the strategy "Left". If both players thus play their strictly dominant strategy, the strategy combination (Top, Left) results as the solution of the game. This strategy combination represents a Nash equilibrium: by definition, in a Nash equilibrium, each player chooses the best response to a *given* strategy of the other player. If a player now has a strictly dominant strategy, then this strategy represents the optimal strategy for *every* strategy of the opponent (for a player with a strictly dominant strategy, it does not matter how the other player decides). A Nash equilibrium that is based solely on dominant strategies of the players is referred to as an "equilibrium in dominant strategies".

We see from the underlining of the payoffs that the strategy combination (Top, Left) is also the only Nash equilibrium of the game, i.e., this Nash equilibrium is *unique*. This is inevitably the case with strictly dominant strategies: if both players have a strictly dominant strategy, then they will play this strategy, regardless of how the other player decides. Under no circumstances they will change their strategy, which can only lead to a unique game result. This is a very convincing solution concept, as there is no strategic uncertainty. Unfortunately, this case is very rare.

The desirable property of a unique game result is already no longer guaranteed if the relevant strategies are only *weakly dominant*. If player i has a *weakly dominant strategy* $s_i\prime$, then this strategy achieves for every strategy of the opponent s_{-i} a higher or equal payoff compared to the choice of any other strategys_i''. However, while this strategy can also be as good as any other strategy, in at least one case a strictly higher payoff must be present. Formally expressed, weakly dominant strategies are thus defined as

$$u_i(s_i\prime, s_{-i}) \geq u_i\left(s_i'', s_{-i}\right) \forall s_{-i} \in S_{-i} \text{ and}$$

$$u_i(s_i\prime, s_{-i}\prime) > u_i\left(s_i'', s_{-i}\prime\right) \text{ for at least one } s_{-i}\prime \in S_{-i} \tag{3.2}$$

In contrast to the case of a strictly dominant strategy, nearly all payoffs of a weakly dominant strategy can be identical to the payoffs of the next best alternative, as long as there is at least one payoff that is strictly greater. The right matrix of Table 3.1 illustrates this: Again, the first strategy of each player is a dominant strategy, but it is only a *weakly dominant strategy*. To highlight the difference from the strictly dominant strategies discussed earlier, equality of two payoffs in the matrix is marked with a double underline.[1] The combination of dominant strategies (Up, Left) again leads to an *equilibrium in dominant strategies*, which has the payoff combination (4, 4). In contrast to the case of strictly dominant strategies, another Nash equilibrium can now be identified in the right matrix of Table 3.1, which results from the strategy combination (Down, Right) with the payoff combination (2, 2). Both payoffs here only have a double underline, which means that both players here are merely *indifferent* between remaining in this Nash equilibrium and deviating to the (next best) alternative. Such an equilibrium is referred to as a *weak Nash equilibrium* because it is, so to speak, not "strong enough" to clearly prevent deviations.

An equilibrium in dominant strategies is therefore unique if and only if both players have a strictly dominant strategy. If, on the other hand, there is just one weakly dominant strategy involved, the uniqueness is no longer guaranteed and there can be additional Nash equilibria. The same applies to the mixed variant, where one of the players has a strictly dominant strategy, but the other only has a weakly dominant strategy. Table 3.2 presents two examples of this mixed variant.

In the left matrix, the equilibrium in dominant strategies is found in the strategy combination (Up, Left) with the payoff combination (4, 4). In the right matrix,

[1] For simplification purposes, we will later switch back to the solid underline; in this section, however, the double underline should serve for better visual distinguishability.

Table 3.2 Equilibrium in dominant strategies: Mixed cases

Alice, Bob	Left	Right
Top	(4, 4)*	(2, 2)
Bottom	(2, 2)	(0, 2)

Alice, Bob	Left	Right
Top	(4, 4)*	(2, 4)*
Below	(2, 0)	(0, 2)

"Right" is the weakly dominant strategy of Bob, so the equilibrium in dominant strategies represents the strategy combination (Up, Right) with the payoffs (2, 4). In the left matrix, the equilibrium in dominant strategies is unique, in the right one it is not: The strategy combination (Up, Left) also leads to a Nash equilibrium with the payoffs (4, 4). However, this second Nash equilibrium is not *an equilibrium in dominant strategies* (Bob does not play a dominant strategy) and not *a weak Nash equilibrium* (only Bob is indifferent as opposed to Alice).

The previously discussed rules for the uniqueness of Nash equilibria can basically also be applied to the cases where of N players $N-1$ players have a dominant strategy and only one player does not have a dominant strategy available. In 2×2 matrix games, this would mean that only one of the two players has a dominant strategy. If this player now chooses the dominant strategy, the other player will optimally adapt to this strategy choice, i.e., choose the *best response* to it. Table 3.3 presents two examples of this situation.

In the left matrix, Alice has the strictly dominant strategy "Up". Bob responds optimally to this with the answer "Left". The strategy combination (Up, Left) is thus a Nash equilibrium. Since only Alice has a dominant strategy available and this is also strictly dominant, the optimal reaction to this always leads to a unique Nash equilibrium: If Alice has a strictly dominant strategy, she will under no circumstances play anything other than "Up". A second Nash equilibrium can only occur if Bob does not have a clear answer to the strategy choice "Up", i.e., he is indifferent between "Left" and "Right". If Bob were indifferent between "Left" and "Right", then Bob would automatically have a weakly dominant strategy. However, we excluded this by assumption (we would have one of the situations from

Table 3.3 Dominant strategy and best response

Alice, Bob	Left	Right
Top	(4, 2)*	(2, 0)
Bottom	(2, 0)	(0, 2)

Alice, Bob	Left	Right
Top	(4, 2)*	(2, 0)
Below	(2, 0)	(2, 2)*

Table 3.4 Dominant strategy and best response

Alice, Bob	Left	Center	Right
Top	(3, 2)	(1, 1)	(2, 3)
Bottom	(4, 0)	(0, 2)	(1, 1)

Table 3.2 and no longer from Table 3.3). If, therefore, Bob is not supposed to have a dominant strategy (neither strictly nor weakly), his reaction will always be clear, and it must therefore be a unique Nash equilibrium. The situation is different if Alice only has a weakly dominant strategy (right matrix in Table 3.2). In this constellation, it is possible that a second equilibrium is found in the matrix. In the example, such a second Nash equilibrium is at the strategy combination (Down, Right) with the payoffs (2, 2). Again, as soon as weakly dominant strategies are in play, the uniqueness of the first, dominance-derived, Nash equilibrium is no longer guaranteed. It is then possible (though not mandatory) that another equilibrium exists.

Dominated strategies and the method of iterated elimination. In many games there are no dominant strategies, so we cannot—at least at first glance—apply the solution methods discussed so far. However, it may be that in the game in question there are at least *dominated strategies*. Dominated strategies are, in a way, the counterpart to the dominant strategies.

A *strictly dominated strategy* $s_{i'}$ gives player i for each strategy of the opponent s_{-i} a genuinely lower payoff/utility than at least one alternative strategy s_i'',

$$u_i\left(s_i', s_{-i}\right) < u_i\left(s_i'', s_{-i}\right) \forall s_{-i} \in S_{-i} \tag{3.3}$$

As expression (3.3) illustrates, we now change the perspective from the "superior" strategies to the "inferior" strategies. We thus proceed according to the exclusion principle and ask which strategies do *not* qualify as equilibrium strategies. Analogous to our considerations with dominant strategies, we assume that a rational player will *not* play a strictly dominated strategy.

In the above definition, the phrase "at least one" is particularly important: While a strictly dominant strategy always performs genuinely better than any alternative strategy, it is sufficient for a strictly dominated strategy that it performs strictly worse than a single alternative strategy. The matrix from Table 3.4 can illustrate this.

In Table 3.4, Alice has no dominant strategy and—since she only has two strategies—consequently also has no dominated strategy.[2] Bob also does not have a dominant strategy. However, Bob's strategy "Left" always (that is, for both strategies of Alice) yields lower payoffs than the "Right" strategy, (2 instead of 3

[2] If a player only has two strategies, then the concepts of "dominant strategy" and "dominated strategy" do not differ: If one of two strategies is dominant, then the other is automatically dominated and vice versa.

Table 3.5 Iterated elimination of strictly dominated strategies

Alice, Bob	Left 1)	Center 2)	Right
Top	(3, 2)	(1, 1)	(2, 3)*
Bottom	(4, 0)	(0, 2)	(1, 1)

for "Top" and 0 instead of 1 for "Bottom"). The "Left" strategy is therefore a strictly dominated strategy of Bob. However, the "Left" strategy does not have to be dominated by *all* alternative strategies: the "Left" and "Center" strategies do not dominate each other, i.e., in case of "Top" "Left" is better, while in case of "Bottom" "Center" is better. So, it is enough to find a single superior strategy. This is sufficient for obvious reasons: A decision-maker who compares the "Left" and "Right" strategies can only improve if he always chooses "Right" instead of "Left". He can therefore make no mistake if he henceforth renounces the "Left" strategy. More technically, we refer to this procedure as the *elimination of dominated strategies*. If this procedure is applied not only once but repeatedly to a matrix, this is referred to as *iterated* elimination. The elimination of strictly dominated strategies is particularly helpful for solving complex matrix games, as the removal of irrelevant strategies reduces the dimension of the matrix and thus makes it easier to handle. This will be briefly illustrated using the game from Table 3.4. To make the procedure easier to understand, the solution steps are separately shown in Table 3.5.

As Bob will not play the "Left" strategy, we can eliminate this strategy, that is, the strategy is—as shown in step (1) in the table—simply crossed out in the matrix. For the further solution of the game, we can now focus on the remaining, smaller 2 × 2 matrix, which is highlighted in the table. In this smaller matrix, Alice's strategy "Top" is now a strictly dominant strategy for her—the underlying assumption, of course, is that she also knows that Bob will never choose a dominated strategy. We can now again use our solution techniques for cases with dominant strategies. Since Bob also does not have a dominant strategy in the smaller matrix, it remains only to clarify what his best reaction to "Top" looks like: If Alice chooses her strictly dominant strategy "Top", then it is optimal for Bob to respond with "Right", i.e., the strategy combination (Top, Right) emerges as the Nash equilibrium of the game. The solution of the reduced game with the 2 × 2 matrix thus corresponds to the already known problem from Table 3.3, in which only one player has a strictly dominant strategy and the other optimally adapts to it. The Nash equilibrium (Top, Right) is also unique: The eliminated strategy of Bob is a *strictly* dominated strategy, which by definition cannot be an equilibrium strategy. Since in the reduced game only Alice has a dominant strategy and this is *strictly* dominant, there can be no further Nash equilibrium.

In general, strategies that represent the best response to the opponent's strategy choice survive the iterated elimination of *strictly* dominated strategies. Such strategies are also referred to as *rationalizable strategies*, since these are the only strategies considered by rational players. On the other hand, strategies that do not represent the best response to the opponent's strategy are not rationalizable, as no rational player will play such a strategy. In the matrix from Table 3.5, the "Center" and "Right" strategies are rationalizable strategies: "Center" is the best response of Bob to "Bottom" and "Right" is his best response to "Top". An equilibrium strategy, which by definition is a rationalizable strategy, is therefore never removed by the elimination procedure. Moreover, the order of elimination does not matter in the iterated elimination of strictly dominated strategies. This means that in the case where we can eliminate multiple strategies, we do not have to worry about which elimination we start with.

However, the latter does not apply to the iterated elimination of *weakly* dominated strategies. A strategy $s_i\prime$, which is weakly dominated by another strategy$s_i^{\prime\prime}$, has either the same or lower payoffs than the comparison strategy$s_i^{\prime\prime}$:

$$u_i(s_i\prime,\ s_{-i}) \leq u_i\left(s_i^{\prime\prime},\ s_{-i}\right) \forall s_{-i} \in S_{-i} \text{ and}$$

$$u_i(s_i\prime,\ s_{-i}\prime) < u_i\left(s_i^{\prime\prime},\ s_{-i}\prime\right) \text{ for at least one } s_{-i}\prime \in S_{-i} \tag{3.4}$$

Now, if we apply the method of iterated elimination to a game where weakly dominated strategies exist, we do indeed obtain a Nash equilibrium as the solution to the game, but we cannot be sure whether we have really found all Nash equilibria. If there are multiple weakly dominated strategies in such a game, then even the order of elimination can be decisive for the resulting equilibrium. Let us consider the following matrix as an example:

Alice, Bob	Left	Center	Right
Top	(2, 2)	(2, 0)	(0, 2)
Bottom	(1, 1)	(2, 2)	(0, 2)

In this game, Alice has the weakly dominated strategy "Bottom" (or the weakly dominant strategy "Top"). Bob has a total of two weakly dominated strategies: Both "Left" and "Center" are weakly dominated by the strategy "Right". We now have a total of three options for initiating the elimination of weakly dominated strategies. If we first eliminate the strategy "Bottom" of Bob, the matrix simplifies to

Alice, Bob	Left	Center	Right
Top	(2, 2)*	(2, 0)	(0, 2)*
~~Bottom~~	~~(1, 1)~~	~~(2, 2)~~	~~(0, 2)~~

Bob chooses as the best response either "Left" or "Right", as he is indifferent between these two strategies. As a result, we get two Nash equilibria: One with the strategy combination (Top, Left) and one with the strategy combination (Top, Right). Alternatively, we can also start the elimination procedure with one of the strategies of Bob. If we eliminate the strategy "Left" in the first step, then we get the matrix:

Alice, Bob	Left	Center	Right
Top	(2, 2)	(2, 0)	(0, 2)*
Bottom	(1, 1)	(2, 2)	(0, 2)*

Now the strategy "Center" is weakly dominated, so we can also eliminate this in the second step. Finally, with (Top, Right) and (Bottom, Right) two identical payoff combinations remain, which thus both represent Nash equilibria. While we have already found (Top, Right) before, (Bottom, Right) remained undiscovered when we started the elimination with "Bottom". In contrast, the previously found Nash equilibrium (Top, Left) was first eliminated here. Finally, as a third option, we can start by eliminating the strategy "Center":

Alice, Bob	Left	Center	Right
Top	(2, 2)*	(2, 0)	(0, 2)*
Bottom	(1, 1)	(2, 2)	(0, 2)

If we eliminate the strategy "Center" because it is weakly dominated by "Right", we again get a reduced 2 × 2 matrix. In this, both players now have a weakly dominated strategy: For Bob, the strategy "Left" and for Alice, the strategy "Bottom" is weakly dominated. If we now eliminate the strategy "Left", then we would have the situation as before. But if we instead eliminate the strategy "Bottom" in the second step (as shown), we get the result as in the first case, when we started with the elimination of the strategy "Bottom". This clearly demonstrates that the order of elimination is also important in the second elimination round for the question of which Nash equilibria are found with the procedure and which are not. Probably the most interesting point of the preceding analysis is that we have overlooked the fourth Nash equilibrium at the strategy combination (Bottom, Center) until the end, despite the three separately pursued approaches. The method of eliminating weakly dominated strategies has always found only two of the total four Nash equilibria per elimination procedure.

Figure 3.1 gives an overview of how Nash equilibria can be systematically tracked down with the methods discussed in this section and in which cases it is ensured that the equilibria found are also unique and complete solutions.

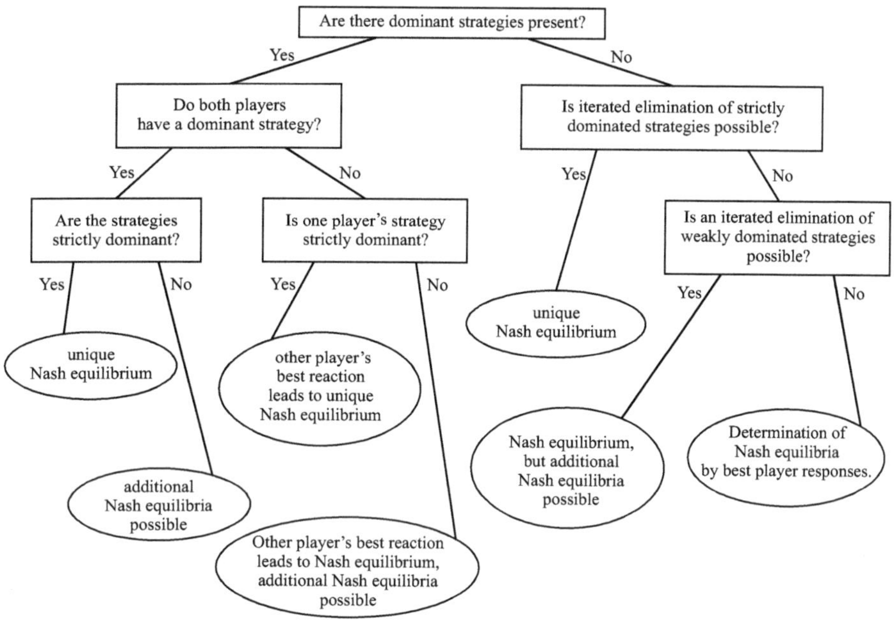

Fig. 3.1 Procedure for dominance considerations

3.2 Nash Equilibrium[3]

In this section, we want to deal with the concept of the Nash equilibrium in more detail to better understand its prerequisites and significance as a solution concept. The Nash equilibrium goes back to the mathematician and laureate of the Alfred Nobel Memorial Prize in Economic Sciences John F. Nash (1928–2015), who is known to many from the movie "A Beautiful Mind" (2001). We have already repeatedly considered the basic idea of his concept: Each player chooses the optimal strategy, given all other players also choose their optimal strategy. More precisely: A strategy combination s^* is a Nash equilibrium if each player by choosing the optimal strategy s_i^* receives a payoff that is at least as high as an alternative strategy s_i under the condition that all other players have chosen their optimal strategy s_{-i}^*:

$$u_i\left(s_i^*, s_{-i}^*\right) \geq u_i\left(s_i, s_{-i}^*\right) \ \forall s_i \in S_i \tag{3.5}$$

[3] This section deals with the prerequisites of a Nash equilibrium. Due to the mathematical background of game theory, this may appear somewhat abstract to some readers. In most cases, however, knowledge of the conditions presented is not absolutely necessary for a basic understanding or practical application of this solution concept, so this section may be skipped.

A prerequisite for the existence of the Nash equilibrium is that all players are rational and behave accordingly. Of course, this also includes the common-knowledge assumption (see Sect. 2.3). If a player does not behave rationally, i.e., does not choose the best response, the player's actual best response does not lead to the optimal result.

Excursion: Number selection game. In a so-called number selection game, we can illustrate how important it is for the Nash equilibrium that all players behave "correctly". In a number selection game, several players participate and each of them is asked to name an integer number between 1 and 100. The average of all the numbers mentioned is then calculated. Whoever's number is then closer to 80% of this value wins the game. Before you continue to read, ask yourself what number you would choose!

Let us now take an analytical look at the game. First, you should form expectations about the behavior of all other players, i.e., about which numbers they will choose, and thus which number you should pick to win the game. Let us assume that all players choose the highest possible number, i.e., 100, an average of 100 would result. 80% of this value would be 80, which means you should not choose a number between 80 and 100, as you would not win with it. So, let us choose 80. However, if we assume that all players are rational, we should be aware that the other players have also made this consideration and will therefore never choose a number from this range. That means they will also choose 80. This would make the average 80 and 80% of it is 64. Thus, we should not choose a number in the range between 64 and 80. So, what about choosing 64? Again, we must assume that the others would come to the same conclusion. So, the number to choose would not be 64, but 51.2 or, in integer numbers, 51. We can continue with these considerations until we reach the smallest possible selectable number, which is 1. Thus, in the Nash equilibrium every player will choose 1.

But what happens if not all players are rational? If you, as a completely rational player, were to choose the Nash equilibrium strategy 1, you would almost certainly not win. If all the other players only came to the first conclusion and would choose 80, they would have won, as 80 is closer to 64 (or a slightly lower value) than 1 with a relatively high number of players. So, if you doubt that most other players do not reason that far, you should better choose a number other than 1. However, this implies that the Nash equilibrium is not realized.

However, if the basic prerequisites of rationality and common knowledge are met, each player forms correct expectations regarding the behavior of the other players and chooses the best response accordingly. In general, the following reaction correspondence formally results for each player:

$$r_i(s_{-i}) = \left\{\hat{s}_i \in s_i \middle| u_i\left(\hat{s}_i, s_{-i}\right) \geq u_i(s_i, s_{-i}) \; \forall \, s_i \in S_i\right\}, \tag{3.6}$$

where $\hat{s}_i$ denotes the strategy that ensures a higher utility compared to all other strategies s_i. The Nash equilibrium is then determined by the mutually best responses of the players: The expectations of all players are only fulfilled when the strategies assumed for the other players s_{-i} are in turn the best responses, that is, $s^* \in r(s^*)$.

For a solution concept to be truly convincing, it makes sense if it can always be applied. Hence, we will now briefly show the conditions under which a Nash equilibrium exists.

Existence of Nash Equilibrium

A Nash equilibrium exists if it holds for a game $\Gamma(N, S, u)$ that

(i) the strategy space S is compact and convex and
(ii) the payoff function $u_i(s)$ of each player is continuous and bounded in $s \in S$ and quasi-concave in s_i.

Let us start with (i), i.e., the strategy set must be compact and convex. *Compact* means that the set is *bounded* and *closed*. Bounded means that there is an upper and a lower limit and closed means that these limit points must be included in the set. Thus, the number space of all natural numbers N is not compact, since it is not limited (after every natural number, there is always another natural number). The interval $[1, 100]$ would be bounded, but not closed, as the lower limit is included in the interval (represented by the square, inclusive bracket), but not the upper limit (the round bracket symbolizes that this number is not included in the interval). A set is *convex*, if all points on a connecting line between any two elements of the set are also included in the set. The simplest example of a non-convex set is a donut: if we connect any two points, there is a risk that this connection goes through the hole in the center and is therefore outside of the donut. However, this should never be the case with a convex set. Condition (ii) is more of a technical nature and ultimately implies that the payoff functions of the players must be differentiable and that convex indifference curves result from this. For economic applications or models, however, this condition is de facto always ensured.

The inspection game in Table 3.6 shows a game which has obviously no Nash equilibrium, as the conditions on the strategy space are not met. The situation here is as follows: An employee can decide to work in the interest of the company, which is exhausting, or to shirk and engage in more enjoyable activities. The manager, on the other hand, has the option to monitor the work performance, which consumes her working time and therefore causes costs, or to trust that everything is fine (laissez-faire). The incentive structure is as follows: If the employee works, it makes little sense for the manager to monitor him. If the employee is not monitored, he would rather shirk than work. But if the employee shirks, he should be monitored and confronted by the manager. But if the employee is monitored, he will work because he does not want to lose his job. Thus, we are now back at the start of our analysis—as we can see, there is no resting point here, which could consequently be a Nash equilibrium.

What exactly is the problem with the inspection game? Specifically, it shows that the strategy space is discrete and therefore not convex: While the employee's

Table 3.6 Inspection game

Employee (A), Manager (B)	s_{B1} (Monitor)	s_{B1} (laissez-faire)
s_{A1} (Work)	($\underline{1}$, 1)	(1, $\underline{2}$)
s_{A2} (Shirk)	(–1, $\underline{2}$)	($\underline{2}$, 0)

strategy space is indeed compact—there are exactly two strategies that can be chosen—it is not convex—he can either shirk or work. Thus, he can do nothing between these extremes, such as just a little bit of shirking or working. In all previous games with a discrete strategy space, we were simply lucky that a Nash equilibrium existed—in general, we cannot assume that we find a Nash equilibrium in such a strategy space. However, we can modify the game to obtain a Nash equilibrium by transforming the discrete strategy space into a continuous strategy space. The employee could then decide, for example, what arbitrary proportion between 0 and 100% of his working time he wants to spend on work, i.e., $0 \leq s_A \leq 100$ or $S_A = [0, 100]$. The manager, for her part, could use any amount of her available budget of 50,000 euros to set up a monitoring system, $0 \leq s_B \leq 50.000$ or $S_B = [0, 50.000]$. Since the strategy spaces of both players are now compact and convex,[4] a Nash equilibrium will exist.

Another way to overcome the problem of a non-convex strategy space, without fundamentally changing the game as just happened, is the use of mixed strategies. We will therefore look at these in more detail in the next section.

3.3 Mixed Strategies

So far, we have always considered pure strategies, in which a player i chooses the strategies from a discrete strategy set $s_i \in \{s_{i1}, s_{i2}\}$. Accordingly, the resulting Nash equilibria were always *equilibria in pure strategies*, which can be read directly from the matrix. However, it is also possible that a player plays the pure strategies only with a certain probability and thus—somewhat technically, but more precisely formulated—*randomizes* over the pure strategies. Since the player in this case combines or "mixes" the pure strategies over the probabilities, we speak of mixed strategies.

Let us assume, p_i is the probability with which player i chooses the first strategy s_{i1}, i.e., with the counter-probability $1 - p_i$ the remaining second strategy s_{i2} is chosen. The strategic variable of the player is now the probability p_i that is picked from the continuous strategy set $p_i \in [0, 1]$, since probabilities can only range between 0 and 100%. The two pure strategies are special cases of the mixed strategies, i.e., we find the discrete strategy set of pure strategies as special cases $p_i \in \{0, 1\}$ of the mixed strategies again. Here the choice of probability $p_i = 1$ corresponds to the pure strategy s_{i1} and accordingly $p_i = 0$ corresponds to the strategy choice s_{i2}.

[4] Since the sets are specified in each case, the set is compact. Convexity is also quite easy to check: A connecting line between "monitor" and "laissez-faire" would include, e.g., 25% "monitor" and 75% "laissez-faire", but this does not exist as a strategy and is therefore outside the strategy set. However, if we take any two expenditures for the monitoring system, e.g., 4,000 euros and 25,000 euros, then 25%·4,000 + 75%·25,000 = 19,750 euros would be expenditures that are within the strategy set and can therefore be selected.

Table 3.7 Penalty shootout

Shooter (A), Goalkeeper (B)	s_{B1} (left)	s_{B2} (right)
s_{A1} (left)	$(0, \underline{1})$	$(\underline{1}, 0)$
s_{A2} (right)	$(\underline{1}, 0)$	$(0, \underline{1})$

The concept of mixed strategies is not necessarily easy to understand intuitively. After all, each player can only choose one pure strategy, so it is not clear how to imagine the strategy choice "randomizing" exactly. Even if the concept seems somewhat abstract, we often encounter the underlying logic in everyday life. Mixed strategies become relevant whenever there is a conflict of interest between two players, through which *tactical behavior* becomes an option (see also the inspection game in the previous section). Take, for example, the penalty shootout game shown in Table 3.7 from Chap. 2.

The tactical element in penalty shootouts lies in the fact that the shooter and the goalkeeper have exactly opposite interests: For the goalkeeper, it would be best if the shooter's action was already recognizable to him before the shot; for the shooter, on the other hand, it is precisely important that he remains inscrutable in his intention. The same applies in reverse: The shooter would like to know beforehand which corner the goalkeeper will jump to, but for the goalkeeper, it is optimal to remain unpredictable in his defensive reaction. The optimal strategy for both is therefore the strategy in which their opponent remains uncertain about their intention. The opponent, for example, the goalkeeper, can only be uncertain if he expects the shooter to play both strategies with a positive probability until the execution of the game. This makes it clear what exactly is meant by "randomizing": The shooter must behave as if he were making his decision dependent on a random mechanism (such as a coin or a dice), and he must choose the probability for his strategy in such a way that the goalkeeper is indifferent between "left" and "right". Thus, the expected payoffs[5] must be the same ("$\stackrel{!}{=}$"). The following equation shows how we can calculate the probability p_A^* of this optimal strategy of player A (shooter):

$$\underbrace{p_A \cdot 1 + (1 - p_A) \cdot 0}_{E[u_B(p_A, p_B=1)]} \stackrel{!}{=} \underbrace{p_A \cdot 0 + (1 - p_A) \cdot 1}_{E[u_B(p_A, p_B=0)]} \Leftrightarrow p_A^* = \frac{1}{2} \tag{3.7}$$

Here p_A denotes the probability with which player A, the shooter, chooses his first strategy "left", which implies that he chooses his second strategy "right" accordingly with the counter-probability $(1 - p_A)$.[6] On the left side of the equation is the expected payoff of player B (goalkeeper) if he chooses "left" ($p_B = 1$) and

[5] The calculation here is analogous to Sect. 1.3.

[6] A better and more explicit notation for player A choosing the first strategy would be $p_{A1} = 1$ and correspondingly $p_{A2} = 1$ for the second and so on. However, since we only have two strategies here and therefore $p_{A2} = 1 - p_{A1}$, we choose a simple notation here.

on the right side of the equation is the corresponding payoff of player B if he chooses "right" ($p_B = 0$). The shooter now chooses his strategy p_A in a way that both sides of the equation are identical, thereby keeping the goalkeeper indifferent. We can easily calculate the optimal probability $p_A^* = 1/2$, i.e., the shooter chooses both directions with a 50% probability each. What if he were to shoot to the left with a higher probability than 50%, e.g., with 60%? For this, just look at Eq. (3.7): If the shooter chooses $p_A > 1/2$, then the left side of the equation is greater than the right. The goalkeeper thus achieves a higher expected payoff $E[u_B]$, if he jumps to the left ($p_B = 1$). If the shooter tries to shoot to the left more often, the goalkeeper will also always jump to the left, which is not optimal for the shooter. If, on the other hand, he was to shoot to the right more often, i.e., choose $p_A < 1/2$, then in Eq. (3.7) the right side is greater than the left side. The goalkeeper will then have a higher expected payoff $E[u_B]$ if he jumps to the right, which he would do as a rational player. We can therefore see that a deviation of the shooter from $p_A^* = 1/2$ in both directions only leads to the goalkeeper having a clearly best response to the shooter's strategy choice, which worsens the shooter's situation.

Now let us consider the problem from the goalkeeper's perspective. For him, too, it is optimal to keep the shooter indifferent:

$$\underbrace{p_B \cdot 0 + (1 - p_B) \cdot 1}_{E[u_A(p_A=1, p_B)]} \stackrel{!}{=} \underbrace{p_B \cdot 1 + (1 - p_B) \cdot 0}_{E[u_A(p_A=0, p_B)]} \Leftrightarrow p_B^* = \frac{1}{2} \tag{3.8}$$

Since the game "Penalty Shootout" is symmetric, the same optimal probability for the strategy "left" results for the goalkeeper, i.e., $p_B^* = 1/2$. We can easily understand with the same considerations as above that it is also not optimal for the goalkeeper to deviate from the probability of 50%.

We refer to the method presented here for determining equilibrium in mixed strategies as *expected utility comparison.* The procedure is to equate the expected payoffs/utilities of player i depending on the probability choice of player $-i$. This allows us to determine the probability of player $-i$ making player i indifferent. This procedure for determining an equilibrium in mixed strategies can be carried out relatively quickly, it is only necessary to ensure that the correct payoffs are equated. An alternative way to determine an equilibrium in mixed strategies is the *optimization approach.*

Optimization is based on the player's expected payoff. For player i this is generally

$$\begin{aligned} E\big[u_i(p_i, p_j)\big] = {} & p_i \cdot p_j \cdot u_i\big(p_i = 1, p_j = 1\big) \\ & + p_i \cdot \big(1 - p_j\big) \cdot u_i\big(p_i = 1, p_j = 0\big) \\ & + (1 - p_i) \cdot p_j \cdot u_i\big(p_i = 0, p_j = 1\big) \\ & + (1 - p_i) \cdot \big(1 - p_j\big) \cdot u_i\big(p_i = 0, p_j = 0\big). \end{aligned} \tag{3.9}$$

The expected payoff of player i consists of four summands, which describe the four possible strategy combinations: With the probability $p_i \cdot p_j$ both players play their first strategy; in this case, player i receives the payoff $u_i(p_i = 1, p_j = 1)$. With a probability of $p_i \cdot (1 - p_j)$ player i plays the first and player j the second strategy; in this case, player i receives a payoff of $u_i(p_i = 1, p_j = 0)$. The two remaining summands result in the same way. This expression thus represents the expected payoff of player i for all strategy combinations (p_i, p_j). Next comes the optimization: We partially derive this expression according to the strategic variable of player i, p_i, and then set the derivative equal to zero.[7] For player A, the expected payoff $E[u_A(p_A, p_B)]$ is derived with respect to p_A and for player B, the expected payoff $E[u_B(p_A, p_B)]$ is derived with respect to p_B. For player A, the expected payoff in the penalty game is thus represented as

$$E[u_A(p_A, p_B)] = p_A \cdot p_B \cdot 0 + p_A \cdot (1 - p_B) \cdot 1 + (1 - p_A) \cdot p_B \cdot 1 + (1 - p_A) \cdot (1 - p_B) \cdot 0 \tag{3.10}$$

This is to be partially derived with respect to his strategic variable p_A

$$\frac{\partial E[u_A(p_A, p_B)]}{\partial p_A} = p_B \cdot 0 + (1 - p_B) \cdot 1 - p_B \cdot 1 - (1 - p_B) \cdot 0 = 1 - 2 \cdot p_B. \tag{3.11}$$

As we can see, this expression only depends on p_B. Thus, setting it equal to zero yields the equilibrium probability:

$$1 - 2 \cdot p_B \stackrel{!}{=} 0 \quad \Leftrightarrow \quad p_B^* = \frac{1}{2}. \tag{3.12}$$

Let us now look at the first-order condition (this is what the first derivative of an objective function is called) from Eq. (3.12) in more detail. The expression $1 - 2 \cdot p_B$ is the (expected) marginal utility of player B with respect to his first strategy. The expression describes the amount by which the expected payoff of player A changes if probability p_A is marginally increased. We can see that the answer to this question depends solely on player B, as the marginal utility of player A depends only on the strategy variable p_B (and not on p_A). If player B (goalkeeper) chooses his first strategy ("left") with a probability greater than 0.5 (e.g., $p_B = 0.8$), then the marginal utility of player A becomes negative ($1 - 2 \cdot p_B < 0$). This means that player A *decreases* his expected utility if he increases p_A, i.e., plays the "left" strategy with a higher probability. In this case, player A is better off if he decreases p_A, i.e., shoots more often to the "right". If, on the other hand, player B chooses his first strategy with a probability less than 0.5 (e.g., $p_B = 0.3$), i.e., the goalkeeper tends to jump to the right, then the marginal

[7] If you need the basics or a refresher for mathematical optimization, take a closer look at Chap. 12 and especially 12.3.

utility of player A becomes positive ($1 - 2 \cdot p_B > 0$). In this case, player A will *increase* his expected utility by increasing p_A, i.e., shooting to the "left" with a higher probability. If player B chooses exactly $p_B = 0.5$, the marginal utility of player A is zero, which simply means that player A is indifferent between his two (pure) strategies—he can neither improve by increasing p_A nor by reducing it. The optimization approach thus clearly shows that the player's indifference is indeed also a best response to the opponent's strategy, since the marginal utility is zero in this case. Thus, in a mixed strategy equilibrium, it is optimal for player i to keep player j indifferent between his pure strategies and optimal for player j to remain indifferent, since no improvement arises for him out of this equilibrium.

Let us summarize the considerations made so far as a reaction correspondence:

$$p_A^*(p_B) = \begin{cases} 0 & p_B \in \left(\frac{1}{2}, 1\right] \\ [0, 1] & p_B = \frac{1}{2} \\ 1 & p_B \in \left[0, \frac{1}{2}\right) \end{cases} \qquad p_B^*(p_A) = \begin{cases} 0 & p_A \in \left[0, \frac{1}{2}\right) \\ [0, 1] & p_A = \frac{1}{2} \\ 1 & p_A \in \left(\frac{1}{2}, 1\right] \end{cases} \tag{3.13}$$

As we see, the strategy combination $\left(p_A = \frac{1}{2}, p_B = \frac{1}{2}\right)$ is a Nash equilibrium of the game. The reaction correspondences in mixed strategies are to be interpreted in the same way as those for pure strategies (see Chap. 2):

- If player B (goalkeeper) chooses his first strategy ("left") with a probability from the interval $(\frac{1}{2}, 1]$, i.e., it is played with a probability of more than 50% (50% itself is not in the interval), the best response of player A is to choose his first strategy with probability $p_A^* = 0$, i.e., to play "right" with certainty.
- If player B chooses exactly $p_B = \frac{1}{2}$, player A is indifferent between "left" and "right" and will therefore choose any p_A^* from the interval $[0, 1]$ as the best response.
- If player B chooses p_B from the interval $[0, \frac{1}{2})$, the best response of player A is $p_A^* = 1$, i.e., he will choose "left" with certainty.

Accordingly, the optimal reaction of player B to the probability choice of player A can be read from the right expression of (3.13). The reaction correspondences are symmetric and only the probabilities $p_A = p_B = \frac{1}{2}$ represent strategies which are consistent with each other. At this probability, the strategies of the players represent the best responses, i.e., this is a Nash equilibrium.

Note and Tip When you set up the reaction correspondences for the players, make sure that you always specify the best response of a player for the opponent's *complete choice* of strategy. You must, e.g., specify the best response of player A for all strategies of player B from the entire spectrum $p_B \in [0, 1]$. So, there should never be a "gap"; every player i must have a best response for every strategy choice of every player j.

Every Nash equilibrium is an intersection of the reaction correspondences, as we can see from the graphical representation in Fig. 3.2.

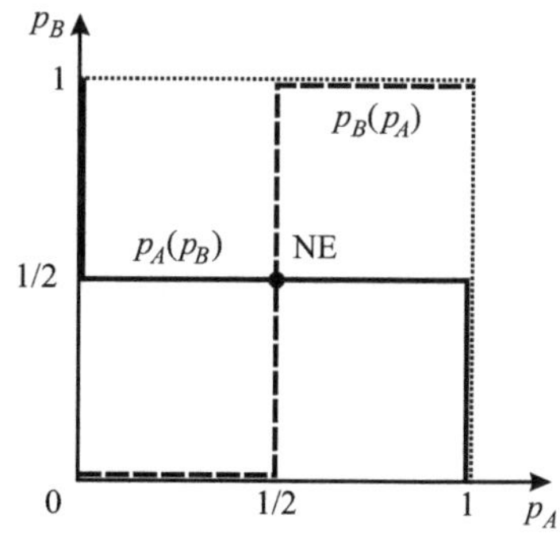

Fig. 3.2 Reaction correspondences for the penalty shootout game

The graphical representation is a direct transfer of the reaction correspondences from (3.11). Let us consider the best response from the shooter (A) as an example: In the p_B -interval $(\frac{1}{2}, 1]$ the solid line is a vertical line that runs exactly over $p_A =$ 0. At $p_B = \frac{1}{2}$, $p_A(p_B)$ is a horizontal line, which means that the best response from A can be found across the entire spectrum $p_A \in [0, 1]$. If the goalkeeper (B) chooses a lower probability than $\frac{1}{2}$, the reaction correspondence of A becomes vertical again, and runs over$p_A^* = 1$. The reaction correspondence of player A is thus vertical in those probability ranges of p_B where Player A has a best response and is therefore not indifferent. Similarly, we obtain the reaction correspondence of player B, which is represented by the dashed line labeled $p_B(p_A)$ in the figure and stems from (3.11). The Nash equilibrium (NE) results as the intersection of the reaction correspondences.

As we see from the reaction correspondence there is only one intersection point, $\left(p_A^* = \frac{1}{2},\ p_B^* = \frac{1}{2}\right)$, in the penalty shootout game which implies that there does not exist any further equilibrium. We can easily confirm this finding in the matrix from Table 3.7: If we again underline the deviation directions row by row in this game, we find that there is no equilibrium in pure strategies in this game. Thus, the penalty shootout game has only one Nash equilibrium in mixed strategies. A helpful guideline is that *the number of all Nash equilibria is always odd.*[8]

Hint and Tip Every game has an odd number of Nash equilibria. If the number of Nash equilibria in pure strategies is even, then you know that there must be one (but only one!) additional equilibrium in mixed strategies.

If we find no equilibrium in pure strategies—like in the penalty shootout game—we can be sure that there must be exactly one equilibrium in mixed strategies. All the matrix games we analyzed always have only one equilibrium in mixed strategies (alongside an even number of Nash equilibria in pure strategies).

[8] You can read about it in Lemke, C.E./ Howson, J.T. Jr. (1964): Equilibrium Points of Bimatrix Games, Journal of the Society for Industrial and Applied Mathematics, Vol. 12, No. 2, 413–423.

We can now return to our problem from Sect. 3.2, more specifically the inspection game in Table 3.6, in which we could not find a Nash equilibrium in pure strategies either. As we now know, by extending the strategy space to include mixed strategies, we should be able to find a Nash equilibrium in this game. Specifically, it turns out that the working probability of the employee is $p_A = 2/3$ while the monitoring probability of the manager is $p_B = 1/3$. Due to a similar incentive structure, the corresponding reaction correspondences resemble those in Fig. 3.2. By considering mixed strategies, a discrete strategy space thus becomes a compact and convex strategy set, thereby ensuring the existence of a Nash equilibrium.

Finally, we can also determine the *equilibrium payoffs* of the players. While we can see them directly in the matrix for pure strategies, we must calculate them in the case of mixed strategies. This is because they result from the expected profit both players receive when they play their equilibrium strategies. For this, we refer to the general formula (3.9), which gives us the expected payoff of the players, and insert the equilibrium strategies. In general, we calculate the (expected) equilibrium payoff of player i as

$$\begin{aligned} E\left[u_i(p_i, p_j)\right] = \; & p_i^* \cdot p_j^* \cdot u_i(p_i = 1, p_j = 1) \\ & + p_i^* \cdot \left(1 - p_j^*\right) \cdot u_i(p_i = 1, p_j = 0) \\ & + (1 - p_i^*) \cdot p_j^* \cdot u_i(p_i = 0, p_j = 1) \\ & + (1 - p_i^*) \cdot \left(1 - p_j^*\right) \cdot u_i(p_i = 0, p_j = 0). \end{aligned} \tag{3.14}$$

For player A (shooter) and player B (goalkeeper), this results in

$$E\left[u_A\left(p_A = \frac{1}{2}, p_B = \frac{1}{2}\right)\right] = \frac{1}{2} \cdot \frac{1}{2} \cdot 0 + \frac{1}{2} \cdot \frac{1}{2} \cdot 1 + \frac{1}{2} \cdot \frac{1}{2} \cdot 1 + \frac{1}{2} \cdot \frac{1}{2} \cdot 0 = \frac{1}{2}$$

$$E\left[u_B\left(p_A = \frac{1}{2}, p_B = \frac{1}{2}\right)\right] = \frac{1}{2} \cdot \frac{1}{2} \cdot 1 + \frac{1}{2} \cdot \frac{1}{2} \cdot 0 + \frac{1}{2} \cdot \frac{1}{2} \cdot 0 + \frac{1}{2} \cdot \frac{1}{2} \cdot 1 = \frac{1}{2}.$$

If both players play their equilibrium strategies, i.e., both, the goalkeeper and the shooter, choose "left" or "right" with a probability of 50%, both receive the expected payoff of 1/2.

Excursion: Penalty Kicks—Why not just shoot in the center? In the 1974 World Cup final, the Dutchman Johan Neeskens was the first penalty shooter to aim for the center of the goal and was successful. Sports journalists refer to the paradigm shift resulting from this event as the Neeskens effect, as this penalty strategy was used more frequently from then on. Previously, this option seemed to be taboo, because for the shooter an unsuccessful shot in the center was considered a disgrace. But a goalkeeper who expects a shot in the center also takes a risk. Since he does not move out of the center in this case, his "inaction" appears quite idle to observers. In case of a failure, the criticism could equally be made against the player or the goalkeeper that they did not show enough effort.

The German economists Wolfgang Leininger and Axel Ockenfels (2007) examined the strategic behavior of the shooter and the goalkeeper for about 450 penalties in the Bundesliga and were able to empirically substantiate that after 1974 not only the frequency of penalties shot into the center increased, but also the success rate of penalties overall. The game-theoretical explanation for this is simply that the strategy set for the shooter and the goalkeeper has been expanded. The fact that the shooter effectively has one more option available, shifts the duel situation in his favor. As can be easily calculated, in a modified penalty shootout game, which is expanded by the strategy "center", the expected payoff of the player increases to 2/3 whereas the expected payoff of the goalkeeper decreases to 1/3 (zero-sum game). Although the strategy "center" is now considered a firmly calculable option, it is still significantly less chosen than the two conventional shooting directions "left" and "right".

Let us now consider as a second application the matrix game in Table 3.8. The game is called game of chicken, as it represents a conflict situation like the one found in a duel among teenagers. The game is said to have been originally inspired by the 1955 film "Rebel Without a Cause" starring James Dean. The players can each choose between a soft and an aggressive strategy. "Aggressive" means in the context of a duel or conflict situation that the player in question seeks confrontation with the opponent. "Soft", on the other hand, means rather evasive, conflict-avoiding behavior. If the players choose asymmetric strategies, then the player who plays "aggressive" is always better off: The "aggressive" player receives a payoff of 3, while the "soft" player only gets a payoff of 1. If both play "soft", each receives a payoff of 2. However, if both play "aggressive", a catastrophe results and both receive a payoff of −1. In the film, the face-off situation consists of two teenagers driving towards each other and the one who sidesteps first loses. In this context, playing "aggressive" means that the player holds on, resisting the fear of evasion, and the catastrophe consists of the life-threatening collision of the vehicles. From the matrix we see that there are two Nash equilibria in pure strategies: the two asymmetric strategy combinations (aggressive, soft) and (soft, aggressive). According to our guideline, if there is an even number of equilibria in pure strategies (here two), there must be another one in mixed strategies. This can easily be found by comparing expected utilities,

$$\underbrace{p_A \cdot 2 + (1 - p_A) \cdot 1}_{E[u_B(p_A, p_B=1)]} \stackrel{!}{=} \underbrace{p_A \cdot 3 + (1 - p_A) \cdot (-1)}_{E[u_A(p_A, p_B=0)]} \Leftrightarrow p_A^* = \frac{2}{3}$$

In this case, the optimal probability for Alice to choose her first strategy ("soft") is 2/3, or 66.6%. Why is it that Alice does not randomize with 50%, as was the case in the penalty shootout game? The answer comes from the players' payoffs. If Alice were to play her first strategy with a 50% probability, Bob would switch to "soft" as he would have a higher payoff (1.5 > 1). In this case, Alice would be too aggressive for Bob, so he would prefer to avoid her. To keep Bob indifferent, Alice

Table 3.8 Game of Chicken

Alice, Bob	s_{B1} (soft)	s_{B2} (aggressive)
s_{A1} (soft)	(2, 2)	($\underline{1}$, $\underline{3}$)
s_{A2} (aggressive)	($\underline{3}$, $\underline{1}$)	(−1, −1)

must ensure that the yielding strategy loses attractiveness for Bob. She achieves this by being somewhat softer herself: The more Alice takes a step back, the more attractive it becomes for Bob to play aggressively himself.

Since the game is symmetric, we get the same optimal probability for Bob as well. The reaction correspondences for the game of chicken are

$$p_A^*(p_B) = \begin{cases} 0 & p_B \in \left(\frac{2}{3}, 1\right] \\ [0, 1] & p_B = \frac{2}{3} \\ 1 & p_B \in \left[0, \frac{2}{3}\right) \end{cases} \qquad p_B^*(p_A) = \begin{cases} 0 & p_A \in \left[0, \frac{2}{3}\right) \\ [0, 1] & p_A = \frac{2}{3} \\ 1 & p_A \in \left(\frac{2}{3}, 1\right] \end{cases}$$

and are graphically represented in Fig. 3.3. We can see in this figure that the reaction correspondences have three intersection points. At the intersection points $\left(p_A^* = 1, p_B^* = 0\right)$ as well as $\left(p_A^* = 0, p_B^* = 1\right)$ these are the already known equilibria in pure strategies. From this, we can easily understand that the concept of mixed strategies is a more general concept that always includes the pure strategies as special cases.

Finally, let us determine the Nash equilibrium in mixed strategies for a general payoff structure. This is helpful to see how the payoffs affect the result, i.e., how the players' payoffs contribute to the optimal probabilities. You can also use the result in the future to check that your own calculations are correct. The starting point is the matrix from Table 3.9, in which we represent the payoffs of both players in general form, i.e., without numerical values. One result that we would like to anticipate already is that the optimal probability of player i depends exclusively on the payoffs of the opposing player j (and vice versa). For generalization as well as for further practice, we will determine equilibrium strategies in detail using our two methods.

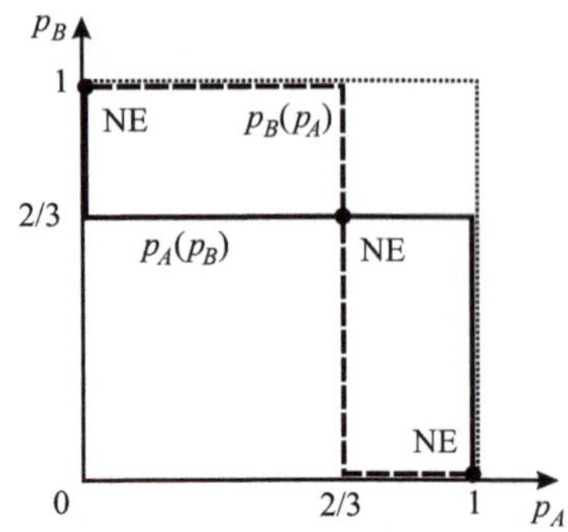

Fig. 3.3 Reaction correspondences for the game of chicken

Table 3.9 General approach to determine a Nash equilibrium in mixed strategies

Alice, Bob	s_{B1}	s_{B2}
s_{A1}	(a, A)	(b, B)
s_{A2}	(c, C)	(d, D)

Optimization Approach: The basic idea of this method is that each player chooses the probability(ies) with which they play their strategies in such a way that their expected payoff is maximized. This is done by replacing the actual discrete strategies of the players with the probabilities of these strategies as strategic variables creating a continuous strategy space. The (expected) payoff arises from the respective payoffs being weighed by their probability of occurrence. This probability, in turn, results from the joint probability choice of both players, i.e., from the choice of strategic variables. The procedure is the same as with any optimization.

Procedure for Optimization Approach

1. Determination of the payoff function depending on the own strategic variables and those of the co-player(s).
2. Partial derivative of the function with respect to the strategic variable of the player.
3. Determination of the first-order condition, i.e., setting the first derivative equal to zero.
4. Solving for the strategic variable.

Accordingly, the procedure for Alice generally looks like this:

$$\begin{aligned} E[u_A(p_A, p_B)] &= p_A \cdot p_B \cdot a + p_A \cdot (1 - p_B) \cdot b \\ &\quad + (1 - p_A) \cdot p_B \cdot c + (1 - p_A) \cdot (1 - p_B) \cdot d \\ \frac{\partial E[u_A(p_A, p_B)]}{\partial p_A} &= p_B \cdot a + (1 - p_B) \cdot b - p_B \cdot c - (1 - p_B) \cdot d \stackrel{!}{=} 0 \\ &\Rightarrow p_B = \frac{d - b}{a - b - c + d}. \end{aligned}$$

The procedure for Bob is analogous:

$$\begin{aligned} E[u_B(p_A, p_B)] &= p_A \cdot p_B \cdot A + p_A \cdot (1 - p_B) \cdot B \\ &\quad + (1 - p_A) \cdot p_B \cdot C + (1 - p_A) \cdot (1 - p_B) \cdot D \\ \frac{\partial E[u_B(p_A, p_B)]}{\partial p_B} &= p_A \cdot A - p_A \cdot B + (1 - p_A) \cdot C - (1 - p_A) \cdot D \stackrel{!}{=} 0 \\ &\Rightarrow p_B = \frac{D - C}{A - B - C + D}. \end{aligned}$$

Expected utility comparison: The basic idea of this method is to find the strategy with the highest expected payoff. This procedure is thus comparable to the decision between lotteries (Sect. 1.3), but with the difference that the probability of the individual payoffs is not exogenously chosen by "nature", but endogenously by an opponent. This method takes advantage of the fact that it deals with discrete strategies (and thus only a manageable number of options exists). It is then to find that probability of the opponent that makes the considered player indifferent between their two strategies, i.e., both give them the same expected payoff.

Procedure for Expected Utility Comparison

1. Determination of the expected payoff for each strategy (the payoffs of the considered player are weighed by the probability chosen by the opponent).
2. Comparison of the expected payoffs.
3. Determination of the probability for which the expected payoffs are equal.

Note and tip This methodology is particularly recommended for 2 × 2 matrices because there is less computational effort (and thus less susceptibility to errors) and the reaction correspondence is easier to read. For matrices of higher order, the optimization approach is recommended instead.

For Alice we need to check for which values of p_B (probability that Bob chooses his first strategy, s_{B1}) it is better (or equally good) for her to choose s_{A1}, i.e., $p_A = 1$ rather than s_{A2}, i.e., $p_A = 0$. This is the case if

$$\begin{aligned} E[u_A(p_A = 1, p_B)] &\geq E[u_A(p_A = 0, p_B)] \\ a \cdot p_B + b \cdot (1 - p_B) &\geq c \cdot p_B + d \cdot (1 - p_B) \\ p_B &\geq \frac{d - b}{a - b - c + d}. \end{aligned}$$

From this, Alice's reaction correspondence can now be easily read: If p_B is strictly greater than the calculated value, then Alice will choose s_{A1}, i.e., $p_A = 1$, since the expected payoff is higher. At the calculated value, Alice is indifferent, i.e., any probability for s_{A1} is possible, and for a smaller p_B she will choose s_{A2}, i.e., $p_A = 0$. This formally results in

$$p_A^*(p_B) = \begin{cases} 0 & p_B \in \left[0, \frac{d-b}{a-b-c+d}\right) \\ [0, 1] & p_B = \frac{d-b}{a-b-c+d} \\ 1 & p_B \in \left(\frac{d-b}{a-b-c+d}, 1\right] \end{cases}$$

Similarly, for Bob we need to check at which values of p_A (probability that Alice chooses her first strategy, s_{A1}) it is better (or equally good) for him to choose s_{B1}, i.e., $p_B = 1$ rather than s_{B2}, i.e., $p_B = 0$. This is the case if

$$\begin{aligned} E[u_B(p_A, p_B = 1)] &\geq E[u_B(p_A, p_B = 0)] \\ A \cdot p_A + C \cdot (1 - p_A) &\geq B \cdot p_A + D \cdot (1 - p_A) \\ p_A &\geq \frac{D - C}{A - B - C + D}. \end{aligned}$$

Bob's reaction correspondence is therefore

$$p_B^*(p_A) = \begin{cases} 0 & p_A \in \left[0, \frac{D-C}{A-B-C+D}\right) \\ [0, 1] & p_A = \frac{D-C}{A-B-C+D} \\ 1 & p_A \in \left(\frac{D-C}{A-B-C+D}, 1\right] \end{cases}$$

Note and Tip We can always conclude the number of Nash equilibria from the appearance of the reaction correspondences (and vice versa), as the following figure illustrates:

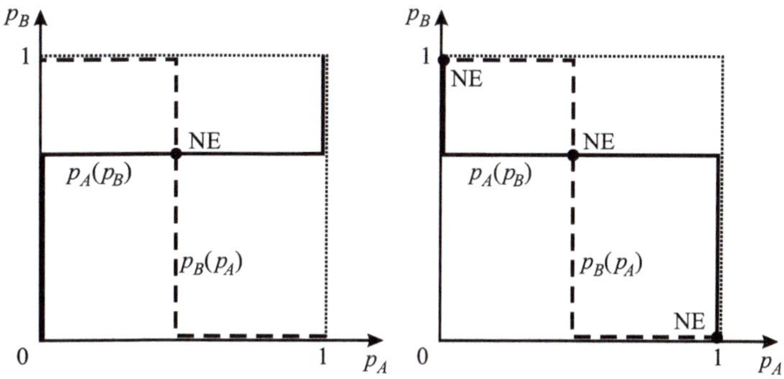

On the left, there is only one Nash equilibrium in mixed strategies, whereas on the right there are three equilibria (one in mixed and two in pure strategies). The curves can also be drawn in mirror image depending on how the probabilities were chosen. When drawing the reaction diagrams, always check whether the number of Nash equilibria you have found is also reflected in the number of intersections of the reaction diagrams.

At the end of this section, we also want to briefly establish a connection to the beginning, i.e., to Sect. 3.1, and show that a (strictly) dominant strategy does not necessarily have to be a pure strategy but can also be a mixed strategy. We can illustrate this using the two matrices in Table 3.10. On the left you see the basic game. Since we are not interested in the payoffs of Bob, these are merely designated by a placeholder. As we can see, Alice does not have a dominant strategy: If Bob chooses his first strategy, then s_{A1} is better, if Bob chooses his second strategy, then s_{A2} is better. However, we also see that Alice will never choose her third strategy s_{A3}.

Now let us consider a mixed strategy where Alice plays her first strategy s_{A1} with probability p_A, her second strategy s_{A2} with the counter-probability $(1 - p_A)$, and her third strategy s_{A3} with probability 0. Since we are now considering probabilities, Alice does not receive a certain payoff for a certain strategy combination,

Table 3.10 Mixed strategy as a dominant strategy

Alice, Bob	s_{B1}	s_{B2}
s_{A1}	$(4, \cdot)$	$(0, \cdot)$
s_{A2}	$(0, \cdot)$	$(4, \cdot)$
s_{A3}	$(1, \cdot)$	$(1, \cdot)$

Alice, Bob	s_{B1}	s_{B2}
$p_A \cdot s_{A1} + (1 - p_A) \cdot s_{A2}$	$(4 \cdot p_A, \cdot)$	$(4 \cdot (1 - p_A), \cdot)$
s_{A3}	$(1, \cdot)$	$(1, \cdot)$

but only an expected one—depending on which pure strategy is realized by the chosen random mechanism. Thus, when Alice chooses the mixed strategy, her expected payoff depends on Bob's strategy as follows: If he chooses s_{B1} she expects $p_A \cdot 4 + (1 - p_A) \cdot 0 = 4 \cdot p_A$ while if he chooses s_{B2} she expects $p_A \cdot 0 + (1 - p_A) \cdot 4 = 4 \cdot (1 - p_A)$ as a payoff. This strategy dominates s_{A3} strictly if it generates a higher payoff than s_{A3} for both of Bob's strategies, i.e., $4 \cdot p_A > 1$ or $p_A > 0.25$ and $4 \cdot (1 - p_A) > 1$ or $p_A < 0.75$. Thus, if p_A is chosen in the range $(0.25, 0.75)$ this mixed strategy dominates a pure strategy. For example, by choosing her first and second strategy with equal probability, Alice could create a dominant strategy.

3.4 Efficiency, Equilibrium Analysis, and Conflict Structure in Simple Matrix Games

So far, our focus has been on suitable methods of equilibrium *search*. However, equilibrium *selection*, on the other hand, has only played a subordinate role so far. This section now takes a closer look at the economic evaluation of the possible outcomes of a game, especially regarding the criteria of efficiency and plausibility of the solution. We will consider six matrix games that differ in terms of their payoff or conflict structure. These games are paradigmatic for different types of conflicts of interest and are known in game theory under striking names, such as "prisoner's dilemma".

Efficiency. Before we start with the equilibrium analysis of matrix games, let us first discuss three basic efficiency concepts. In a broader sense, the three concepts represent economic welfare measures and are therefore also of great importance for economic analyses formulated in terms of game theory. After all, we not only want to understand which game outcome ("solution") is to be expected, but also, which game outcomes would be desirable for all players. The first step of analysis is *positive analysis*, which focuses on the actual conditions (current state, "what is?"). The second step of analysis, on the other hand, is *normative*, because it seeks potential for improvement (desired state, "what should be?"). The systematic comparison of current and desired state can then provide starting points for the use of economic policy instruments, which a regulating, social planner, who aims for the most efficient game outcome in the sense of all players, would implement.

The most important efficiency criterion of economics is *Pareto optimality* (or the *Pareto criterion*). A situation in a multi-person game is referred to as Pareto-optimal when it is not possible to improve the situation of one decision-maker without simultaneously making another worse off.

Let us consider the game in Table 3.11. In this game, there are three Pareto-optimal payoff combinations, which are marked with the superscript P. If you switch between the three Pareto-optimal payoff combinations (2, 2), (3, 0), and (0, 3), one of the players always improves, but the other one loses. Therefore, it is not possible to improve the situation of one player *without* making the other player

Table 3.11 Prisoner's dilemma

Alice, Bob	s_{B1}	s_{B2}
s_{A1}	$(2, 2)^P$	$(0, \underline{3})^P$
s_{A2}	$(\underline{3}, 0)^P$	$(\underline{1}, \underline{1})^*$

Table 3.12 Inspection game

Potential delinquent ($\underline{A}$), Inspector ($\underline{B}$)	control (s_{B1})	no control (s_{B2})
no breach of rules (s_{A1})	$(1, 2 - C)$	$(1, 2)$
breach of rules (s_{A2})	$(2 - S, S - C)$	$(2, 0)$

worse off. A Pareto-optimal payoff combination is thus *efficient* if any change to this allocation would only allow unilateral and consequently conflict-prone improvements (Table 3.12).

If we speak of an efficient constellation, we must always refer to the applied criterion. We should therefore describe the payoff combinations (2, 2), (3, 0), and (0, 3) as "*efficient according to the Pareto criterion*". The Pareto criterion is a relatively weak efficiency concept, as it is based on a minimal consensus. In many cases, the set of efficient solutions can be very large, leaving a social planner with little guidance for decisions.

The remaining, fourth payoff combination (1, 1), constitutes the Nash equilibrium in dominant strategies, which is not Pareto-optimal: If you switch from here to (2, 2), then *both* players would improve. This leads us to the concept of *Pareto dominance*: In general, the statement can be made that one payout combination (A, B) *strictly Pareto-dominates* another (a, b) for $A > a$ and $B > b$. Alternatively, this fact can also be formulated in the passive form: The payoff combination (a, b) is *strictly Pareto-dominated* by (A, B) (for $a < A$ and $b < B$). A payoff combination that is Pareto-dominated by another is always *inefficient*. In the example from Table 3.11, the payoff combination (1, 1) is Pareto-dominated by (2, 2) *and is therefore inefficient*. Finally, there is also the *weak Pareto dominance* of a payoff combination (A, b) over an alternative (a, b) for $A > a$. In a weakly Pareto-dominant situation, no player worsens, but at least one improves compared to the alternative.

Figure 3.4 shows the payoff space for the prisoner's dilemma. The Pareto-optimal payoff combinations can be identified by the fact that the line connecting them is concavely curved and represents the upper edge of the payoffs. This line is referred to as the *efficiency line*, the *efficient edge*, or the *Pareto boundary*. The Nash equilibrium at (1, 1) is located below the Pareto boundary, indicating that it is a Pareto-dominated payoff combination. If we now draw a vertical and a horizontal line from point (1, 1) (dashed lines in the figure), we can identify a payoff

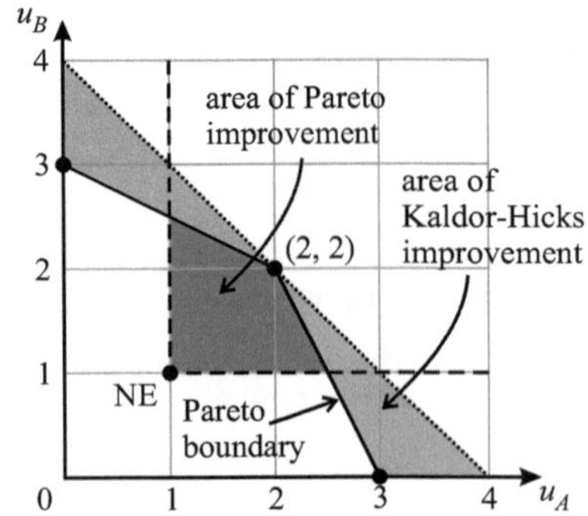

Fig. 3.4 Pareto boundary and improvement possibilities according to the Pareto and Kaldor-Hicks compensation criterion

area between these two lines and the Pareto boundary where each of the two players can improve without worsening the other. This dark gray area in the figure thus marks the area of Pareto improvements.

The third efficiency criterion is the *Kaldor-Hicks compensation criterion.* This criterion assumes interpersonal comparability and transferability of the individual utilities of the players. While it is usually assumed that the utility of two players is not (directly) comparable, this criterion allows both comparisons and possible redistributions between the players. In the prisoner's dilemma from Table 3.11, only the payoff combination (2, 2) *is efficient according to the Kaldor-Hicks compensation criterion:* For (2, 2) the total benefit is 4, while for the others it is either 3 or even only 2. Thus, starting from e.g. (3, 0), it would be possible when switching to (2, 2) that Alice receives 3 from the total sum, i.e., she is not worse off, but 1 remains, which can be given to Bob, who improves as a result. However, (3.5, 0.5) would also be possible, so that both improve compared to (3, 0). Starting from (2, 2), redistributions are thus possible that strictly or weakly pareto-dominate all other payoffs. In Fig. 3.4 the areas are highlighted in light gray, which according to the Kaldor-Hicks compensation criterion allow for further improvements and therefore are located outside the Pareto boundary. The compensation criterion is, due to the assumptions, a much stronger and significantly less consensual efficiency criterion than Pareto efficiency. It provides a social planner in most cases with a smaller number of efficient solutions and is thus a better decision aid. Nevertheless, it is debatable from a welfare economic point of view since it is assumed that the losses of one player can simply be offset with the gains of the other and a redistribution is fundamentally possible. Please note that only the possibility of improvement is sufficient, an actual redistribution does not have to take place.

Following the description of the three efficiency concepts, the question arises under what conditions *Nash equilibria are efficient* (in terms of the Pareto criterion). There is a quite general, but still clear answer to this question: A Nash equilibrium is always efficient if *no external effects* are present.

Externality. To understand this statement, we first need to clarify what an external effect or an *externality* in economics is. Simply put, an external effect refers to the impact of decision-maker i on an uninvolved person j, which is not considered in the decision-making of i. Because decision-maker i does not consider these effects, the result brought about by the decision is not efficient from a collective point of view (i.e., from the joint perspective of decision-maker i and person(s) j). Thus, in the case of external effects, from the perspective of economic policy, individual optimal behavior deviates from the socially optimal solution (the so-called *social optimum*). We distinguish *positive* and *negative* external effects. In the case of positive external effects, the action of decision-maker i has a positive effect on the utility of j, which is not considered by decision-maker i. Formally expressed, the marginal utility of player j is positive, i.e., their utility U_j increases if player i increases activity x_i:

$$\text{positive external effect:} \quad \frac{\partial U_j}{\partial x_i} > 0. \tag{3.15}$$

Examples of positive externalities include investments in research and development (R&D) or vaccinations. The R&D investments of a company are usually too low, as the company only considers the effects of these investments on its own profit. From a wider economic perspective, however, a higher level of R&D activity would be desirable, as the benefits of innovations ultimately benefit the entire society. The example of vaccinations should make the connection even clearer: A vaccination not only provides everyone with protection against infectious diseases, but also reduces the risk of infection for others who may not be able to get vaccinated, and thus the danger of an epidemic spreading (herd immunity). This is the case with smallpox. A higher willingness of decision-maker i to get vaccinated thus increases not only the own utility U_i, but also the utility U_j of other people.

In the case of a negative externality, the action of decision-maker i has a negative effect on the utility of j, which is not considered by decision-maker i. Here, the marginal utility of player j is negative, i.e., their utility U_j increases (decreases) if player i decreases (increases) activity x_i:

$$\text{negative external effect:} \quad \frac{\partial U_j}{\partial x_i} < 0. \tag{3.16}$$

Examples of negative externalities include emissions or the consumption of non-renewable resources. The production of goods or road traffic is associated with high emissions of pollutants and noise, which have a detrimental effect on the well-being of all participants. On a smaller scale, smoking, for example, represents a negative externality, as non-consumers are also exposed to the odor and health risks. To take these external effects into account (*internalization of an external effect*), every producer or every driver would have to reduce the harmful activity or at least adjust it in an equivalent manner by installing appropriate countermeasures such as noise protection or pollutant filters. In the example of finite resources, inefficient activity results from excessive consumption, which can lead to accelerated exhaustion of resources (such as oil, fishing grounds). If this negative external effect were internalized by each player, this would lead to a lower total consumption and enable society to use the resource sustainably.

Constant sum games and anti-coordination games. We begin our analysis of (matrix) games with the most conflict-prone variant, a constant-sum game or zero-sum game. As already explained in Sect. 2.2, there is no game outcome that pareto-dominates another game outcome (one player's gains always come at the expense of the other player) and could thus enable both players to gain efficiency. This implies that in a constant-sum game, every game outcome is efficient according to both the Pareto criterion and the Kaldor-Hicks compensation criterion (try to figure out why!). However, this does not mean that the players are indifferent to the game outcome. On the contrary: A constant-sum game shows a strong conflict of interest between the players, especially when there is a high payoff difference between the players in each cell. In such a case, the game offers the players no compromise solution: every possible game outcome would be good for one player but bad for the other. Examples of this are the penalty game already discussed in detail, the game of rock paper scissors, and the game of "heads or tails" (Matching Pennies). In each cell of these games, one of the players always has an incentive to deviate, so there is no Nash equilibrium in pure strategies. For this reason, these games can be considered as *anti-coordination games*, as each player always wants to achieve exactly the opposite of what the other is interested in. As we have already seen in the penalty game, these games have an equilibrium in mixed strategies, which reveals the strong tactical component in this type of games.

A special category of anti-coordination games, which is very important for economic applications, are *inspection games*—we have already briefly discussed one of these in Table 3.6. In terms of incentive structure, inspection games correspond to the penalty game or "heads or tails", but primarily address the interaction between an inspector and a potential delinquent.[9] There are many possible applications for this type of interaction: The interaction between the tax office and the taxpayer, ticket inspector and passenger, auditing company and company, health department and restaurant owner, exam supervisor and examinee, or police and criminal. All these examples demonstrate the high relevance and ubiquity of inspection games.

Table 3.12 shows the general incentive structure of an inspection game. For this example, we assume that the inspector realizes a (gross) utility of 2 if the potential delinquent does not commit a rule violation, but 0 if the delinquent commits a rule violation. A control is associated with costs C. If the inspector catches the delinquent, the latter must pay a penalty of S to the inspector. For the delinquent, a non-sanctioned rule violation is attractive; in this case, she receives 2 instead of 1 if she refrains from violating the rule. For $0 < C < S$ and, e.g., $C = 1$ we get the classic anti-coordinative incentive structure of an inspection game—the payoff structure in Table 3.6 results for $S = 3$ and $C = 1$. Let us assume that the potential delinquent (A) refrains from a rule violation and is still controlled (upper left cell

[9] Occasionally, inspection games are presented as zero-sum games, although strictly speaking they are not (the above example illustrates this). They are primarily anti-coordination games and thus have *the same properties* as the zero-sum games, penalty shootout or matching pennies.

in the matrix). In this case, the control becomes unnecessary, so the inspector (B) will refrain from it (upper right cell in the matrix). However, without control, it is interesting for player A to break the rules (bottom right in the matrix). If player A does this, it is better for the inspector to control (bottom left in the matrix). This in turn makes the rule violation too costly for the delinquent, so A will refrain from it, and the cycle begins anew. In Problem 3.7.2 at the end of this chapter, we will determine the equilibrium in mixed strategies for an inspection game with general payoffs

Prisoner's Dilemma Next, we want to take a closer look at the *prisoner's dilemma*. We have already presented the payoff structure in Table 3.11. The prisoner's dilemma is by far the most famous game, which is used in many economic, political, and related contexts to illustrate conflicts of interest. According to its name, it describes the situation of two prisoners who are confronted by the prosecutor with a leniency policy. Although the prosecutor has enough evidence against both (suspect) offenders to charge them for a minor offense, she relies on the statements of both, to uncover a larger crime. The situation is symmetrical for both prisoners, as they can choose between a cooperative strategy s_{i1} ("remain silent") and a fraudulent or opportunistic strategy s_{i2} ("betray"). If both cooperate and remain silent, they will only have to expect a minor penalty (payoff 2). If only one of them behaves opportunistically and takes advantage of the leniency policy (while the other remains silent), the informant goes unpunished (payoff 3) while the other receives the maximum penalty (payoff 0). If both decide to testify, they both receive the mitigated maximum penalty (payoff 1), as they were at least willing to testify. For the prisoner's dilemma, we have already seen that there are three efficient payoff combinations according to the Pareto criterion and one according to the Kaldor-Hicks compensation criterion. The payoff combination (2, 2) is efficient according to both criteria and therefore presents itself as a desirable goal for the players—the prisoners are best off if they both remain silent. The dilemma lies in the fact that neither the efficient solution to be aimed for, (s_{A1}, s_{B1}), nor one of the other two Pareto-optimal payoff combinations are a Nash equilibrium of the game. Both players have a strictly dominant strategy with s_{i2}, i.e., they will both opt for "betray". The strategy combination $(s_{A2}, s_{B2})^*$ is therefore the unique Nash equilibrium of the game. This equilibrium is also *stable*, as the choice of a strictly dominant strategy for both players is a clear and risk-free decision: Regardless of their opponent's choice, they are always best off if they choose this strategy.[10] The problem of the prisoner's dilemma is that strong incentives of the players lead

[10] Occasionally, descriptions of the prisoner's dilemma include the additional assumption that the prisoners are kept in separate cells and have no possibility to communicate. This assumption is needless from a game theory perspective: Since both prisoners have a strictly dominant strategy, their optimal decision is independent of the strategy choice of the opponent. The two prisoners could therefore also have a coffee together first and agree on a common strategy choice. Each will then hope the other will honor the agreement so they can look forward to the higher payoff of 3. The game theoretical prediction of their decision remains unaffected by this: Both will choose "betray" for sure.

to an inefficient result. It can therefore be characterized as a *strong cooperation problem.* More precisely, we speak of a *strong cooperation problem* if the Nash equilibrium of the game is Pareto-dominated by a payoff combination that (1) is *efficient* and (2) has *largely symmetric payoffs,* so that players could agree on this payoff combination without conflict if *binding agreements* were possible.

The payoff combination (2, 2) meets these conditions: It pareto-dominates the equilibrium solution (1, 1), it is pareto-optimal, and "fair" in the sense that both players would agree on this solution if they could make binding agreements. The prisoner's dilemma could therefore be overcome by extending the game with the possibility of binding agreements, such as a contract that can be enforced by a third party (e.g., the legal system) if necessary. Both players would have a clear interest in signing such a contract, as this contract removes the opportunistic incentive from both and thus improves their situation. Whenever players resort to such an external, institutional solution, it is also referred to as a *commitment* ("self-binding") of the players. Another possibility to overcome the prisoner's dilemma is that the players in the subsequent round could expect credible rewards or punishments for their behavior. If, for example, both prisoners could be sure that they would be considered "men of honor" in their community for remaining silent (reward through a positive reputation) or rightly fear that they would be expelled from the community for betrayal (punishment through condemnation), then this would also eliminate the opportunistic incentive. The proposed solutions to overcome the prisoner's dilemma, whether through contracts, rewards, or punishments, however, always represent *extensions* of the original game. To explicitly show how this can overcome the opportunistic incentive, one would have to transform the simultaneous game "prisoner's dilemma" either into a sequential game (see Chap. 4) or into a repeated game (see Chap. 5) and add the corresponding rewards and punishments.

The prisoner's dilemma is representative of a whole range of applications that can be described by the same incentive structure and thus also represent a strong cooperation problem. One application, which also originates from the illegitimate area, is that of *collusion* or *cartel agreement.* Here, two companies decide whether they will "cooperate" in their pricing policy as a cartel, e.g., they set a high price, (s_{i1}) or whether they opportunistically "deviate" from the cartel agreement, e.g., they offer cheaper (s_{i2}). Since both have the strictly dominant strategy of breaking the cartel agreement (s_{i2}), the cartel solution albeit advantageous for both companies does not come about. The example of the cartel agreement is interesting insofar as it highlights an important limitation of our concept of efficiency: If both companies were to overcome the dilemma and adhere to their cartel agreement, then the resulting cooperative solution would be efficient for the companies. Nevertheless, we know from the theory of competition policy that the cartel solution is *inefficient from a societal perspective,* as the production quantities are too low, and the prices are too high—only perfect competition guarantees a pareto-efficient result for society. If we were to expand the cartel game to include the group of consumers as a third player, then the strictly dominant, opportunistic strategy of the companies would lead us directly to the societal optimum and not into a dilemma.

The normative implications of a game, i.e., the question of what the desirable, efficient solution looks like, therefore cannot be generalized without further ado, but depend on the definition of the game. In the next section, in which we solve a game with continuous strategies, we will take a closer look at duopoly competition again.

Further applications of the prisoner's dilemma with implications of high societal relevance are the *Public Good Game* and the *Tragedy of the Commons.* In the public good game, a public good (such as education, medical care, internal security, public infrastructure, dike, etc.) is jointly financed and used by a group, with the financing being voluntary and no one can be excluded from the use (*non-excludability* of a public good). Take the dike, for example—whether you paid for it or not, it protects you from high tides. This results in the opportunistic incentive for each group member to free ride, again as a strictly dominant strategy: Regardless of whether the others "contribute" (s_{i1}) or not, it is optimal for the individual member to "not contribute" (s_{i2}). Since everyone behaves this way in equilibrium, the provision of the public good fails due to lack of funds. The tragedy of the commons refers to the problem of overexploitation of natural resources, such as raw materials or fishing grounds. Here again, we can consider the two strategies "sustainable cooperation" and "opportunistic behavior", where "sustainable resource use" (s_{i1}) in this context means that the players consider each other when using resources and only take as much as they absolutely need. "Opportunistic behavior" (s_{i2}) accordingly implies the strategy of appropriating as much of the resource as possible. In equilibrium, the players have the incentive to choose the opportunistic strategy, which ultimately leads to them not using the resource sustainably and thus harming themselves.[11]

In all applications of the prisoner's dilemma, the external effect that reveals the inefficiency of the Nash equilibrium is evident. Whether it is a positive or a negative external effect depends solely on whether the *cooperative* or the *opportunistic* strategy is assumed. The cooperative strategy (s_{i1}) would correspond in the previously listed examples to the strategies "remain silent", "contribute", or "Sustainable resource use", whereas opportunistic behavior (s_{i2}) here includes the strategies "betray", "not contribute ", or "overexploitation". Regarding the cooperative strategies, there is a *positive* external effect: It would be better for all players, i.e., total welfare increases, if everyone cooperates *more* than they are willing to do in equilibrium (deviation in the *positive* direction). If, on the other hand, we focus on the opportunistic strategy, then there is a *negative* external effect: Total welfare increases if everyone behaves *less* opportunistically than is the case in equilibrium. Which variant one prefers to prove the inefficiency of the Nash equilibrium is irrelevant; both views are completely equivalent. After all, it doesn't matter whether

[11] The term "tragedy of the commons" dates to the eighteenth century. It describes the problem observed at the time that farmers and shepherds tended to overuse the pasture and farmland they jointly cultivated on behalf of the crown.

Table 3.13 Hawk-dove and weak cooperation problem

(a) Hawk-dove

Alice, Bob	s_{B1}	s_{B2}
s_{A1}	$(2, 2)^P$	$(\underline{1}, \underline{3})^{P*}$
s_{A2}	$(\underline{3}, \underline{1})^{P*}$	(0,0)

(b) Weak cooperation problem

Alice, Bob	s_{B1}	s_{B2}
s_{A1}	$(\underline{2}, 2)^P$	$(\underline{1}, \underline{3})^{P*}$
s_{A2}	$(0, \underline{0})$	$(0, \underline{0})$

you say, "more of the good is better" or "less of the bad is better". What is important is that the designations "positive" or "negative" in the external effects refer exclusively to the "more" or "less" (and not to "good" or "bad"). Regardless of whether it is a positive or a negative external effect, from an economic point of view, both are "bad" in the sense that they cause a loss of efficiency and represent a (allocative) market failure.

Hawk-Dove. Let us consider the game in Table 3.13(a). This game is called the *hawk-dove game* or also the *game of chicken*. Both variants describe the same conflict structure; we have already encountered the latter in Table 3.8 in connection with mixed strategies.

The term hawk-dove describes, in analogy to the food fight in the animal world, a distribution conflict that can easily escalate and then harm both parties. Here, "hawk" (s_{i1}) stands for an aggressive or offensive strategy and "dove" (s_{i2}) for a soft or peaceful strategy.[12] If two decision-makers play the "dove" strategy, they peacefully agree on an equal division (2, 2). However, if only one plays the "dove" strategy and one the "hawk" strategy, then the more aggressive player prevails and receives the correspondingly larger share with the payoff of 3, whereas the peacefully acting player only gets a payoff of 1. If both players play the aggressive "hawk" strategy, a fight over the resource ensues, causing both to inflict such high damage on each other that they only end up with (0, 0). A common feature of this game with the prisoner's dilemma is the "fair" payoff combination (2, 2) in the top left, which again offers itself as a compromise for the players and which they would also choose as rational decision-makers if they could make binding agreements. The first important question is therefore whether this potential compromise solution results in this game without binding agreements. For this, the strategy combination (s_{A1}, s_{B1}) would have to constitute a Nash equilibrium. As we easily recognize, there are two Nash equilibria in pure strategies with (s_{A2}, s_{B1})

[12] The terms "hawk" and "dove" stand for strategies. In evolutionary game theory, a part of game theory grounded in evolutionary theory, these terms, however, refer to animal populations.

and (s_{A1}, s_{B2}). However, the interesting strategy combination (s_{A1}, s_{B1}) *is not* a Nash equilibrium. Thus, there is also a cooperation problem in this game but in contrast to the prisoner's dilemma, the two Nash equilibria with the payoff combinations (3, 1) and (1, 3) are Pareto-optimal. Although the compromise solution (2, 2) cannot be realized again, the two realizable solutions (3, 1) and (1, 3) are efficient—both according to the Pareto criterion and the Kaldor-Hicks compensation criterion. From the perspective of the social planner, the problem in the hawk-dove game is therefore significantly mitigated, as the players—starting from the Pareto-optimal Nash equilibria—at least theoretically have the possibility to reach the compromise solution through a subsequent redistribution of resources ("side payments"). For this reason, we refer to the situation of the hawk-dove game as a *weak cooperation problem.* One way to overcome a weak cooperation problem, in addition to binding agreements, is the use of *correlated strategies,* in which both players make their strategy choice dependent on a jointly observed random mechanism (for example, a coin toss or the realization of the weather). For example, with heads, the more advantageous equilibrium for Alice is played and with tails the one for Bob.

However, in the hawk-dove game, another problem arises: Since there are two Nash equilibria here, the players must be able to agree on one of the equilibria *implicitly.* Such a communication problem between two equilibria is called a *coordination problem.*

We refer to "implicit" communication because it is a simultaneous game in which, by definition, both sides have *imperfect* information about the (intended) action of the opponent during the game. The players must therefore derive a clue about the possible action of the opponent solely from what they already know about each other at this point. Such a clue is called a *focal point.* In this context, we distinguish between *exogenous* and *endogenous focal points.* An endogenous focal point can be read directly from the payoffs of the game. For example, if one of the Nash equilibria pareto-dominates the other, it is likely that both will orient themselves towards this more attractive equilibrium without the possibility of communication. If such an endogenous focal point is not present (which is the case here), an exogenous focal point could still help. Such a point of orientation is called "exogenous" because it does not result from the payoffs or the payoff structure but results from factors outside the game. For example, the players might simply follow their habits or be guided by traditions and social conventions, e.g., that you pass each other on the right (or left). The power structure and hierarchy at the level of society (such as the family) or an organization (such as a company) can also provide clarity in the selection of an equilibrium. Statements about an exogenous focal point can therefore only be derived from the specific (external) context of the game.

If there are no focal points, there is a risk of miscoordination, where both players miss the efficient Nash equilibria and end up in the worst constellation (0, 0). One circumstance that makes miscoordination likely in the hawk-dove game is the direct conflict of interest over the two Nash equilibria: Each player wants to enforce exactly the Nash equilibrium that the other wants to prevent. So, if both

Table 3.14 Battle of the sexes and assurance game

(a) Battle of the sexes

Alice, Bob	s_{B1}	s_{B2}
s_{A1}	(0, 0)	$(\underline{1}, \underline{3})^{P*}$
s_{A2}	$(\underline{3}, \underline{1})^{P*}$	(0, 0)

(b) Assurance game

Alice, Bob	s_{B1}	s_{B2}
s_{A1}	$(\underline{3}, \underline{3})^{P*}$	(0, 2)
s_{A2}	(2, 0)	$(\underline{1}, \underline{1})^{*}$

players try to act as aggressively as possible at the same time to force the other to give in, the confrontation results in the worst outcome for both (0, 0). The hawk-dove game illustrates the danger of a conflict escalation that ends tragically for all parties ("catastrophe").

Both problems discussed in this game, the (weak) cooperation problem and the coordination problem, can be considered independently of each other. In the second game in Table 3.13(b), the hawk-dove game was slightly modified (the cell in the bottom left was "neutralized" by the payoffs (0, 0)). With this change, we have removed the coordination problem, so now only the (weak) cooperation problem remains: Alice has a dominant strategy and Bob chooses the result. If, on the other hand, one isolates only the pure coordination problem of the hawk-dove game, one arrives at the left game in Table 3.14(a), which is also known as the "Battle of the Sexes".

Battle of the Sexes. This (somewhat overly dramatic) name of the game describes the situation of a married couple who want to agree on a common leisure activity, but both have different wishes. One variant would be, for example, that Alice prefers to go to the cinema, while Bob wants to go to a football match. The strategies here—similar to the hawk-dove game or the game of chicken—would either be "comply"/ "give in" (s_{i1}), or "enforce" (s_{i2}). This game is in some ways even riskier than the hawk-dove game, as here miscoordination is possible in both directions. In the hawk-dove game, a rather cautious, risk-averse player will rather opt for the yielding or peaceful strategy: He can never assert himself against the other player with this strategy, but he can at least prevent the catastrophe alone, which would be the higher priority for such a player. In the battle of the sexes, there is no such safe fallback strategy. Since there is no endogenous focal point in this game (there is as much or as little to be said for (3, 1) as for (1, 3)), only an exogenous focal point could serve as a coordination aid. For example, it could be that the couple has been to the football together the last few times, so now it is time for a visit to the cinema again.[13]

Assurance game The second game in Table 3.14(b) also involves a coordination problem. This game has several names. In the literature, it is known as the *assurance game, trust dilemma, stag hunt,* and *arms race.* In contrast to the coordination problem in Table 3.14(a), there is an endogenous focal point at the strategy combination (s_{A1}, s_{B1}) in the top left. The payoff combination (3, 3) Pareto-dominates all others, meaning it is also the only efficient result of the game, and it simultaneously emerges as a Nash equilibrium. Consequently, there is much to suggest that the players will agree on this equilibrium. However, there is also a residual risk for both players in this game, which consists of misjudging the other. If, despite all the plausibility of (s_{A1}, s_{B1}) there is a miscoordination, the players either end up at (s_{A2}, s_{B1}) with the payoff combination (2, 0) or at (s_{A1}, s_{B2}) with the payoffs (0, 2). So, there is also a conflict of interest here, which, however—unlike the hawk-dove game or the battle of the sexes—occurs away from the equilibria. The term "stag hunt" for this game is based on the story of two hunters who can either kill a hare individually ((s_{A2}, s_{B2}) with (1, 1)) or a stag together ((s_{A1}, s_{B1}) with (3, 3)). The conflict arises when one relies on the other to support him in the stag hunt, but the latter abandons the former to secure a hare (asymmetric payoffs (0, 2)). In the variant of the "arms race", the same fear of a unilateral deviation by the co-player (in this case it would be a country with the strategies "rearm", s_{i1}, and "disarm", s_{i2}) comes into play: The endogenous focal point (3, 3) stands for general disarmament (s_{A1}, s_{B1}), which would clearly be advantageous for both countries. Nevertheless, a country can be deterred from taking the sensible path of disarmament solely by the residual risk that the other country does not pursue the same interests.

The fact that a conflict of interest *aside from equilibria* can influence the outcome of the game is illustrated by the game in Table 3.15(a). This is also a version of the assurance game, but the payoffs were chosen so that the risk of miscoordination is particularly evident. Again, the epsilon (ε) stands for the smallest possible positive value; in terms of money units, this would be one cent. From the perspective of Alice, it would be very plausible for both players to choose the first strategy. However, it is conceivable that Bob does not notice the marginal difference between $2 + \varepsilon$ and 2 at all and simply randomly chooses the second strategy, as he is indifferent between both strategies. A second possibility would be that Bob is competitively minded and is willing to accept the loss of the small ε to

[13] An example of an endogenous focal point would be that the football game promises little excitement, whereas the movie is a big blockbuster that Alice has been looking forward to for a long time and with which Bob can also arrange. In this case, the players would have different payoffs: The payoff combination for the movie could be (6, 2), for example, where the high number 6 expresses Alice's anticipation. For football, the payoff combination (1, 2.5) would be conceivable, i.e., Bob would only have a slightly greater interest in football than in the movie in this case. Both payoff combinations would still be Pareto-optimal, but it is very plausible that the variant (6, 2) would prevail as an endogenous focal point.

Table 3.15 Risky assurance game and low-risk coordination game

(a) Risky assurance game

Alice, Bob	s_{B1}	s_{B2}
s_{A1}	$(\underline{2+\varepsilon}, \underline{2+\varepsilon})^{P\,*}$	$(-10, 2)$
s_{A2}	$(2, -10)$	$(\underline{1}, \underline{1})^{*}$

(b) Low-risk coordination game

Alice, Bob	s_{B1}	s_{B2}
s_{A1}	$(\underline{2}, \underline{2})^{P\,*}$	$(0, 0)$
s_{A2}	$(0, 0)$	$(\underline{1}, \underline{1})^{*}$

perform better than Alice. Or Bob is simply afraid that the first two possibilities apply to Alice, so he prefers to play it safe with the strategy choice s_{B2}. It is also legitimate that none of these three considerations come into question if strict rationality and common knowledge of all players are assumed. However, since we are also interested in stable or robust game results, the sensitivity of the common-knowledge assumption is central in this case[14]: In this game, each player must be completely sure that the other player is acting strictly rationally. The same must also apply to the other player. And both must also be equally aware that the other is aware of this, etc. The slightest doubt can immediately justify switching to the second strategy. For this reason, this game is also called the *assurance game* and *trust game*: Each player must assure themselves of the other's rationality or trust in their rationality.

Using the concept of *risk dominance*, we can determine for a Nash equilibrium whether the associated risk is too high for the players and whether they prefer to switch to another equilibrium that seems safer to them. With two players and two strategies, a Nash equilibrium is *risk-dominant* if each player prefers the equilibrium strategy assuming that the other player is equally likely to play both strategies. This criterion implies for coordination problems that a rational and risk-averse player chooses the strategy that keeps the maximum loss within limits. For the risky assurance game in Table 3.15(a), the Nash equilibrium (s_{A1}, s_{B1}) is risk-dominant if

$$(2+\varepsilon)\cdot 0.5 - 10\cdot 0.5 \overset{!}{>} 2\cdot 0.5 + 1\cdot 0.5 \quad \Leftrightarrow \quad \varepsilon \overset{!}{>} 11$$

For this, the left side of the inequality, which indicates the expected payoff for the first strategy (s_{i1}), must be greater than the right (expected payoff of s_{i2}). Since it is a symmetric game, this test is thus simultaneously performed for both players. The probability with which the opponent chooses the strategies is assumed to be 0.5—a plausible assumption if no further knowledge about the other player is

[14] See also the discussion of the centipede game (Fig. 2.5) in Sect. 2.3.

available. We see that for this $\varepsilon > 11$ would have to apply, which according to our initial assumption (ε smallest possible amount) is not permissible. From the high required value for ε, we can see that the first strategy in the game in Table 3.15(a) is far too risky for the players. The risk-dominant Nash equilibrium is therefore clearly in the second strategy (s_{A2}, s_{B21}).

Finally, we consider the coordination problem in Table 3.15(b). Here too, as with any coordination problem, there is the risk of miscoordination, but there is no conflict of interest, neither *between* the equilibria nor *aside* the equilibria. The fact that there is no conflict of interest in this game causing additional risk can also be seen from the fact that the first strategy is risk-dominant in this game. Such a conflict-free coordination problem can therefore be relatively easily overcome by the endogenous focal point at (2, 2).

3.5 Games with Continuous Strategies

So far, we have limited our explanations to matrix games or at least outlined the most important basics using the example of matrix games. Matrix games are games with discrete strategies, where each player can choose from a limited number of options (such as "cooperate" or "cheat", shoot "left" or "right", etc.). In a 2×2 matrix, each player chooses the strategies from the two-element set $s_i \in \{s_{i1}, s_{i2}\}$. However, in some cases, it makes more sense to represent game situations with a continuous strategy space. In contrast to discrete strategies, continuous strategies encompass a continuous spectrum of actions on a cardinal scale—this also implies a convex strategy set, so the existence of a Nash equilibrium is generally ensured. In relation to the penalty game, there would then no longer be just the two strategies "left" and "right", but every point on the goal line would be an element of the strategy set of shooter and goalkeeper. If we were to mentally lay a tape measure on the goal line, then the tape measure covers the continuous strategy set of shooter and goalkeeper. With continuous strategies, each player therefore chooses the strategy from the compact interval $s_i \in [s_i, s_i]$, i.e., between the left and right goalposts. If this interval is infinitely divisible, then player i has an infinitely large strategy set. The formulation of a game in continuous strategies is reasonable when the *quantifiability* of selected actions is in focus, such as quantity or price strategies in corporate competition. In price competition, for example, each company chooses a certain value of the product price from a given price range.

Games with continuous strategies have the advantage that they can be solved with the methods of calculus (see Chap. 12). If we want to determine a Nash equilibrium for these games, we must—as before in the matrix games—maximize the utility of each player given the strategy choice of the opponent. The maximization is now done by setting the first derivative of the objective function of each player to zero, i.e., by determining the mathematical "first-order condition". When determining a Nash equilibrium in mixed strategies, we proceeded similarly in the optimization approach, while there is an important difference: In the mixed strategies, we did not derive with respect to the strategies, but with respect to the probability with which a player plays a discrete strategy. Since each probability is also a continuous variable, we were able to determine the equilibrium using calculus. In the case of continuous strategies, however, these are always pure strategies, because they are directly and clearly chosen by the players and do not emerge as a "mixture" or expected value from a random process.[15]

As an example, let us now consider a market with only two companies (i.e., a *duopoly*), which produce a certain good and compete. The companies decide how much of the product they want to produce and supply to the market. The imperfect competition situation described here is referred to as *Cournot competition.* If the companies were to compete over the price instead, it would be referred to as *Bertrand competition.* Furthermore, we assume that the good in question is a *differentiated* good. In the case of a differentiated good, the product sold by the companies is not considered equivalent for all consumers (*imperfect substitutes*), but a part of the consumer has a certain preference for the product of company A and another part of the customers has a preference for the product of company B. The relevant price, which company i can achieve depending on the supplied quantity x_i, is

$$p_i(x_i, x_j) = 120 - 2 \cdot x_i - x_j. \tag{3.17}$$

The price achievable by company i is inversely related to the quantity: The higher (or lower) the output quantities of both companies i and j, the lower (or higher) is the price. This is a standard property of a "normal" market demand. Note, however, that this is strictly speaking the *inverse demand*, as the actual demand is the quantity resulting from the prevailing price.[16] Since we consider *quantity strategies*, the companies choose their output quantity (somewhat more realistic you can also speak of "production capacity") and the price then adjusts as a result. With a very high output quantity, the product is abundantly available on the market, so the customers' willingness to pay decreases. With a very low output quantity, on the other hand, the product is very scarce on the market, which

[15] Let us consider the following example to clarify the difference between mixed strategies and continuous strategies: The strategy is how much of a cake we want to eat. With two discrete strategies $s \in \{0, 1\}$, we can decide whether to eat the entire cake ($s = 1$) or nothing of it ($s = 0$). With mixed strategies, these two discrete strategies are randomized, meaning that with any probability $p \in [0, 1]$, we eat the entire cake or nothing of it. For example, for $p = 0.5$, a coin can be tossed: For "heads" we eat the whole cake ($s = 1$), for "tails" we do not eat it ($s = 0$). With continuous strategies, we determine the exact proportion (between 0 and 100%) of the cake that we want to eat, meaning that the extreme cases "eat nothing" (0%) or "eat everything" (100%) only specify the minimum and maximum amount. In the entire interval $s \in [0, 1]$, we can choose any proportion of the cake—"down to the smallest crumb" (like 13.7%, 81.294%, etc.). In the end, with mixed strategies the entire cake is gone with a certain probability, while with continuous strategies a part of the cake remains for sure.

[16] Think about your own everyday experience: when you go shopping, you see the prices and then decide how much you want to buy. The shops do not set the price according to the quantity you buy.

is reflected in a correspondingly high willingness to pay—thus in a high price. In the extreme case, where none of the companies offer products on the market ($x_i = x_j = 0$), the price would be maximum ($p = 120$). This maximum price is the intercept of our price function and is also called *prohibitive price.* The different quantity coefficients 2 and 1 mean that the price reaction to a change in x_i is twice as strong as with a change in x_j. Since this is a differentiated good, the own output quantity of company i always has a stronger effect on its own price p_i than the output quantity of the competitor. For example, if company i increases its quantity x_i, then both the willingness to pay of its own group of customers and the willingness to pay of the common group of customers decreases. On the other hand, if company j increases its quantity x_j, then for company i this only has the consequence that only the willingness to pay of the common group decreases.[17]

The objective function to be optimized by company i is its profit. For simplification, we disregard costs here, so the profit π_i corresponds to the turnover of company i. The turnover or revenue of a company results as the product of the selling price and the sales quantity, $R_i = p_i \cdot x_i$. The profit function of company i is then given by

$$\pi_i\left(x_i, x_j\right) = p_i\left(x_i, x_j\right) \cdot x_i = \left(120 - 2 \cdot x_i - x_j\right) \cdot x_i =$$

$$= 120 \cdot x_i - 2 \cdot x_i^2 - x_j \cdot x_i. \tag{3.18}$$

To determine the profit-maximizing output quantity, the profit π_i is partially derived with respect to x_i, the resulting first-order condition is set to zero, and then solved for i's strategic variable x_i, $\partial\pi_i/\partial x_i \overset{!}{=} 0$. The result is the optimal output quantity $x_i^*\left(x_j\right)$ of company i, for which—given the output quantity x_j of its competitor—the profit is maximized. $x_i^*\left(x_j\right)$ is nothing else than the reaction function of company i.[18] Specifically, the reaction function of company i is

$$\frac{\partial\pi_i}{\partial x_i} = 120 - 2 \cdot 2 \cdot x_i - x_j \Rightarrow 120 - 4 \cdot x_i - x_j \overset{!}{=} 0$$

$$\Leftrightarrow \; x_j^*\left(x_j\right) = 30 - \frac{1}{4} \cdot x_j. \tag{3.19}$$

[17] If company j increases its quantity, then of course the willingness to pay of its own group of consumers also decreases. In the case of a fully differentiated good (quantity coefficient of $j = 0$), there would no longer be a common group of customers, meaning that there would then be two separate markets, each with one of the companies as a monopolist. In the case of a purely homogeneous good, on the other hand, the quantity coefficients for both companies i and j are identical, meaning that the effects of a chosen output quantity on both the own and the competitor's price would be equally high.

[18] So far, we have been talking about reaction correspondence, which is more general. Here we deliberately speak of a reaction function since we have a clear assignment $x_i \rightarrow r_j(x_i) = x_j$. This is not the case with mixed strategies—consider here the vertical segment $p_i \rightarrow p_j(p_i) \in [0; 1]$.

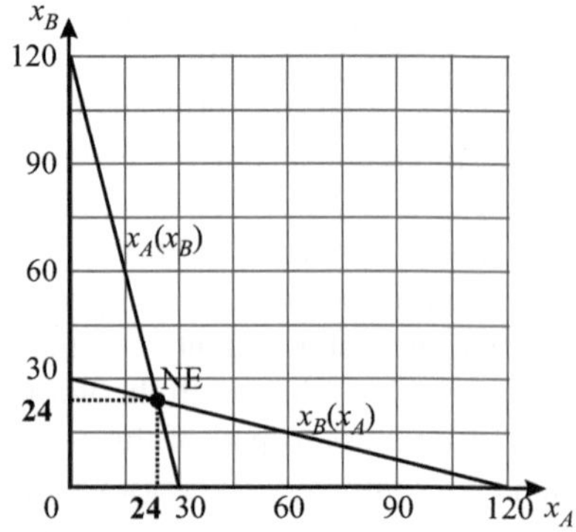

Fig. 3.5 Reaction correspondences for the duopoly with differentiated goods

So far, we have viewed the game from the perspective of a representative company i that is in competition with company j. This more general approach is always useful when a game is symmetrical, i.e., when the outcome of the game does not change due to a (hypothetical) role swap between the players. The advantage of this more general representation is that we obtain a reaction function with a single optimization that is valid for all players. Thus, we can specify the reaction function of the two companies with (3.19).

$$r_A(x_B) : x_A^*(x_B) = 30 - \frac{1}{4} \cdot x_B \Leftrightarrow x_B\left(x_A^*\right) = 120 - 4 \cdot x_A$$

$$r_B(x_A) : \ x_B^*(x_A) = 30 - \frac{1}{4} \cdot x_A \tag{3.20}$$

The reaction function $r_A(x_B)$ represents the best response of company A to the chosen quantity of company B and $r_B(x_A)$ correspondingly the best response of company B to the quantity of company A. Figure 3.5 graphically represents the reaction functions.

When transferring the companies' reaction functions into the graphical representation, it should be noted that depending on the chosen coordinate system (i.e., which variable is on the horizontal and which on the vertical axis) the reaction function of one player must be rearranged. In our case, we must rearrange the reaction function of company A after x_{B2} which is also presented in Eq. (3.20).

The *Nash equilibrium* (NE) of the game results from the intersection of the reaction curves. At this point, the quantities chosen by the companies represent mutually best responses, i.e., none of the companies has an incentive to deviate from this. Mathematically, the solution can be found by substituting the reaction functions into each other.

$$x_A^*\left(x_B^*\right) = 30 - \frac{1}{4} \cdot \left[30 - \frac{1}{4} \cdot x_A\right] \Leftrightarrow x_A^* = 24, \ x_B^* = 24.$$

In equilibrium, the companies will each produce and sell 24 units. By substituting the equilibrium quantities into Eqs. (3.18) and (3.19), we obtain the corresponding prices and profits of the companies in equilibrium. Specifically, we get the prices $p_A^* = p_B^* = 48$ and profits $\pi_A^* = \pi_B^* = 1.152$. The following scheme summarizes the procedure for determining Nash equilibria in continuous strategies.

General Procedure for Continuous Strategies

1. Determination of the strategic variables of the players. (e.g., *price, quantity, …*).
2. Setting up the payoff function depending on the strategic variable of the considered player and that of his/her opponent(s) (e.g., *profit function*).
3. Partial derivative of the payoff function with respect to the strategic variable of the considered player and setting the resulting condition equal to zero (first-order condition).
4. Resolve the condition according to the strategic variable of the considered player to obtain the player's reaction function.
5. Repeat steps 3 and 4 for all other players.
6. Mutual substitution of the reaction functions of all players to determine the Nash equilibrium.

As with matrix games, we can also examine the efficiency of the solution in games with continuous strategies. There is a complication and a convenience at the same time. The complication is that we simply cannot recognize a Pareto-optimal solution immediately in games with continuous strategies, as the strategy sets of the players are infinitely large and there are therefore infinitely many Pareto-optimal solutions. The convenience, however, is that we can determine efficient, i.e., socially optimal, solutions directly via the same derivation procedure that we use to find the Nash equilibria. For this, we only need to optimize the common objective function of all players together. Applied to the duopoly game, we find the socially optimal solution where the sum of the profits of both companies is maximum. We can imagine that the two companies would merge and then ask ourselves which quantities the merged company would choose to maximize its total profit $\pi_{\text{total}} = \pi_A + \pi_B$ with respect to products x_A and x_B. Since there is no possibility to reach a higher profit, this is the best achievable solution.[19]

In general, a social planner or the hypothetical merged company will consider both quantities equally when optimizing the total profit, $\pi_{\text{total}}(x_i, x_j)$, i.e., we must

[19] Of course, this is only the "best achievable solution" for the players considered in this game, i.e., exclusively for the duopoly. In particular, the welfare of consumers is not considered here; they do not appear explicitly as players. The joint profit maximization of a duopoly is identical to the behavior of a cartel. From an overall economic perspective, which also considers the situation of consumers, this leads to an inefficiently low quantity at too high prices.

partially derive the profit with respect to both, x_i and x_j, so that we obtain two first-order conditions that we can solve as a system of equations, $\partial\pi_{\text{total}}(x_i, x_j)/\partial x_i \overset{!}{=} 0$ and $\partial\pi_{\text{total}}(x_i, x_j)/\partial x_j \overset{!}{=} 0$.

For our duopoly game, we obtain a total profit of

$$\begin{aligned}\pi_{\text{total}}(x_A, x_B) &= \pi_A(x_A, x_B) + \pi_B(x_A, x_B) = \\ &= (120 - 2 \cdot x_A - x_B) \cdot x_A + (120 - 2 \cdot x_B - x_A) \cdot x_B\end{aligned}$$

from which the two partial derivatives

$$\begin{aligned}\frac{\partial\pi_{\text{total}}(x_A, x_B)}{\partial x_A} &= 120 - 4 \cdot x_A - x_B - x_B \quad \text{and} \\ \frac{\partial\pi_{\text{total}}(x_A, x_B)}{\partial x_B} &= 120 - 4 \cdot x_B - x_A - x_A\end{aligned} \tag{3.21}$$

result. If we set both derivatives to zero, we obtain a system of equations of the first-order conditions:

$$\begin{aligned}120 - 4 \cdot x_A - 2 \cdot x_B &\overset{!}{=} 0 \\ 120 - 4 \cdot x_B - 2 \cdot x_A &\overset{!}{=} 0\end{aligned}$$

The solution of this system of equations gives the efficient output quantities $x_A^* = x_B^* = 20$. With this, we calculate the prices $p_A^* = p_B^* = 60$ and total profit $\pi_{\text{total}}^* = 2{,}400$.

If we now compare the socially efficient solution with the previously determined competitive solution, we see that the profit of a company in competitive equilibrium with 1,152 is lower than an equal distribution of the total profit, which would yield 1,200 for each company. The comparison of quantities and prices clarifies why this is the case: In the Nash equilibrium, the output quantities are larger and therefore the prices are lower than in the efficient solution. If the companies were to merge and act together in this sense, they would produce less at higher prices and thus achieve higher profits.

Since the equilibrium solution obviously deviates from the efficient solution, there must be an *external effect* present. We can quantify this external effect exactly by comparing the partial derivatives from expression (3.19) and (3.21):

$$\frac{\partial\pi_i(x_i)}{\partial x_i} = 120 - 4 \cdot x_i - x_j \tag{3.22}$$

$$\frac{\partial\pi_{\text{total}}(x_i, x_j)}{\partial x_i} = 120 - 4 \cdot x_i - x_j \underbrace{\boxed{-1 \cdot x_j}}_{\substack{\text{negative} \\ \text{external} \\ \text{effect}}} \tag{3.23}$$

Equation (3.22) describes the individual optimization approach of company i and Eq. (3.23) the joint optimization approach of both companies together (that is, the optimization of the social planner or the merged company). Obviously, the derivatives match each other except for one term. The additional term, which only appears in the joint optimization approach and is highlighted, has a negative sign. How can this term be interpreted? The partial derivative of the profit with respect to the quantity x_i tells us by how much the profit increases when the quantity x_i marginally increases. This profit increase is apparently smaller in the case of joint profit maximization than in individual profit maximization and this marginal profit loss is greater, the higher the quantity x_j of the competitor is. In joint profit maximization, company i will therefore produce less, as it considers that its output quantity x_i also reduces the price p_j and thus the profit π_j of company j.[20] The profit π_j is finally a part of the total profit $\pi_{total} = \pi_i + \pi_j$. In individual maximization, however, company i will not consider this effect, as it is not its own profit. Since both companies ignore this effect, their output quantity is inefficiently high and the resulting prices correspondingly too low (the companies are effectively undercutting each other with high quantities). They could both achieve higher profits if they produced less, to thereby enable higher sales prices. This is a case of a *negative external effect*, as the relevant activity of the players (here the choice of production quantities) *negatively* affects their payoffs (which is also indicated by the sign of the term). So, they could both be better off if they limit this activity, i.e., produce less.

Cournot competition with many companies. In classes on game theory, the case of homogeneous goods is very often considered. This is a special case of differentiated goods in which the quantity coefficients of the firms are identical. With homogeneous goods, no company has its own group of customers anymore, but there is only one jointly shared demand. We now want to take a closer look at this competition and expand our analysis by looking more generally at n companies that compete on quantity. In this case, the game-theoretical tools allow us to link important results of microeconomics regarding the market forms monopoly (on supplier), oligopoly (several suppliers), and polypoly (perfect competition; many suppliers) in a comprehensive approach.

The general specification of a (inverse) linear demand is

$$p(X) = a - b \cdot X.$$

Here, a and b are parameters, where a is the prohibitive price (the price above which there is zero demand) and b is the amount by which the price must fall

[20] The external effect can only occur in relation to the common group of customers. You can see from the quantity coefficient that only the reaction of this group is involved in this problem: The quantity coefficient 1 corresponds exactly to the influence of one's own quantity decision on the *common* group of consumers; 2 was the influence of one's own quantity decision on the *own* group of customers. Regarding the latter, there can be no external effect, since one's own group of customers does not react to the competitor's quantity decision (and vice versa).

for another unit to incentivize the demand of one further unit. X represents the total supply of all n companies in the market, i.e., $X \equiv x_1 + x_2 + \cdots + x_n$ ("$\equiv$" denotes an identity) or expressed more formally $X \equiv \sum_{j=1}^{n} x_j$. We assume that all companies are symmetric and have the same cost function:

$$C(x_i) = c \cdot x_i + f.$$

The first part, $c \cdot x_i$, represents the variable costs, i.e., the costs that depend on the quantity produced. We assume constant marginal costs c, i.e., each additional unit produced costs the same. The second part f are the production-independent fixed costs. Now let us consider any company i whose profit function is given by

$$\pi_i(x_i) = p(X) \cdot x_i - C(x_i) = \left(a - b \cdot \sum_{j=1}^{n} x_j\right) \cdot x_i - (c \cdot x_i + f).$$

As we can see, the profit depends on many players, depending on the exact value of n. As in the simple case with only two companies/players, we need to determine the reaction function of company i to finally obtain the maximum profit. This results from the first-order condition. However, the sum expression, which depends on all output variables and thus also on x_i, is "irritating" here. Thus, let us reformulate this by separating the quantity of company i from the sum:

$$\pi_i(x_i) = \left[a - b \cdot \left(x_i + \sum_{j \neq i}^{n} x_j\right)\right] \cdot x_i - c \cdot x_i - f.$$

The sum now includes all other players except i. In principle, we could therefore also write x_{-i} instead of the sum expression. Since this expression does no longer depend on x_i, it can be considered as a constant in the partial derivative. Thus, we obtain as a first-order condition:

$$\frac{\partial \pi_i}{\partial x_i} = a - 2 \cdot b \cdot x_i - b \cdot \sum_{j \neq i}^{n} x_j - c \stackrel{!}{=} 0.$$

With this, we can determine the reaction function of company i,

$$x_i(x_{-i}) = \frac{a - c}{2 \cdot b} - \frac{1}{2} \cdot \sum_{j \neq i}^{n} x_j. \tag{3.24}$$

As in the case of only two players, we obtain n reaction functions, which we would have to substitute into each other to determine the Nash equilibrium. However, we can take advantage of the symmetry of the game. We know that the companies are symmetric both in terms of their demand and their cost structure, which means that all companies will supply the same quantities in equilibrium: $x_1 = x_2 = \cdots = x_i = \ldots = x_n$. Thus, the total supply is $X = n \cdot x_i$ and accordingly $\sum_{j \neq i}^{n} x_j = (n-1) \cdot x_i$ (it is $(n-1)$ because we excluded x_i). If we consider this in the reaction function (3.24), we can determine the Nash equilibrium:

$$x_i = \frac{a - c}{2 \cdot b} - \frac{1}{2} \cdot (n-1) \cdot x_i \;\Leftrightarrow\; x_i = \frac{a - c}{b \cdot (n+1)} \tag{3.25}$$

Note that we have made use of the symmetry property only in the first-order condition. If we had tried to apply this in the profit function, we would have altered the demand function and thus would have arrived at incorrect results.

If we insert this into the demand function, we obtain the price

$$p = a - b \cdot n \cdot x_i \;\Leftrightarrow\; p = \frac{a + c \cdot n}{n + 1} \tag{3.26}$$

With this, we are now able to immediately determine the quantity and price for any number of companies. Finally, the profit is

$$\pi = \frac{(a - c)^2}{b \cdot (n + 1)^2} - f \tag{3.27}$$

An interpretation of the results shows that, unsurprisingly, as the number of companies increases, the output quantity of each company decreases, the price falls, and less profit can be generated.

Finally, we would like to discuss the existence of a Nash equilibrium in continuous strategies. With discrete strategies, we saw that this is not always guaranteed due to the non-convex strategy sets. The following general result is helpful for this:

If the payoff functions are continuously differentiable and strictly concave, and the strategy space S_i is a compact and convex interval in R^1 (one-dimensional strategy space), then there is always an internal solution of $\partial u_i / \partial s_i = 0$ for all i as a Nash equilibrium.

This statement is helpful because the assumptions made here are all met in the case of linear demand functions. That is, for most of the problems discussed in microeconomic classes, a Nash equilibrium can be found by optimization (as in Fig. 3.5) and is therefore not a boundary solution, for which more complicated solution methods and/or case distinctions would be necessary. Thus, this statement may seem very frightening, but for us it means that we do not need to worry too much about the existence of an equilibrium!

3.6 Comprehension Questions

1. A strictly Pareto-dominated strategy combination …
   A Can never be an equilibrium in dominant strategies
   B Can sometimes not be realized, as a player can improve by deviating
   C Always makes both players worse off than any other strategy combination
   D Does not allow any statement about whether an equilibrium in dominant strategies exists

2. In a 2×2 matrix game, a Nash equilibrium has already been found. This Nash equilibrium is unique if …
   A Both players have a strictly dominant strategy

B Both players have a weakly dominant strategy
C If only one of the players has a dominant strategy, which is also strictly dominant
D If it is a Nash equilibrium in mixed strategies

3. Consider the following payoff combinations of a two-person game: (2, 1), (1, 0), (0, 5), (4, 0). Which of the following statements is correct?
A Only (2, 1) and (0, 5) are Pareto-optimal
B Only (2, 1), (0, 5), and (4, 0) are Pareto-optimal
C (1,0) is Pareto-dominated by (2, 1) and (4, 0)
D (0, 5) dominates (2, 1), (1, 0), and (4, 0) according to the Kaldor-Hicks compensation criterion

4. Consider a symmetric duopoly game with continuous strategies, where the companies choose their R&D investments. If there is a positive external effect in a Nash equilibrium of this game, then …
A The investments have a positive effect on the companies' profits
B The Nash equilibrium is inefficient because the companies invest too much in R&D
C The Nash equilibrium is efficient because the companies invest a lot in R&D
D The Nash equilibrium is inefficient because the companies invest too little in R&D

5. A weak cooperation problem can be overcome …
A By a mechanism of correlated strategies
B By binding agreements
C By an endogenous focal point
D If the strategy that leads to an efficient compromise solution is risk-dominant

6. Which of the following statements is correct?
A In a Nash equilibrium, it is never possible to make one player better off without Making another worse off
B A Nash equilibrium is always efficient
C A payoff combination is Pareto-optimal if none of the players has an incentive to deviate
D If it is not possible to make one player better off without making another player worse off, then this situation is efficient according to the Pareto criterion

1B,D; 2A,C; 3B,C,D; 4D; 5A,B,D: Explanation for D: If you change in the hawk-dove game (Table 3.10(a)) the payoffs of the efficient compromise solution from (2, 2) to (2.5, 2.5), the first strategy becomes risk-dominant for both players. If both players fear a miscoordination on (0, 0), they play the risk-dominant, first strategy to be on the safe side and thus achieve the cooperative solution—by mutual caution, so to speak; 6D.

3.7 Problems

3.7.1 Dominance Considerations

a. Determine the *rationalizable* strategies in the following matrix! Then indicate for the statements (i) and (ii) whether these are correct or incorrect and refer to the matrix if necessary for the justification of your answer!

Alice, Bob	Left	Center	Right
Top	(3, 2)	(1, 0)	(2, 1)
Bottom	(4, 0)	(0, 2)	(1, 1)

i. A rationalizable strategy always survives the iterative elimination of strictly dominated strategies.
ii. A strategy that survives the iterative elimination of strictly dominated strategies is always also a rationalizable strategy.

b. Solve the following matrix game using the method of iterative elimination of *strictly* dominated strategies! If you find a Nash equilibrium in pure strategies: Is this unique? Justify!

Alice, Bob	s_{B1}	s_{B2}	s_{B3}	s_{B4}
s_{A1}	(5, 2)	(1, 4)	(−1, 1)	(2, −1)
s_{A2}	(4, 0)	(2, 1)	(3, 0)	(3, −1)
s_{A3}	(−2, −2)	(1, −1)	(2, 3)	(2, 2)
s_{A4}	(1, 1)	(1, 3)	(1, 2)	(4, 1)

c. Solve the following matrix game using the method of iterative elimination of *weakly* dominated strategies! If you find a Nash equilibrium in pure strategies: Is this unique? Justify your answer!

Alice, Bob	s_{B1}	s_{B2}	s_{B3}
s_{A1}	(1, 0)	(0, 1)	(2, 3)
s_{A2}	(−1, 0)	(0, 1)	(3, 4)
s_{A3}	(2, 5)	(5, 2)	(5, 5)
s_{A4}	(2, 2)	(2, 2)	(1, 3)

Solution in Sect. 3.8.1.

3.7.2 Mixed Strategies

Consider the following decision situation between a customer and the producer of an experience good.[21] The producer sells the good at price p and can produce it in high or low quality. High quality incurs additional effort costs of e. The customer cannot directly recognize the quality at the time of purchase but has the option to carry out a targeted check by, for example, consulting an expert or using special technical aids. The control measure costs the customer k. Without control, the customer has no choice but to trust the producer. A high-quality good provides the customer with a high utility of u_h, a low-quality good only provides a low utility u_l. If the customer checks and finds low quality, he refrains from buying, which gives him a (gross) utility of u_m. It also applies $p - e > 0$ (the producer does not make losses at high quality) and $u_m - k > u_l$ (the consumer prefers to refrain from buying a checked product than to get low quality).

a. Represent this inspection game in matrix form! Show that there is no equilibrium in pure strategies in this game!
b. Determine the Nash equilibrium in mixed strategies both via (i) the expected utility comparison and via (ii) the optimization approach! Interpret the calculated probabilities and explain to which extent the optimal probability of the producer and that of the customer depend on the parameters!
c. Present the result formally and graphically in the form of reaction correspondences! Determine the expected equilibrium payoff of the players!

Solution in Sect. 3.8.2.

3.7.3 Nash Equilibrium in Games with Continuous Strategies: Price Strategies in Duopoly

Two symmetric companies offer a differentiated good and are in price competition with each other, thus choosing the price as a strategic variable. The individual demand for good i is

$$x_i(p_i, p_j) = a - p_i + \beta \cdot p_j.$$

This demand for x_i is negatively dependent on its own price p_i, but positively on the competitor's price p_j, i.e., if the competitor raises its prices, some customers will switch to the considered company because it is cheaper. Since the goods are differentiated, the influence of the competitor's price is assumed to be weaker, that is, $\beta < 1$.

[21] A good is an experience good if its quality (without further help) can only be determined after purchase. Examples of experience goods include the contents of canned food or the durability of an electronic device.

a. Set up the profit function from the perspective of company i, determine the reaction functions, as well as the equilibrium prices and quantities!
b. Represent the reaction functions graphically!
c. Examine the reaction functions to see if there is an external effect in duopoly price competition!

Solution in Sect. 3.8.3.

3.7.4 Competition Among Many Companies

In a market with market demand $X = 15 - 0.5 \cdot p$, where $X \equiv x_1 + x_2 + \cdots + x_n$, n companies compete in quantities. The cost function of company i is given by $C(x_i) = 6 \cdot x_i$.

a. Determine the quantity $x_i(n)$, the price $p(n)$, and the profit of a company $\pi_i(n)$ in oligopoly equilibrium for $n = 2$, 5, and 11 companies!
b. Show for a general specification of the inverse demand $p(X) = a - b \cdot X$ and the cost function $C(x_i) = c \cdot x_i + f$, that for $n \to \infty$ (perfect competition) the price corresponds to the marginal costs!

Solution in Sect. 3.8.4.

3.8 Solutions

3.8.1 Dominance Considerations

Problem Text in Sect. 3.7.1.

Part a)

Alice, Bob	Left	Center	Right
Top	(3, 2)	(1, 0)	(2, 1)
Bottom	(4, 0)	(0, 2)	(1, 1)

Alice has only two strategies; since neither of the two is strictly dominant, both strategies are rationalizable strategies: If Bob chooses "Left", then Alice's best response is "Bottom"; if Bob chooses "Center" or "Right", then Alice's best response is "Top". Both "Top" and "Bottom" are the best response to at least one strategy choice of Bob; therefore, both are rationalizable strategies.

If Alice chooses "Top", then "Left" is Bob's best response; if Alice chooses "Bottom", then "Center" is Bob's best response. The rationalizable strategies of Bob are therefore "Center" and "Left". "Right" is not a rationalizable strategy for Bob, as it does not represent the best response to any strategy choice by Alice.

To (i): The statement is correct. A rationalizable strategy represents by definition the best response to (at least) one strategy of the opponent, i.e., a rationalizable strategy is never strictly dominated.

To (ii): The statement is false. A counterexample to the statement is provided by the strategy "Right" in the matrix: As shown, "Right" is not a rationalizable strategy for Bob. It is also not a dominated strategy, so it would also survive the process of iterative elimination of strictly dominated strategies. Thus, we must be aware that there are strategies which, on the one hand, are not the best response but, on the other hand, are not dominated strategies.

Part b)

If you start the elimination with Alice, the strategies can be eliminated as follows: Alice's s_{A3} is strictly dominated by s_{A2}, Bob's s_{B1} by s_{B2}, Alice's s_{A1} by s_{A2}, and finally Bob's s_{B3} as well as s_{B4} by s_{B2}.

The remaining strategies are s_{A2}, s_{A4}, and s_{B2}. The best response from Alice is s_{A2}, resulting in the Nash equilibrium $(s_{A2}, s_{B2})^*$.

If you start the elimination with Bob, the strategies can be eliminated as follows: Bob's s_{B4} is strictly dominated by s_{B3}, and s_{B1} by s_{B2}, then Alice's s_{A1}, s_{A3}, and s_{A4} by s_{A2}. Thus, the strategies s_{A2}, s_{B2}, and s_{B3} remain. The best response of Bob is s_{B2}, which results in the Nash equilibrium again $(s_{A2}, s_{B2})^*$.

In the iterative elimination of strictly dominated strategies, it does not matter whether the elimination procedure starts with Alice or Bob. Both must lead to the same, unique solution. If a Nash equilibrium can be determined through the iterative elimination of strictly dominated strategies, then it is always unique, as none of the eliminated strategies can be part of a Nash equilibrium (a rational player would always deviate from a strictly dominated strategy).

Part c)

If the elimination starts with Alice, then strategies can be eliminated in this order: Alice's s_{A1} and s_{A2} are strictly dominated by s_{A3}, while s_{A4} is weakly dominated by s_{A3}. Thus, only strategy s_{A3} remains. In his best response, Bob is indifferent between s_{B1} and s_{B3}. The Nash equilibria $(s_{A3}, s_{B1})^*$ and $(s_{A3}, s_{B3})^*$ result.

If the elimination starts with Bob, the strategies can be eliminated as follows: Bob's s_{B2} is strictly dominated by s_{B3}, while s_{B1} is weakly dominated by s_{B3}. Thus, only strategy s_{B3} remains. Alice's best response is then s_{A3}, resulting in the Nash equilibrium $(s_{A3}, s_{B3})^*$.

When eliminating weakly dominated strategies, we are at risk of not finding all Nash equilibria. The elimination starting with Alice leads to two Nash equilibria, while the one starting with Bob leads to only one (proof by contradiction).

3.8.2 Mixed Strategies

Problem Text in Sect. 3.7.2.

Part a)
Inspection game as a matrix:

Producer (A), Customer (B)	s_{B1} (Trust)	s_{B2} (Control)
s_{A1} (high quality)	$(p - e, \underline{u_h})$	$(\underline{p - e}, u_h - k)$
s_{A2} (low quality)	$(\underline{p}, u_l)$	$(0, \underline{u_m - k})$

Underlining indicates that in each cell a player has an incentive to deviate, thus, there is no Nash equilibrium in pure strategies.

Part b)
Let p_A be the probability with which s_{A1} and p_B be the probability with which s_{B1} is chosen.

Producer (player A):
I. Expected Utility Comparison

$$\begin{aligned} E[u_A(p_A = 1, p_B)] &\geq E[u_A(p_A = 0, p_B)] \\ p_B \cdot (p - e) + (1 - p_B) \cdot (p - e) &\geq p_B \cdot p + (1 - p_B) \cdot 0 \\ p_B &\leq \frac{p - e}{p}. \end{aligned}$$

II. Optimization Approach

$$\begin{aligned} E[u_A(p_A, p_B)] &= p_A \cdot p_B \cdot (p - e) + p_A \cdot (1 - p_B) \cdot (p - e) \\ &\quad + (1 - p_A) \cdot p_B \cdot p + (1 - p_A) \cdot (1 - p_B) \cdot 0 \\ &= (p_A + p_B - p_A \cdot p_B) \cdot p - p_A \cdot e \end{aligned}$$

Partial derivative and determination of the first-order condition:

$$\begin{aligned} \frac{\partial E[u_A(p_A, p_B)]}{\partial p_A} &= (1 - p_B) \cdot p - e \stackrel{!}{=} 0 \\ &\Rightarrow p_B^* = \frac{p - e}{p}. \end{aligned}$$

At the optimum, the producer is indifferent between its two strategies. For the producer to remain indifferent between high and low quality, the customer must choose the first strategy ("trust") with a probability of $(p - e)/p$. The customer aligns the probability p_B with the incentives of the producer: The higher the costs e for high quality, the stronger is the producer's incentive to deliver low quality.

For the producer to lose the incentive to do so and to be kept indifferent, the customer must counteract this by trusting less (and thus controlling more) as e increases. This is also described by the expression $p_2^* = (p - e)/p$: The greater the costs e (or the smaller the price p), the smaller the customer's probability to "trust".

Customer (player B):

i. Determination of the equilibrium in mixed strategies via the *expected utility comparison*

$$E[u_B(p_A, p_B = 1)] \geq E[u_B(p_A, p_B = 0)]$$
$$p_A \cdot u_h + (1 - p_A) \cdot u_l \geq p_A \cdot (u_h - k) + (1 - p_A) \cdot (u_m - k)$$
$$p_A \leq \frac{u_m - u_l - k}{u_m - u_l}.$$

ii. Determination of the equilibrium in mixed strategies via the *optimization approach*

$$\begin{aligned} E[u_B(p_A, p_B)] &= p_A \cdot p_B \cdot u_h + p_A \cdot (1 - p_B) \cdot (u_h - k) \\ &\quad + (1 - p_A) \cdot p_B \cdot u_l + (1 - p_A) \cdot (1 - p_B) \cdot (u_m - k) \\ &= p_A \cdot (u_h - u_m) - p_B \cdot (u_m - u_l - k) \\ &\quad + p_A \cdot p_B \cdot (u_m - u_l) + u_m - k. \end{aligned}$$

Partial derivative and determination of the first-order condition:

$$\frac{\partial E[u_B(p_A, p_B)]}{\partial p_B} = (u_m - u_l - k) + p_A \cdot (u_m - u_l) \stackrel{!}{=} 0$$
$$\Rightarrow p_A^* = \frac{u_m - u_l - k}{u_m - u_l}.$$

At the optimum, the customer is also indifferent between the two strategies. The producer creates this indifference by considering the customer's incentives. The higher the costs k for control, the stronger is the customer's incentive to forego control and thus to trust. In contrast to an absolutely trusting customer, the producer would always choose low quality; in contrast to a control-obsessed, distrustful customer, the producer should always choose high quality. For the customer to be just indifferent between trusting and controlling, the producer must choose the high quality with a probability of p_A^*: The higher (or lower) the control costs k for the customer, the more (or less) the producer can afford to choose low quality.

Part c)
From our results in b) we can determine the reaction correspondences as follows:

$$p_A^*(p_B) = \begin{cases} 0 & p_B \in \left(\frac{p-e}{p}, 1\right] \\ [0,1] & p_B = \frac{p-e}{p} \\ 1 & p_B \in \left[0, \frac{p-e}{p}\right) \end{cases} \qquad p_B^*(p_A) = \begin{cases} 0 & p_A \in \left[0, \frac{u_m-u_l-k}{u_m-u_l}\right) \\ [0,1] & p_A = \frac{u_m-u_l-k}{u_m-u_l} \\ 1 & p_A \in \left(\frac{u_m-u_l-k}{u_m-u_l}, 1\right] \end{cases}$$

Graphical representation:

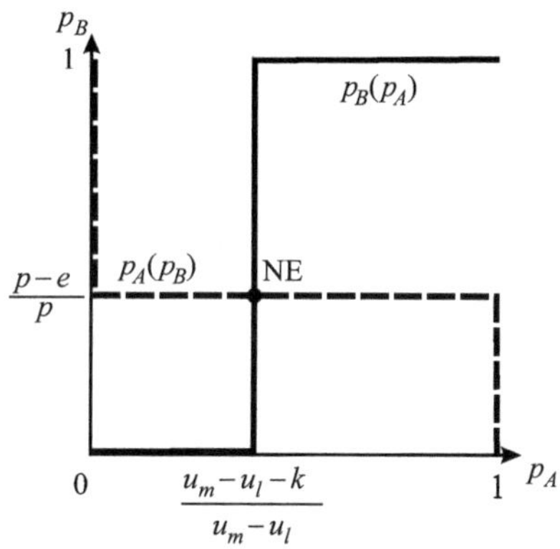

In general, the (expected) equilibrium payoff of player i is calculated as follows:

$$\begin{aligned} E\left[u_i(p_i, p_j)\right] &= p_i^* \cdot p_j^* \cdot u_i(p_i = 1, p_j = 1) \\ &+ p_i^* \cdot \left(1 - p_j^*\right) \cdot u_i(p_i = 1, p_j = 0) \\ &+ \left(1 - p_i^*\right) \cdot p_j^* \cdot u_i \cdot (p_i = 0, p_j = 1) \\ &+ \left(1 - p_i^*\right) \cdot \left(1 - p_j^*\right) \cdot u_i(p_i = 0, p_j = 0). \end{aligned}$$

If both players choose their equilibrium strategies p_A^* or p_B^*, the following expected payoffs for player A (producer) and player B (customer) result:

$$E\left[u_A\left(p_A^*, p_B^*\right)\right] = p - e \quad \text{and} \quad E\left[u_B\left(p_A^*, p_B^*\right)\right] = u_h - \frac{k \cdot (u_h - u_l)}{u_m - u_l}.$$

For player A, this expected payoff is easy to understand: With high quality, the producer (A) always has the same payoff of $p - e$ for each strategy choice of the consumer (B), so that this must also occur with low quality (but as an expected payoff) to keep A indifferent between the two strategies.

3.8.3 Nash Equilibrium in Games with Continuous Strategies: Price Strategies in Duopoly

Problem Text in Sect. 3.7.3.

Part a)

The payoff function or profit is the difference between revenue, $x_i \cdot p_i$, and costs, c_i, of company i, and thus depends on the strategic variable p_i:

$$\pi_i(p_i, p_j) = x_i(p_i, p_j) \cdot (p_i - c_i) = (a - p_i + \beta \cdot p_j) \cdot (p_i - c_i).$$

To determine the profit-maximizing output quantity, the profit π_i is partially derived with respect to the strategic variable p_i. By setting this derivative equal to zero, we obtain the first-order condition, which is then solved for p_i to determine the reaction function of company i, which is its profit-maximizing price $p_i^*(p_j)$ depending on the given price of the competitor. For company i, the reaction function is therefore

$$\frac{\partial \pi_i}{\partial p_i} = a + c_i - 2 \cdot p_i + \beta \cdot p_j \overset{!}{=} 0 \Rightarrow p_i^*(p_j) = \frac{1}{2} \cdot (a + c_i + \beta \cdot p_j). \tag{3.29}$$

The Nash equilibrium of the game is obtained by equating the reaction functions. This results in the equilibrium prices and quantities according to

$$p_i^* = \frac{a \cdot (2 + \beta) + 2 \cdot c_i + \beta \cdot c_j}{4 - \beta^2}$$
$$x_i^* = \frac{a \cdot (2 + \beta) + (\beta^2 - 2) \cdot c_i + \beta \cdot c_j}{4 - \beta^2}.$$

Part b)

To be able to plot the reaction function graphically in a diagram where p_j is on the horizontal axis and p_i is on the vertical axis, we must rearrange the reaction function of company i according to p_j. The reaction function of company j can, however, be used directly, so that we finally obtain both functions for the graphical representation:

$$r_i(p_j) : p_j(p_i^*) = \frac{1}{\beta} \cdot (2 \cdot p_i - a - c_i) \text{ and}$$
$$r_j(p_i) : p_i^*(p_j) = \frac{1}{2} \cdot (a + c_j + \beta \cdot p_i).$$

The following figure graphically represents the reaction functions:

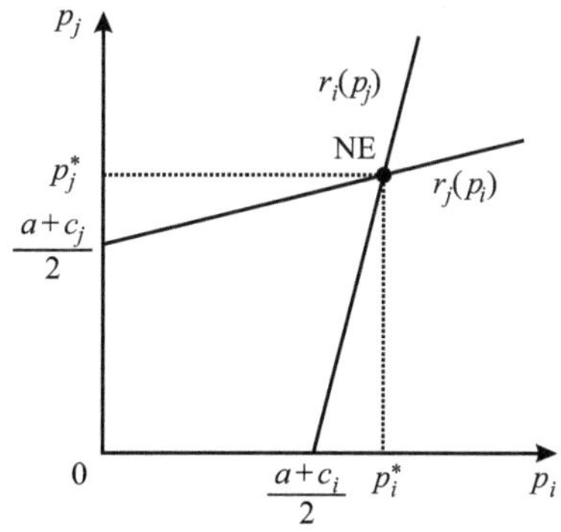

Unlike with quantity strategies, the reaction functions here are increasing. But this should come as no surprise to us: If the other company lowers its price, we will potentially lose customers to our competitor, which is why we will also react by lowering our prices. The Nash equilibrium also results here at the intersection of the two straight lines.

Part c)
We want to show that even with price strategies, a non-Pareto-optimal result is obtained. To do this, we compare the partial derivative from (3.29) with the corresponding partial derivative of the joint profit function $\pi_{\text{total}} = \pi_i + \pi_j = (a - p_i + \beta \cdot p_j) \cdot (p_i - c_i) + (a - p_j + \beta \cdot p_i) \cdot (p_j - c_j)$,

$$\frac{\partial \pi_i}{\partial p_i} = a + c_i - 2 \cdot p_i + \beta \cdot p_j \text{ and}$$
$$\frac{\partial \pi_{\text{total}}}{\partial p_i} = a + c_i - 2 \cdot p_i + \beta \cdot p_j + \beta \cdot (p_j - c_j).$$

The difference between both derivatives is noticeable in the positive last summand $+\beta \cdot (p_j - c_j)$ in the derivative of the total profit function. This term again points to a discrepancy between individual and joint or collective optimization, which is why there is clearly an external effect. For $p_j > c_j$ this external effect is positive and for $p_j < c_j$ this external effect is negative. Since we assume cost-covering prices, there is a positive external effect. The prices of the companies are therefore lower than they would be with joint profit maximization. Both firms could achieve higher profits together through a joint price increase.

3.8.4 Competition Between Many Companies

Problem Text in Sect. 3.7.4.

Part a)
With our knowledge about the general specification, we can directly conclude the solution. Specifically, the following values were assumed: $a = 15$, $b = 0.5$, and $c = 6$. Using these, we get from (3.25) the quantity of a company, from (3.26) the price, and from (3.27) the profit. The results are thus

n	$x_i(n)$	$p(n)$	$\pi_i(n)$
2	4	14	32
5	2	10	8
11	1	8	2

Part b)
Starting from (3.26) we have to determine the limit

$$\lim_{n\to\infty} \frac{a + c \cdot n}{n + 1}.$$

However, the simple "substitution" of "∞" instead of n does not lead to any result, as we would find "∞" in both the numerator and the denominator. Here, a small trick can help: simply divide both the numerator and the denominator by n. The expression remains unchanged, after all, n cancels out. However, we can now substitute "∞" for n and thus derive a clear statement that unequivocally confirms the suspected correlation:

$$\lim_{n\to\infty} \frac{\overbrace{\frac{a}{n}}^{\to 0} + c}{1 + \underbrace{\frac{1}{n}}_{\to 0}} = c$$

Literature Recommendations

Dixit A, Skeath S (2014) Games of strategy, 4th ed. New York (Chaps. 1–3).
Leininger W, Ockenfels A (2007) The penalty-duel and institutional design: is there a Neeskens-Effect? CESifo Working Paper, No. 2187.

Multistage Games

4

The common element of all the games considered in the previous chapter is that they are one-stage simultaneous games: Simultaneous, because the players cannot observe the strategy choice of their fellow players at the time of their decision (imperfect information). One-stage, because the players have only one action choice available at a certain point in time ("stage" of the game). In this chapter, we consider *multistage games*, in which the *temporal* sequence of actions is important. The most important subcategory of multistage games is *sequential games*, in which the players decide one after the other. As a result, the second mover, i.e., the player who decides second, can observe the preceding action of the first mover before making the own decision (perfect and asymmetric information). The central solution concept for multistage games is the method of *backward induction*, in which we determine the solution to a game by conceptually "looking ahead and reasoning backwards". The Nash equilibria determined by this procedure are *subgame-perfect* Nash equilibria. In this chapter, we will deal more closely with the method of backward induction and the concept of subgames and subgame perfection and apply the solution steps to multistage games with discrete and continuous strategies.

4.1 Multistage Games in Discrete Strategies

Temporal Structure of Multistage Games. Before we start with the analysis of multistage games, let us first clarify which variants of multistage games can occur and how these can be categorized, especially regarding the prevailing state of information. Since the analysis with the consideration of multistage games becomes more complex, it is important to always keep an overview to be able to judge safely which game structure exactly exists. Figure 4.1 shows three variants as

F. Bartholomae and M. Wiens, *Game Theory and Applications*, Classroom Companion: Economics, https://doi.org/10.1007/978-3-658-44608-6_4

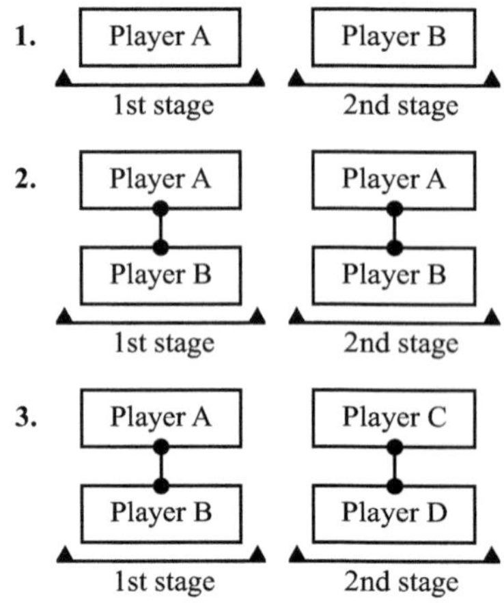

Fig. 4.1 Variants of multistage games

examples of how multistage games can be structured. These three variants already cover most of the multistage games that are usually dealt with in lectures.

Game structure 1 represents the simplest case of a sequential game, in which the players make their decisions one after the other. Player A is the first one to decide, i.e., player A is the *first mover.* Thus, A must form expectations about how player B will react to the chosen action. Since B's decision is not known at this point in time, player A acts under *imperfect information* (B's move is not observable due to the sequence of moves). Player B is the *second mover* and reacts to A's move after observing it. Since B is aware of A's move, B has *perfect information* regarding A's strategy choice. Thus, as player A has imperfect information, but player B has perfect information, the state of information between the two players is *asymmetric.* We have already encountered this simple sequential game structure in Sects. 2.3 and 2.4 in connection with the "Future Energy" game. Here, the company was the first mover with its investment decision and the government was the second mover with its funding decision. Most of the games discussed in this chapter are also simple sequential games of this kind.

Game structure 2 is a multistage game, but not a sequential game in the true sense, as the players do not decide one after the other and thus do not react to each other (which is the characteristic element of a sequential game). Instead, these are two different, consecutive simultaneous games. The players could be, for example, two competing companies. At the first stage, they simultaneously decide on their investments in research and development (R&D) and, at the second stage, they again simultaneously decide on their output quantities.[1] Since the players always decide simultaneously, there is symmetric and imperfect information at each stage with respect to the decision made there (at stage 1 it is the investments, at stage

[1] It is important that the stage games are *different games.* If, instead, the exact same game was played again, it would not be a multistage game, but a *repeated game.* Repeated games are discussed in Chap. 5.

2 the output quantities). Between the stages, there is perfect and symmetric information: At the time of their quantity decision, both players are perfectly informed about the previous investment decision, both their own and that of the competitor. In the last section of this chapter, we will solve the R&D game using a specific example.

In game structure 3, *player groups* can be distinguished: Players A and B act simultaneously at the first stage (first group) and players C and D at the second (second group). This is a sequential game between the player groups with the same asymmetric information as in game structure 1. Within the groups, however, there is again symmetric and imperfect information. An example of such a game structure would be a political election: Players A and B are two lobby groups, e.g., for the economy and the environment, which simultaneously influence policy at the first stage; players C and D are companies, which then simultaneously determine their corporate strategies under the resulting competitive conditions.

Information structure of multistage games. To become more familiar with the characteristics of multistage games, we start with a comparison of simultaneous and sequential games. Here, the first question is whether and why the game solution changes when we transform a simultaneous game into a sequential game. For a specific case, the answer to this question is simple: If both players have a strictly dominant strategy, there is no difference between a simultaneous game and a sequential game. As we showed in the discussion on dominance in Sect. 3.1, a strictly dominant strategy is always the optimal response, regardless of the choice of the opponent. In a strategy choice that does not depend on the behavior of the other player, it does not matter whether the relevant player has perfect or imperfect information (a player with a strictly dominant strategy does not need further information) and thus the game structure has no influence on the game result. From the same considerations, it follows that the game result also does not change if the second mover (e.g., Bob) has a strictly dominant strategy. In the simultaneous game, Bob will choose the strictly dominant strategy and the other player (Alice) will respond to this as best as possible. This best response is exactly what Alice will also choose as the first mover: Since she knows that the second mover Bob has a strictly dominant strategy, she already knows his reaction at the beginning. Alice therefore only needs to choose the strategy that is most favorable to her. The crucial point is that the strictly dominant strategy of the second mover compensates for the information disadvantage of the first mover—not knowing exactly how the second mover will react. Thus, from an information perspective, the sequential and the simultaneous game are ultimately identical. However, if the player with the strictly dominant strategy is in the role of the first mover, then the solution of the game variants can indeed differ. Let us consider the simultaneous game in Table 4.1.

In this game, only Alice has a strictly dominant strategy (s_{A1}) and Bob's best response to this is s_{B2}, so the strategy combination $(s_{A1}, s_{B2})^*$ is the unique Nash equilibrium. Now let us compare this to the sequential game from Fig. 4.2 (left game tree), where Alice is the first mover.

Table 4.1 Simultaneous game with a strictly dominant strategy

Alice, Bob	s_{B1}	s_{B2}
s_{A1}	(3, 3)	(1, 4)*
s_{A2}	(2, 2)	(0, 0)

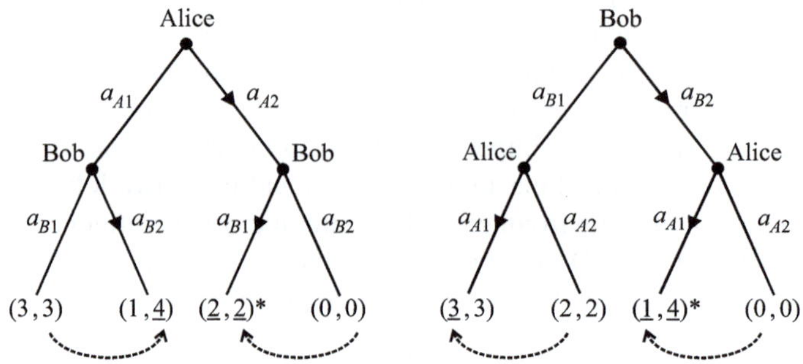

Fig. 4.2 Sequential game with strictly dominant strategy and different order of moves

To solve the game, we use *backward induction*, which means we start with the optimal reaction of the last player, Bob, at the second stage and infer Alice's optimal move in the preceding first stage. If Bob must decide at the left node, it is best for him to choose action a_{B2}.[2] If he is on the other hand at the right decision node, he chooses a_{B1}. Bob's optimal strategy is therefore $s_B^* = \{a_{B2}|a_{A1}, a_{B1}|a_{A2}\}$. The arrows in the game tree illustrate the direction of Bob's decision at each node. If Bob chooses this optimal strategy at the second stage of the game, Alice's best reaction is to choose her second action: If she chooses a_{A1}, Bob chooses a_{B2}, which gives her a payoff of 1; if she chooses a_{A2}, Bob's response a_{B1} grants her a higher payoff of 2 instead. The Nash equilibrium of the sequential game is therefore at $(a_{A2}, \{a_{B2}|a_{A1}, a_{B1}|a_{A2}\})$ with the payoff combination (2, 2). As we can see, this payoff combination in the simultaneous game in Table 4.1 is *not* a Nash equilibrium. As Alice anticipates Bob's reaction and she can decide first, she can influence the decision in her favor by choosing the right branch. Since Alice benefits from the role of the first mover in this case, this is referred to as a *first-mover advantage.* In games where a player benefits from the role of the second mover, there is correspondingly a *second-mover advantage.* The game on the right in Fig. 4.2 is the same game with the only difference being that the players take turns in reverse order, that is, Bob is now the first mover and Alice is the second mover. The solution to this game results in the payoff combination (1, 4), which is also a Nash equilibrium in the simultaneous game, but the equilibrium here is $(a_{B2}, \{a_{A1}|a_{B1}, a_{A1}|a_{B2}\})$ compared to (a_{A1}, a_{B2}) in the simultaneous game. Since Alice has a strictly dominant strategy, it does not matter whether she makes

[2] We refer to "actions" because a "strategy" in a sequential game always consists of conditional actions. See the detailed explanations in Sect. 2.1.

her decision after Bob or at the same time. Her decision is perfectly predictable for Bob in the simultaneous game, so he can optimally adapt to it by choosing his second strategy.

4.2 Backward Induction, Subgame Perfection, and Credibility

With the help of backward induction, we can thus check whether a decision considered optimal by the second mover at the beginning of the game still proves to be optimal when the first mover has already decided. Nash equilibria that meet this further requirement are referred to as *subgame perfect Nash equilibria.* To better understand the concept of *subgame perfection,* let us first explain what a subgame is by using the game tree representation. An independent subgame ("proper subgame") starts at a decision node X if all subsequent nodes are connected to the rest of the game *only* via *this node X*. To illustrate this, look at the game tree in Fig. 4.3. While an independent subgame starts at node d, this is not the case for nodes b and c. Since nodes b and c belong to the same information set (indicated by the dashed connection line), the game progression following b cannot be separated from node c. The subgame to which nodes b and c belong starts at node a and is thus identical to the entire game (the entire game starting at node a also represents a subgame).

A *subgame perfect Nash equilibrium* s^* exists when there is no incentive for any player to deviate from s^* in any subgame that starts at any node of the game tree. The concept of subgame perfection is a criterion for equilibrium selection that allows to test for the plausibility of Nash equilibria in sequential games. Subgame perfect Nash equilibria are plausible insofar as they represent strategies that are consistent over time and therefore *credible.* The important property of the credibility of strategies will be discussed in more detail in the next section.

However, let us first use the example of Fig. 4.4 to illustrate how backward induction is systematically applied in a more complex multistage game.

- Tree I shows the complete game that we want to solve. This game consists of a total of three stages and five subgames (a, b, c, d, e). The number of subgames can easily be identified by the number of decision nodes, each of which lies in its own information set (i.e., there are no dashed lines between the nodes). Alice can decide in three of the five subgames, (a, b, e), and Bob in two of the five subgames, (c, d).

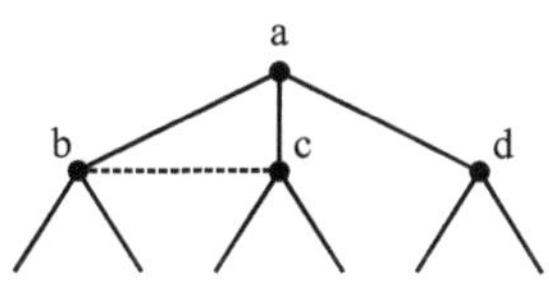

Fig. 4.3 Information set and subgame

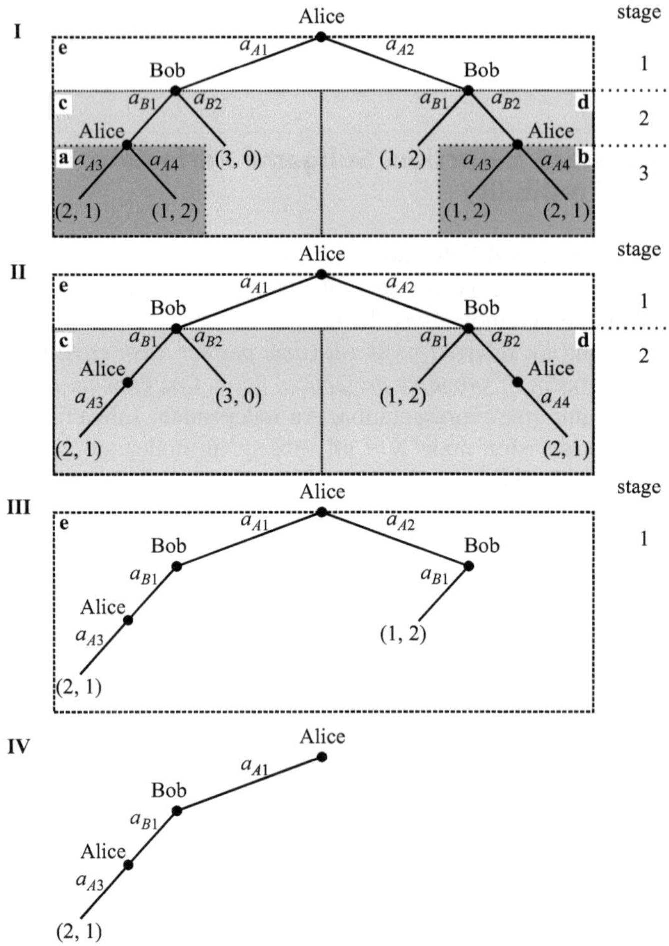

Fig. 4.4 Procedure for backward induction

- As we can see, Alice is the player who can decide at the last stage. Therefore, we must start the backward induction with her. Alice can choose between actions a_{A3} and a_{A4}. In subgame a, this leads to a payoff of 2 for a_{A3} and 1 for a_{A4}. Thus, Alice will prefer a_{A3} over a_{A4}. In subgame b, which is also located at the last stage, Alice will choose a_{A4}, as she can achieve a higher payoff here (2 instead of 1). The actions not chosen (a_{A4} in subgame a and a_{A3} in subgame b) do not need to be considered further in the course of the analysis. We have therefore omitted them in tree II.
- We have thus already solved two subgames at the third stage and can proceed analogously with the remaining stages. In the second stage, in II, two more subgames need to be solved, in which Bob must now decide. Bob can choose between actions a_{B1} and a_{B2}. In subgame c, he will choose a_{B1}, as he can achieve a higher payoff compared to a_{B2}: Since we solve the game backwards,

we know how Alice will decide when Bob chooses a_{B1}. Bob will consider this and therefore knows that his payoff at a_{B1} will be 1. At a_{B2} the game ends with a lower payoff of 0 in comparison. In subgame d, Bob will end the game by choosing a_{B1}, as he can achieve a payoff of 2 instead of 1. Again, we can cross out the actions not chosen (a_{B2} in subgame c and a_{B2} in subgame d).
- In tree III, we have already reached the first stage, which consists of only one subgame in which Alice can choose between her actions a_{A1} and a_{A2}. As we can see from the remaining game tree, she will choose a_{A1}, as she can achieve a payoff of 2 instead of 1 at a_{A2}. If we cross out a_{A2}, this finally leads to the equilibrium path in IV.

Thus, in the subgame perfect equilibrium, the payoff will be (2, 1). To fully specify the equilibrium, we now need to determine the equilibrium strategies. This strategy is the combination of all decisions that the player will make, that is, a player's strategy includes as many elements as there are decisions to be made. Alice can make three decisions (in the subgames a, b, and e), which is why her strategy consists of three elements. In chronological order of the subgames, Alice's equilibrium strategy is: Choose s_{A1} in subgame e, s_{A3} in subgame a, and s_{A4} in subgame b. For Bob, who can decide in the two subgames c and d, the equilibrium strategy is: Choose s_{B1} in subgame c and s_{B1} in subgame d. It is important to note that the *equilibrium path* is not identical to the *equilibrium strategies* but is merely included in them.

Since in general subgames are not named as in our example, the individual elements are given depending on the *history* of the game. This describes how the various subgames were reached. For example, to reach subgame a, Alice must have chosen a_{A1} and Bob must have chosen a_{B1} or in short, the history of subgame a is (a_{A1}, a_{B1}). As we know, in this subgame Alice will choose a_{A3}. Thus, we can use the notation $a_{A3}(a_{A1}, a_{B1})$ to state that Alice will choose a_{A3} in subgame a. The complete strategies of the players depending on the respective history are therefore: Alice's strategy is $\{a_{A1}, a_{A3}(a_{A1}, a_{B1}), a_{A4}(a_{A2}, a_{B2})\}$ and Bob's strategy is $\{a_{B1}(a_{A1}), a_{B1}(a_{A2})\}$. The subgame perfect equilibrium is therefore $\{\{a_{A1}, a_{A3}(a_{A1}, a_{B1}), a_{A4}(a_{A2}, a_{B2})\}, \{a_{B1}(a_{A1}), a_{B1}(a_{A2})\}\}$.

Credibility of Announcements. The focus on subgame perfect Nash equilibria by applying backward induction represents a *test of credibility* of the second-mover strategy. The assessment of the credibility of a strategy is important in contexts in which the first mover attempts to influence the decision of the second mover by announcing a strategy aimed at their own advantage. Let us consider the following case: Bob promises Alice in the left game of Fig. 4.2 to choose a_{B1} if Alice in turn chooses a_{A1}. For Alice, it would naturally be advantageous if Bob kept his promise—she would receive 3 instead of 2. However, Bob's announcement is an incredible promise: If Alice trusts this and chooses a_{A1} so that Bob can decide on the left branch, he will opportunistically deviate to a_{B2} to improve from 3 to 4—hence, Alice will only receive 1. The subgame perfect Nash equilibrium eliminates such incredible or "empty" promises/threats if the second mover threatens with an action choice detrimental to the first mover off the equilibrium path. The reason is that the subgame perfect Nash equilibrium must be a Nash equilibrium in *every*

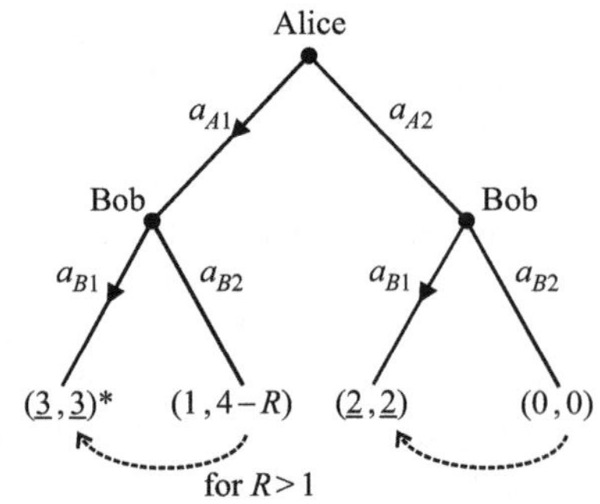

Fig. 4.5 Announcements with Commitment

subgame. In our example, a_{B1} is not the optimal choice and therefore no rational player would choose it.

Let us now assume that Bob would find it uncomfortable to break a given promise, as this would cause him "feelings of remorse" amounting to $R > 1$. Figure 4.5 represents this changed payoff structure. As we can see, this changes the incentives of Bob and he will now prefer his promise (a_{B1}) to opportunistic behavior (a_{B2}), whereby both could reach the payoff combination (3, 3). The credibility problem could also alternatively be overcome by the second mover securing his promise with a pledge of $R > 1$: If he breaks his promise, he loses this amount of money. As can be seen from the payoffs in Fig. 4.5, the second mover has an incentive for such a trust-building measure: Without a pledge, his promise would be incredible, and the payoff combination of the subgame perfect Nash equilibrium (2, 2) would result. With the help of a pledge, the payoff combination (3, 3) would be achievable. Tools of this kind, which enable a player to make an otherwise incredible strategy credible and thus a subgame perfect Nash equilibrium, are referred to as *Commitment* or *Commitment-Device.*

Such a commitment will only be successful if it is credible and irreversible. The latter point is obvious: what would prevent a player from undoing something reversible if it was beneficial to him? Locking a cookie jar doesn't mean you can't open it again. So, the measure must be reversible, if at all, only with a high and therefore not worthwhile use of resources. However, it is much more difficult to make a measure credible. There are two options here: Restricting the freedom of action or changing the payoffs.

There are several conceivable options for *restricting the freedom of action.* We can think of *automatic fulfillment*, i.e., a mechanical device or programming that once activated cannot be switched off. An example would be the fictitious doomsday machine from Stanley Kubrick's legendary 1964 film satire, which was a credible deterrent as it was irrevocably programmed for global destruction. Instead of leaving the decision to a machine, the decision can also be *delegated to a third party* who is not involved, such as a trustee or notary. The British Navy is said to have preferred to recruit soldiers who were non-swimmers in the seventeenth century: A non-swimmer does not need to swear that he will defend the ship under all circumstances, as he already has a considerable self-interest. Another ultimate solution is to *burn the bridge*, making a return impossible, which Alexander the Great literally did to send a credible signal of his determination to the enemy. In the famous Greek saga, Odysseus preferred to tie himself to a mast to resist the temptations of the sirens.

An *interruption in communication*, i.e., after signal transmission "no more communication" can also lend credibility to a strategy (which is difficult to imagine nowadays with permanent availability). If, after an announcement, there is no longer any possibility of additional negotiations, for example, the partners may feel compelled to stick to their agreements. Compared to the other methods, however, this appears to be a weaker form of self-commitment.

As in our previous example, a change in payoffs (secured by a pledge or in the form of contractual agreements) can also lead to a credible self-commitment. *Irrational* behavior can also create credibility if the other players have no clarity about the exact payoffs, i.e., they do not know the true motives and can therefore assume that there may be even worse consequences if they do not accept them. You can think here of a typical movie villain who would really trigger the bomb. Less dramatic, but much more relevant in real life, would be a *loss of reputation*, which can affect future payoffs (we will also look at this in more detail in the repeated games in Chap. 5). Specifically, this means that if you deviate from the announced strategy, you may have an advantage in the short term, but, in the long term, no one will believe you or want to cooperate with you.

In almost all areas of economic policy, such as monetary policy, tax, and fiscal policy, and especially in the case of profound reforms, the credibility of economic policy programs plays a major role. The economists Kydland and Prescott received their Nobel Prize in 2004 for work that was crucially related to the credibility issues considered here. They showed a positive correlation between credible economic policy and stable economic cycles and concluded that economic policy decision-makers should submit to self-commitment (self-commitment) in their own interest for a priori incredible measures. The 1995 Nobel laureate, Robert Lucas, addressed a similar dilemma with the Lucas critique named after him: An economic policy measure becomes ineffective (and thus loses credibility from the start) the moment the target group anticipates the measure and takes it into account in their decisions.

Whether this commitment is designed as a threat, or a promise depends on what is to be achieved. A *threat* is present when other players are negatively impacted if they act against our interests. A *promise*, in turn, should then be implemented to encourage other players to act in our interests. Both have the objective of either persuading the other player to refrain from an action (*deterrence*) or to choose a certain action (*compellence*). There is a big difference between the two methods in terms of cost. A threat is passive and only occurs when the undesired behavior is shown. As this is ideally avoided, the threat does not have to be carried out, meaning that it is potentially more costly than it needs to be. The situation is different with a promise that is actively implemented if the desired behavior is shown and thus causes real costs. It is therefore chosen as efficiently as possible to keep the costs within limits. Both methods are used as *strategic moves* by which the players try to influence the game to their advantage.

4.3 Multistage Games in Continuous Strategies

The subgame perfect Nash equilibrium described in Sect. 4.1 can also be determined in the same way in multistage games with continuous strategies. For this, let us consider the strategic interaction of two companies that each have a monopoly situation at different levels of a supply chain. Company A, a manufacturer, produces goods for company B, a retailer, who sells the product to the end customer. The market demand (end customers) is given by $x^D(p) = a - b \cdot p$. In the chronological sequence, the manufacturer first sets the wholesale price q; on this basis, the retailer determines the selling price p for the end customer from which the sales quantity x results that also corresponds to the demand for the manufacturer's product. This two-stage game thus corresponds to variant 1 from Fig. 4.1. We solve it using backward induction, i.e., again the strategic sequence runs exactly opposite to the chronological sequence: The manufacturer must anticipate the retailer's strategy in its decision about the wholesale price q. If it chooses q too high, it may be that the retailer orders too little of the product; if the manufacturer chooses q too low, it may happen that the retailer takes advantage of the lower purchase price for a higher own margin—in both cases, this is at the expense of the manufacturer's profit. Thus, in the first step, we consider the retailer's decision for a given wholesale price q in the second stage and determine the profit-maximizing commodity price $p^*(q)$ or the profit-maximizing quantity $x^D[p^*(q)]$, which are each dependent on q. Based on this, in the second step, we consider the manufacturer's decision in the first stage by inserting the reaction function $p^*(q)$ into its profit function and determining the profit maximum. As a result, we obtain the profit-maximizing wholesale price q^* of the manufacturer. From q^*, we finally obtain p^* or x^*. The profit-maximizing strategies of both monopolists then represent a subgame-perfect Nash equilibrium. The procedure is outlined again in Fig. 4.6.

As described, we start first with the decision of the retailer (company B). The profit of company B is $\pi_B(p, q) = (p - q) \cdot x^D(p) = (p - q) \cdot (a - b \cdot p)$ and thus depends on the strategic variables of the two players, p and q. The price to be paid to the manufacturer q corresponds to the retailer's variable costs and

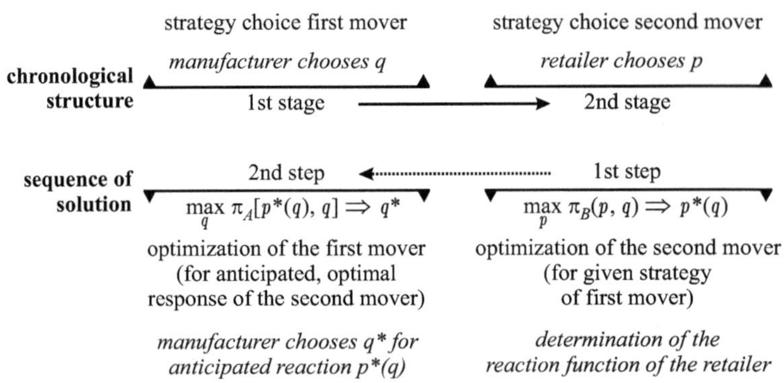

Fig. 4.6 Backward induction in the two-stage game with continuous strategies

the sales quantity is represented by the price-sales function. The maximization of profit by determining the marginal profit (partial derivative of the payoff function with respect to the strategic variable p) and setting the first derivative to zero yields the reaction function $p^*(q)$ of the retailer as well as the corresponding quantity $x^*(q)$:

$$\frac{\partial \pi_B(p,q)}{\partial p} = a + b \cdot q - 2 \cdot b \cdot p \stackrel{!}{=} 0$$
$$\Rightarrow p^*(q) = \frac{1}{2} \cdot \left(\frac{a}{b} + q\right)$$
$$\Rightarrow x^D\big[p^*(q)\big] = \frac{1}{2} \cdot (a - b \cdot q)$$

In the second step, the retailer's reaction function is inserted into the manufacturer's optimization problem. We assume that the manufacturer's production costs per unit are c. The profit function of the manufacturer, already adjusted to the anticipated price strategy of the retailer, is therefore $\pi_A\big[p^*(q), q\big] = (q - c) \cdot x^D\big[p^*(q)\big] = (q - c) \cdot \frac{1}{2} \cdot (a - b \cdot q)$. The profit-maximizing wholesale price q^* is then:

$$\frac{\partial \pi_A}{\partial q} = \frac{a}{2} - b \cdot q + \frac{b \cdot c}{2} \stackrel{!}{=} 0$$
$$\Rightarrow q^* = \frac{1}{2} \cdot \left(\frac{a}{b} + c\right)$$

This tells us what price the manufacturer chooses, and we can include this in the retailer's reaction function. Thus, we obtain the following price and sales quantity:

$$p^*(q^*) = \frac{1}{4} \cdot \left(\frac{3 \cdot a}{b} + c\right)$$
$$x^D\big[p^*(q^*)\big] = \frac{1}{2} \cdot \left(\frac{a}{2} - c \cdot b\right)$$

The interesting phenomenon from an industrial economics perspective in this example is that the selling price p^* is higher than if company B not only distributed the good but also produced it itself *(vertical integration).* We can quickly verify this by substituting the purchase price q against the production costs c. The reason for this is that the two involved monopolists also impose a price markup twice—each firm at its respective level of the value chain. Consequently, the resulting price is higher and the resulting quantity lower than in the case when only one company produces. The resulting inefficiency is referred to as the problem of *double marginalization.* The problem can be mitigated, for example, by different stakeholders in a supply chain cooperating and coordinating their pricing among themselves. In contrast to cooperation in horizontal competition, i.e., cartel formation, cooperation in vertical competition is welfare-enhancing and thus does not pose an antitrust problem.

In a sequential game with *horizontal* quantity competition, i.e., competition at the same value chain level, the procedure is analogous. At the first stage, the first mover determines the quantity, after which the competitor chooses the best response at the second stage. The solution then again follows the procedure described in Fig. 4.6 of backward induction: The optimal reaction of the second mover is first determined and then inserted into the target or profit function of the first mover. The first mover then maximizes the profit considering the anticipated reaction of the competitor. The first mover in this game is also traditionally referred to as *Stackelberg leader* (in the sense of a market leader), the second mover as *Stackelberg follower,* and the subgame perfect Nash equilibrium determined by backward induction accordingly also as *Stackelberg equilibrium.* This designation goes back to the German economist Heinrich Freiherr von Stackelberg (1905–1946), who developed the sequential version of the Cournot model (game with simultaneous quantity competition, Sect. 3.5).

Observability and strategic effect. In the following, we will discuss how important observability is in sequential games and how it can lead to problematic incentives if the first mover knows exactly that the second mover will adapt optimally to his decision. For this, let us consider two companies that are in simultaneous quantity competition with each other (Cournot competition). However, company A can now make an R&D investment before the actual competition takes place, which leads to a reduction in its marginal costs and can be observed by company B. What are the consequences for the subsequent competition? Let us consider the general reaction function, Eq. (3.24), in the case of two companies:

$$x_i\left(x_j\right) = \frac{a - c_i}{2 \cdot b} - \frac{x_j}{2}$$

We see from this function that a reduction in the marginal costs c_i leads to a larger output quantity of company i for any given quantity decision of the other company j—the reduction in marginal costs increases the numerator in the first term. As Fig. 4.7 shows, this shifts the reaction function of company A outward, so that in the new equilibrium at point 3, the now more efficient company A (due to cheaper production) produces more, and the relatively less efficient company B produces less than in the equilibrium without investment (point 1). However, this new equilibrium will only be realized if company B can observe that company A has made the investment, thereby shifting its reaction function and expanding its production. By contrast, if it cannot observe the move of company A, it would still assume the old reaction function of company A and produce its original quantity $\overline{x}_B$ adhering to the old equilibrium.

Overall, we distinguish three effects that result from the R&D investment of company A and are related to the observability of this measure:

1. *Direct effect*: This can be observed at the quantity combination at point 1: Due to the investment, company A now has lower production costs, which increases its profit as it must spend less on the same quantity.

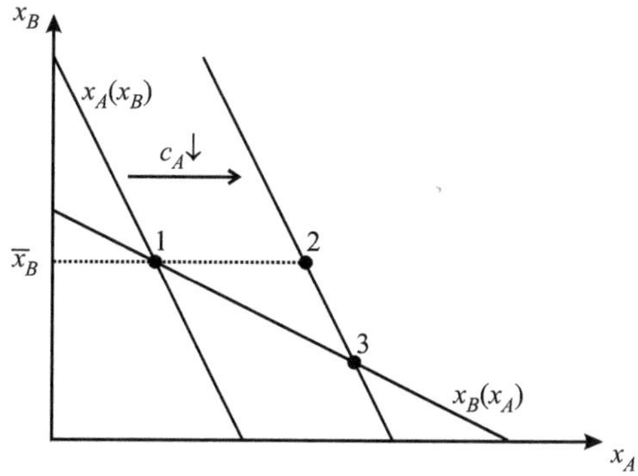

Fig. 4.7 Effects of a cost-reducing investment in quantity competition

2. *Quantity expansion effect*: Company A is now relatively more efficient than company B. Therefore, it makes sense from an overall economic efficiency perspective that company A produces more than company B. As we have already discussed, company A produces more at any quantity of company B, in particular also at the original equilibrium quantity $\overline{x}_2$. Thus, the new combination of quantities is described by point 2. The resulting higher sales increase the profit of company A further.
3. *Strategic effect*: This follows from the fact that company B can observe that company A has become more efficient through the investment and will therefore produce more. This causes company B to produce less, as it is less able to compete with company A given its unfavorable cost structure. This frees up demand and company A will produce more accordingly so that we finally find ourselves in the new equilibrium at point 3.

The strategic effect is particularly interesting because it significantly influences the original investment decision of company A: From an overall economic perspective, there occurs overinvestment, as the investing company anticipates that its competitor will reduce its quantity, thereby increasing its market share. The aim of the investment is therefore not only profit maximization through cost reduction but also partly the displacement of the competitor. We observe the exact opposite effect in price competition: If the opportunity exists to reduce costs through investment before such competition, this will also be undertaken, but there is a tendency to invest too little (underinvestment). After all, any efficiency gains made by a company are passed on to consumers in the form of lower prices, so that the increase in profits for the company is limited.

4.4 Comprehension Questions

1. A sequential game is always also...
   A a repeated game.
   B a multiplayer game.

C a multistage game.
D a game with incomplete information.

2. A Nash equilibrium determined by backward induction in a multistage game…
   A is always Pareto-optimal.
   B is a plausible, because credible, strategy combination.
   C is also always a Nash equilibrium in the corresponding simultaneous game.
   D is always subgame perfect.
3. At decision node X, a proper subgame starts if …
   A all preceding nodes are connected to the rest of the game only via node X.
   B all preceding nodes except for node X are connected to the rest of the game.
   C node X is the origin of the game tree.
   D all subsequent nodes are connected to the rest of the game only via node X.
4. When solving multistage games with continuous strategies, …
   A in the first step, the reaction function of the second mover is determined.
   B in the second step, the reaction function of the second mover is determined.
   C in the first step, the optimal strategy of the first mover is determined.
   D no backward induction is applied.

1b,c; 2b,c,d; 3c,d; 4a

4.5 Problems

4.5.1 Backward Induction

After a successful bank robbery, the five gangsters, Al, Bonnie, Clyde, Donna, and Eddie want to divide their loot of 10 money bags among themselves. They agree on the following procedure: First, Al can make a proposal for the division, which all five gangsters vote on. If it is accepted by a majority of votes, the division takes place as proposed. However, if it is rejected, Al is "eliminated" and Bonnie can make a division proposal, which the remaining four gangsters vote on. If there is a tie or majority of votes, the proposal is accepted. Otherwise, she is also "eliminated", and it is Clyde's turn. The order continues with Donna and Eddie until a division has been accepted. How is the loot divided if a money bag cannot be split and each gangster only has their own profit in mind?

Solution in Sect. 4.6.1.

4.5.2 Sequential Game: Matrix Form Versus Extensive Form

Two companies are in competition with each other in a market with a time horizon of two periods. Company A can choose between the actions "price war" $\left(a_{A1}^{t}\right)$ or "market division"$\left(a_{A2}^{t}\right)$. Company B observes this decision in $t = 0$ and decides at the beginning of period $t = 1$ on the actions "market exit" [a_{B1} with $\pi_{B}^{t}\left(a_{A2}^{t}\right) = 0$]

and "remain in the market" [a^t_{B2} with $\pi^t_B(a^t_{A1}) = -1, \pi^t_B(a^t_{A2}) = 2$]. This decision can be observed by company A. The period profit of company A is $\pi^t_{A,D}(a^t_{A1}) = -1$ and $\pi^t_{A,D} = 2$ if both companies are active in the market (duopoly) and $\pi^t_{A,M} = 6$ if company B has left the market (monopoly of A).

a. Represent the game in period $t = 1$ in strategic form (matrix form) and determine the Nash equilibria in pure strategies! Now represent the game in extensive form (game tree) and solve it! Are all Nash equilibria determined in the matrix equally plausible?
b. Represent the entire game in extensive form! Which method can be used to solve this game? Determine this solution! Why should we not just look at the equilibrium path alone?

Solution in Sect. 4.6.2.

4.5.3 Credibility in Sequential Games

Consider the following game:

Alice, Bob	a_{B1}	a_{B2}
a_{A1}	(3, 3)	$(1 - W, 4)$
a_{A2}	(2, 2)	(0, 0)

a. Transfer this simultaneous game into a sequential game with Bob as the first mover and Alice as the second mover! Represent the sequential game in extensive form! Solve the game using backward induction and determine the subgame perfect Nash equilibrium for $W = 0$!
b. Now assume that the second mover (Alice) issues the following threat to the first mover (Bob) before the game: "Since you are the first to move, I want you to choose a_{B1}. If you still decide for a_{B2}, I will retaliate with a_{A2} as a punishment!" What do you think of this announcement? Justify! What would change in your considerations if the additional payoff W with $W > 1$ comes into play?

Solution in Sect. 4.6.3.

4.6 Solutions

4.6.1 Subgame Perfection and Continuous Strategies

Problem Text in Sect. 4.5.1.

To obtain the optimum division, we apply the method of backward induction. In this game, the distribution choice simultaneously represents the players' payoffs.

The last to move is Eddie (No. 5). When it is his turn, all the others have already been eliminated, so he would claim all 10 sacks for himself. However, he does not get that far: Before him, it is Donna's turn (No. 4). According to the distribution rule, the proposal is accepted in case of a tie, which means she will claim all 10 sacks for herself regardless of Eddie's approval. This results in the payoffs (†, †, †, 10, 0).

Before Donna, Clyde (No. 3) makes a proposal. He knows that if Donna gets her turn, Eddie will get nothing. So, he only needs to ensure that Eddie agrees to his proposal. To do this, it is enough to give him 1 money bag. Thus, his proposal is (†, †, 9, 0, 1).

Let us now consider Bonnie's (No. 2) decision in the second stage. She would have to give Donna (No. 4) 1 money back to persuade her to accept her proposal. She does not care about Clyde and Eddie's approval, as her proposal is accepted in the case of a tie. Thus, she proposes the distribution (†, 9, 0, 1, 0).

Of course, Al (No. 1) is the first to go, who is aware of the other gangsters' distribution considerations and forms corresponding alliances. Since he knows that Clyde and Eddie get nothing in Bonnie's proposal, he can win them over by offering them each 1 money bag. He will therefore submit the distribution proposal (8, 0, 1, 0, 1), which is accepted with 3 to 2 opposing votes and without any elimination.

4.6.2 Sequential Game: Matrix Form Versus Extensive Form

Problem Text in Sect. 4.5.2.

Part a)
It is important to read out the payoffs relevant to the first stage from the many given information. The following matrix shows the payoffs of the two companies:

C_A, C_B	Exit (a^1_{B1})	Remain (a^1_{B2})
Price war (a^1_{A1})	(6, 0)	(−1, − 1)
Division (a^1_{A2})	(6, 0)	(2, 2)

The exponent here denotes the period $t = 1$, in which the action is chosen. We see that company A has a weakly dominant strategy, division (a^1_{A2}), which results in a Nash equilibrium at (a^1_{A2}, a^1_{B2}). However, further verification yields an additional Nash equilibrium at (a^1_{A1}, a^1_{B1}). Thus, there are two Nash equilibria.

In the extensive form of the game, it should be noted that in $t = 1$, company B decides first. This results in the following game tree:

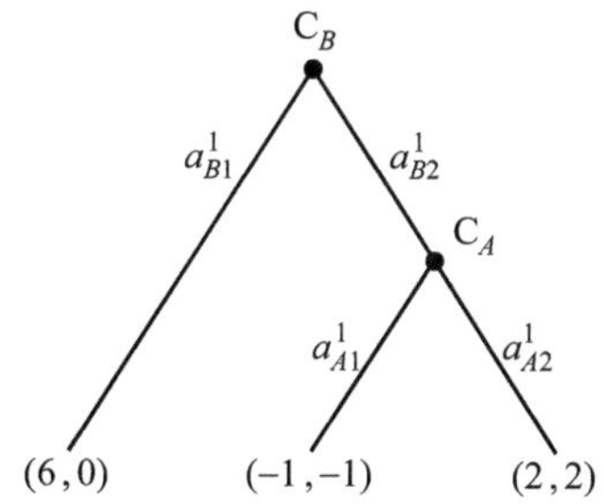

To make it easier to compare the game tree with the matrix (and also part b), the first payoff is that of company A and not, as usual, that of the player who moves first (in this case B). We solve the game by backward induction as follows: In the last stage, company A decides between price war $\left(a^1_{A1}\right)$, which results in a payoff of -1, and division$\left(a^1_{A2}\right)$, with payoff 2. Therefore, it will decide for division$\left(a^1_{A2}\right)$. Company B can in turn decide at the first stage between exit $\left(a^1_{B1}\right)$ with a payoff of 0 or remain $\left(a^1_{B2}\right)$ with a payoff of 2. It will therefore choose to remain$\left(a^1_{B2}\right)$, resulting in the only subgame perfect Nash equilibrium$\left(a^1_{A2}, a^1_{B2}\right)$.

The other equilibrium that we found in the matrix, $\left(a^1_{A1}, a^1_{B1}\right)$, is an incredible threat: Company A threatens to choose a price war, causing company B to leave the market. The reason for this threat is the higher payoff for company A (6 instead of 2) if company B believes the threat and leaves the market. However, if company B stays in the market, company A will divide the market, as it would otherwise be worse off by implementing the threat (it would receive -1 instead of 2). Hence, the threat is not implemented and therefore incredible. Analytically, this equilibrium is not subgame perfect, as it is an equilibrium in the first subgame (if company A chooses a price war, company B should leave the market), but not in the second subgame (company A will never choose a price war here).

Part b)

We know the game in $t = 1$ from a) and now must extend it by the preceding period. Here, company A decides between a price war $\left(a^0_{A1}\right)$ and division $\left(a^0_{A2}\right)$, which also influences the final payoffs of the two players. If company A decides for a price war, it reduces the total payoffs of all players by 1, if it chooses division, the payoff of each player increases by 2. This results in the following game tree.

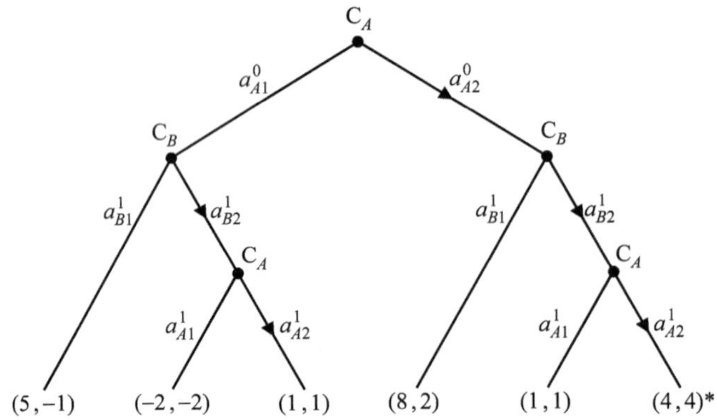

Again, we solve this game by backward induction. Since the subgames of the period $t = 1$ have already been solved in a) and no changes occur due to the uniform change of all payoffs in the subgames, we can immediately consider the subgame in period $t = 0$. Here, company A can choose between a price war $\left(a^0_{A1}\right)$, where it would receive a payoff of 1, and division $\left(a^0_{A2}\right)$ with a payoff of 4. It will therefore choose division $\left(a^0_{A2}\right)$. This can also be well understood by looking at the game tree.

The subgame perfect Nash equilibrium is therefore:

$$\left(\left\{a^0_{A2}, a^1_{A2}\left(a^0_{A1}, a^1_{B2}\right), a^1_{A2}\left(a^0_{A2}, a^1_{B2}\right)\right\}, \left\{a^1_{B2}\left(a^0_{A1}\right), a^1_{B2}\left(a^0_{A2}\right)\right\}\right).$$

From the equilibrium path $a^0_{A2} - a^1_{B2} - a^1_{A2}$ results the payoff (4, 4). The mere indication of the path instead of the strategies is not sufficient, as it cannot be ruled out that this is based on an incredible threat. By specifying the strategies, it can be easily checked whether they really represent an equilibrium in every subgame and thus form a subgame-perfect Nash equilibrium.

4.6.3 Credibility in Sequential Games

Problem Text in Sect. 4.5.3.

Part a)

The representation in extensive form is as follows:

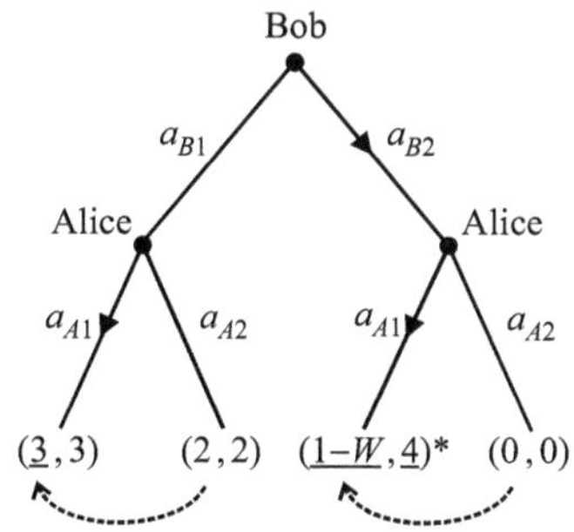

When applying backward induction, we start on the last, i.e., second, stage with the decision of the second mover (Alice). For $W = 0$, she has a strictly dominant strategy, a_{A1}, when it is her turn. Her optimal strategy is therefore $s_A^* = \{a_{A1}|a_{B1}, a_{A1}|a_{B2}\}$. Anticipating this, Bob chooses his second action/ strategy a_{B2} on the first stage. The subgame-perfect Nash equilibrium is therefore $(a_{B2}, \{a_{A1}|a_{B1}, a_{A1}|a_{B2}\})^*$ with the payoff combination (1, 4). The subgame-perfect Nash equilibrium coincides here with the equilibrium of the simultaneous game since Alice as the second mover has a strictly dominant strategy.

Part b)

The announcement of the second mover is an incredible or "empty" threat: If Bob ignores the threat and chooses a_{B2}, then Alice would have no incentive to carry out her threat and would switch to a_{A1}. The parameter $W > 1$ would, however, take away this incentive. For this, Alice would need to find a way to ensure—before the game—that the loss of W *automatically* activates as soon as she must decide at the right node of the game tree (if Alice lands at the right node, she must therefore lose control over W). Only with such a commitment, can she establish a credible threat and persuade Bob to make a different opening move.

Literature Recommendations

Dixit A, Skeath S, Reiley (2014) Games of strategy, 4th ed. New York (Chaps. 6 and 10).
Kydland FE, Prescott EC (1972) Rules rather than discretion: the inconsistency of optimal plans. J Polit Econ 4:103–124.
Tirole J (1988) The theory of industrial organization. MIT-Press (Chaps. 11.1 to 11.3).

Part b)

Repeated Games

5

In the games we discussed so far, the players only interacted with each other once. Therefore, it is often referred to as *one-shot games*. However, in most (everyday) situations, it is the case that one encounters the same "player" repeatedly and must make largely identical decisions. Examples of this are the collaboration between colleagues at the workplace, the regular purchase of a product by a loyal customer, the business relationship between a car manufacturer and its supplier, or the usual squabbles among siblings. In these cases, the players do not play the respective game only once but repeatedly with each other—hence it is referred to as a repeated game or also *dynamic game*. The underlying one-shot game is then referred to as the base game. In this chapter, we want to examine this in more detail and thus also particularly demonstrate how the prisoner's dilemma can be overcome through the possibility to repeat the base game.

5.1 Discount Factor

We can broadly distinguish between two types of repeated games: those with *stationary* structure and those with *non-stationary* structure. In games with a stationary structure, the underlying base game is repeated without any change in the central parameters of this game, i.e., there are no variations in the number of players, the strategy space, the information structure, and the payoffs. In contrast, in games with a non-stationary structure at least one of these parameters changes over time, e.g., because the demand on a market fluctuates with the economic cycle or because a stock of resources is continuously running out, which is reflected in higher extraction costs for the companies. In the following, we will discuss non-stationary games using an example, to illustrate the temporal variation of the payoffs. Another example of this, the Rubinstein bargaining model, will be

F. Bartholomae and M. Wiens, *Game Theory and Applications*, Classroom Companion: Economics, https://doi.org/10.1007/978-3-658-44608-6_5

discussed in the context of negotiation games in problem 7.1.11. Repeated games with a non-stationary structure are rarely covered in teaching, mainly due to the relatively demanding solution methods.[1] Our focus is therefore also on repeated games with stationary structure (Sect. 5.3). These games allow us to depict several interesting phenomena with little formal effort, which we observe in everyday life, especially when opportunities for repeated interaction arise. The most important mechanisms of this kind include reputation and reciprocity.

Regardless of whether it is a game with a stationary or non-stationary structure, in repeated games, the *discount factor* $\delta \in [0; 1]$ is an important preference parameter. The discount factor indicates the value that a decision-maker assigns to the future ("value of the future") and thus tells us what percentage of a future payoff the decision-maker must receive today to be willing to forego a future payoff. The higher the discount factor, the more the decision-maker weighs future payoffs compared to today's payoff. In this case, the decision-maker is more willing to forego today's consumption and wait until tomorrow, thus tending to orientate themselves more long term and behave patiently. For this reason, the discount factor is also often interpreted as the "degree of patience". More technically formulated, the discount factor measures the value of a payoff realized tomorrow (i.e., in period $t+1$) as a proportion of a realized payoff unit today (i.e., in period t). With the help of the discount factor, we are thus able to make payoffs of different periods comparable to each other. To determine the total utility U, which accrues over all considered periods, the utility values of all periods are added and weighed with the discount factor. Starting at the initial period $t = 0$ and considering a time range up to an arbitrary end period T, the total utility is calculated as follows:

$$U_{t=0}^{T} = \delta^0 \cdot u_0 + \delta^1 \cdot u_1 + \delta^2 \cdot u_2 + \delta^3 \cdot u_3 + \cdots + \delta^T \cdot u_T = \sum_{t=0}^{T} \delta^t \cdot u_t. \quad (5.1)$$

Normally, $0 < \delta < 1$, so that the δ^t-term becomes smaller with each additional period. This means that payoffs receive an exponentially higher weight, the closer they are to the current period $t = 0$. The fact that the exponent of the discount factor is determined by the temporal distance to the reference period (here period 0) can be easily illustrated by the following consideration: From the decision-maker's point of view, the following relationship exists between the value of a payoff in period t and in period $t+1$: $u_t = \delta \cdot u_{t+1}$. Similarly, the relationship between periods $t+1$ and $t+2$ is $u_{t+1} = \delta \cdot u_{t+2}$. We can now establish the relationship between periods t and $t+2$ by directly linking the two equations: $u_t = \delta \cdot (\delta \cdot u_{t+2}) = \delta^2 \cdot u_{t+2}$. Hence, if we compare two payoffs with a temporal distance of Δ periods, we get accordingly $u_t = \delta^{\Delta} \cdot u_{t+\Delta}$, , i.e., in the special case that we compare a future payoff with today ($t = 0$), we get $u_0 = \delta^t \cdot u_t$.

[1] In general, knowledge of discrete dynamic optimization is required for this, which cannot be assumed among bachelor students.

The discount factor always relates period $t+1$ to period t, that is $\delta \cdot u_3$ represents the utility of period 3 in period 2. However, we want to consider all utilities in relation to period 0. Therefore, we weigh the utility from period 2, $\delta \cdot u_3$, again, to get the utility in period 1: $\delta \cdot \delta \cdot u_3$ or $\delta^2 \cdot u_3$. This is still a utility accruing in the future from the perspective of period 0, which needs to be further discounted: $\delta \cdot \delta^2 \cdot u_3$ or $\delta^3 \cdot u_3$. . The last step finally sets the utility from period 3 in relation to period 0. Thus, more distant payoffs receive a kind of "valuation discount" (hence the term *discount factor*). If the discount factor reaches its maximum ($\delta = 1$), then the decision-maker assigns the same weight to *all* periods, regardless of whether it is the current period or a period in the distant future. For $\delta = 1$, Eq. (5.1) turns into the simple sum of all (equally weighted) period utilities:

$$U_{t=0}^{T} = u_0 + u_1 + u_2 + \cdots + u_T = \sum_{t=0}^{T} u_t \tag{5.2}$$

If the discount factor is very low, however, current payoffs are assigned a higher value. In this case, the decision-maker is hardly interested in future payoffs but weighs current consumption very highly. A very low discount factor thus stands for a high present preference or a high degree of impatience. With the smallest possible discount factor ($\delta = 0$), the decision-maker is only interested in the present and completely ignores the future. Thus, for $\delta = 0$ from Eq. (5.1), we get $U_{t=0}^{T} = u_0$, since $0^0 = 1$.

Note and Tip The discount factor δ is often confused by students with the discount rate i, the interest rate. There is a clear relationship between the two parameters, as they both are used to compare the payoffs of different periods—but from different perspectives. The discount factor is about the *subjective* valuation of a payoff tomorrow u_1 compared to a payoff today u_0 as a reference, i.e., $u_0 = \delta \cdot u_1$. Suppose the decision-maker receives 100 euros tomorrow and assesses the "value of the future" with $\delta = 0.8$. This means that for the decision-maker, 80 euros today is worth as much as 100 euros tomorrow. The discount rate i on the other hand, is an *objective* parameter, namely the price commonly referred to as interest on the capital market. This price causes the demand for and supply of capital to reach equilibrium. The demand for capital corresponds to the desire for credit and is thus identical to the willingness to shift a payoff tomorrow to today. The supply of capital corresponds to the desire to save and is therefore identical to the willingness to wait until tomorrow to use the available money of today. The demand for capital corresponds to the desire for credit and is thus identical to the willingness to bring forward a payoff from tomorrow to today. The capital market usually demands a positive interest rate i to compensate savers for the opportunity costs of lending money. The equivalent amount of money tomorrow (u_1) is therefore obtained by compounding today's amount u_0 by the factor $(1+i)$, i.e., $u_0 \cdot (1+i) = u_1$. If all decision-makers behave rationally and make their credit or savings decisions via a frictionless capital market, then the two equations $u_0 = \delta \cdot u_1$ and $u_0 \cdot (1+i) = u_1$ represent two sides of the same coin. Under these conditions, there is a clear relationship between the discount factor δ and the discount rate i, namely $\delta = i/(1+i)$. If you use this relationship in Eq. (5.1), you get the well-known net present value (NPV) formula used in finance.

5.2 Repeated Games with Non-stationary Structure

Repeated games with non-stationary structure have the property that certain parameters of the game, usually certain components of the payoff, change over time. The decision that a player makes at a certain time t is therefore not comparable with the decision that is made later in $t + 1$, since optimization calculus may have changed by this time.

Consider, for example, a company that produces a so-called experience good, that is, a good whose quality can only be determined after purchase, i.e., only through consumption. For simplification, we assume only two periods. This good is sold by the company in both periods, $t = 1$ and $t = 2$, to n_1 and n_2 consumers, respectively. The strategic parameter is the quality x which is chosen from the interval [0; 1]. Here, $x = 0$ means the lowest quality and accordingly $x = 1$ the highest quality. To produce a good with the quality x, the company incurs costs of $c \cdot x^2$; c is a fixed cost parameter with $c > 1$. The quality has an influence on the loyalty of the first-period consumers beyond the direct cost aspect: The higher the chosen quality level in period 1, the larger the proportion of customers who buy the good again in period 2; specifically, we assume that in the second period $x_1 \cdot n_1$ of the previous customers buy the good again. In addition, in period 2, n_2 new customers are added. The (constant) price of the good is p_1 in the first and p_2 in the second period.

The intertemporal profit function of the company depends on the quality x_1 and x_2 which is chosen in $t = 1$ and $t = 2$ and is thus given by

$$U(x_1, x_2) = p_1 \cdot n_1 - c \cdot x_1^2 + \delta \cdot \left[p_2 \cdot (x_1 \cdot n_1 + n_2) - c \cdot x_2^2\right]. \tag{5.3}$$

The first term $p_1 \cdot n_1 - c \cdot x_1^2$ corresponds to the profit in period 1. The second term $\delta \cdot \left[p_2 \cdot (x_1 \cdot n_1 + n_2) - c \cdot x_2^2\right]$ describes the discounted profit in period 2. For the goods produced in period 2, the company makes its quality decision again, which is again associated with the corresponding costs $\left(c \cdot x_2^2\right)$. Since the profit from period 2 represents a future payoff from the company's current perspective, it must be discounted (δ).

Since we only consider a time horizon of two periods, the quality of period 2 has no consequences for the company, so it will choose $x_2 = 0$. The chosen quality in period 1 (x_1) on the other hand, influences the profit of both periods, as can be clearly seen in (5.3). This interconnection of period payoffs through the strategies of the players is a characteristic of dynamic games with non-stationary structure. If we partially derive (5.3) by x_1, $\partial U(x_1, x_2)/\partial x_1$, and set the resulting derivative equal to zero, we obtain the profit-maximizing quality x_1^*, which the company will choose in the first period:

$$x_1^* = \frac{\delta \cdot p_2 \cdot n_1}{2 \cdot c}.$$

The first-period quality is thus *ceteris paribus* higher, the lower the costs c, the higher the patience δ of the company, and the higher the "revenue of customer loyalty" $p_2 \cdot n_1$ are. The different time indices of the expression $p_2 \cdot n_1$ clearly show that customer retention is an intertemporal decision problem of the company: The price of period 2 can only be realized to the extent that it can retain the customers from period 1.

5.3 Repeated Games with Stationary Structure

In contrast to games with non-stationary structure, in games with stationary structure the period game, i.e., the base game, remains unchanged. This property greatly simplifies the formal solution of such games, as the respective period is only reflected in the discounting.[2] Let us consider as an example the prisoner's dilemma with the general payoffs $0 < N < C < D$ (with C = cooperation payoff, D = defection or deviation payoff and N = Non-cooperation or Nash equilibrium payoff):

A, B	s_{B1}	s_{B2}
s_{A1}	$(C, C)^P$	$(0, D)^P$
s_{A2}	$(D, 0)^P$	$(N, N)^*$

The question now is to what extent new strategies and thus new Nash equilibria can arise solely through the repetition of this basic game. The players could, e.g., try to achieve the cooperative equilibrium (C, C) by playing a *reciprocal retaliation strategy*: They both play cooperatively (s_{i1}) until (at least) one of them deviates (s_{i2}); from then on, they play the uncooperative strategy until the end of the game. Such a strategy is referred to as a trigger strategy,[3] as the other player activates a—in this case relentless—punishment program by deviating for the first time. However, in a *finitely* often repeated game with a predetermined final period T, such a strategy cannot be maintained according to the following theorem.

Theorem on finitely often repeated games ($T < \infty$) *Let s^* be the only Nash equilibrium of the stage game Γ. Then the only subgame perfect Nash equilibrium of the finitely often repeated game $\Gamma(T)$ consists of the T-fold repetition of the Nash equilibrium of the stage game.*

We can also easily explain why this must be the case if we assume that strictly rational players will use *backward induction* in a finitely repeated game: In the last period, it is no longer beneficial for any of the players to cooperate, since there is no possibility to punish such behavior, i.e., both will certainly cheat here. To

---

[2] This is in contrast, as seen in the previous section, to games with non-stationary structure where the period payoff directly depends on the period t.

[3] The term "trigger" symbolizes the start of a program that can no longer be influenced, just as pulling the trigger of a pistol causes a shot to be fired.

benefit from the cheating strategy, each player would have to try to outdo the other and cheat one period earlier. Since the opponent also thinks this way, they will successively move the start of the cheating phase forward, so that in the end they both cheat already in the first period (and thus also in all subsequent periods). It should be noted that the above theorem provides a relatively poor prediction of the actual strategy choice of players for games with a higher number of rounds (about ten or more rounds): In laboratory experiments, in which this game is played under anonymous conditions over ten rounds, the experiment participants do not use backward induction until the last three rounds at the earliest.[4] The cognitive conclusion over several periods thus seems to be too demanding for real players, so they cooperate in the prisoner's dilemma at least in the first periods.

The theoretical prediction for the repeatedly played prisoner's dilemma becomes more realistic when we consider *infinitely repeated* games, which are also referred to as *supergames*. The infinitely long-time horizon of a supergame can be interpreted in two ways: The players assume a certain but unspecified end period (i.e., they know that the game has an end but not exactly when) or they assume in each period with a constant probability that the game will end at some point.[5] Since there is no last period in a supergame, the problem of backward induction disappears—it is simply not applicable—which allows any individually rational payoff combination to be achieved as an equilibrium of the infinitely repeated game as is stated by the following theorem (Folk theorem):

Folk theorem on infinitely repeated games ($T = \infty$) *Any feasible, individually rational payoff vector $u(s)$ can be achieved as a subgame perfect Nash equilibrium of the infinitely repeated game $\Gamma(\infty)$ if the players weigh future payoffs sufficiently strongly (that is, δ close to 1).*

Based on this theorem, the players in the prisoner's dilemma can indeed achieve a cooperative solution, as this represents an individually rational payoff combination. We see this in Fig. 5.1, in which the individually rational payoff combinations for the matrix can be found in the shaded area (including the border). Starting from the Nash equilibrium of the base game (N, N), the cooperative solution (C, C) represents a clear Pareto improvement. In general, individually rational payoffs

[4] Cf. the experiment by Andreoni and Miller (1993).

[5] Even though the first interpretation most closely resembles the perspective we are familiar with from everyday life (we know for sure that at some point our studies or professional activity will end, but usually not exactly when that will be), strictly speaking only the second variant is permissible for infinite games with a stationary structure: Since the probability is the same in each period, the future of the game looks the same from each period. This is not the case with the first interpretation: The temporal distance from today to the uncertain final period becomes smaller and smaller with time, which changes the perspective on the future from period to period (time must not become "scarce" in the infinite game).

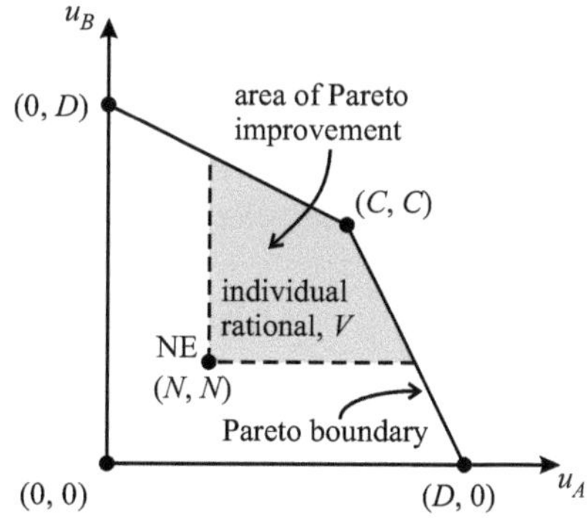

Fig. 5.1 Pareto boundary and possibilities for improvement

are given by

$$V = \{u(s)|s \in S, u_i \geq \hat{u}_i \; \forall \, i \in N\},$$

where $\hat{u}_i$ denotes the payoff that a player can at least secure. In our case, this would be N.

In addition to the potential for improvement, the Folk theorem demands another condition, without which the Pareto improvement cannot be enforced. The players must be sufficiently patient, i.e., they must have a discount factor that does not fall below a critical threshold $\tilde{\delta}$. Let us consider the trigger strategy we have already described, where the players cooperate until one of them deviates. For this, we can calculate the critical discount factor that ensures that a balance is reached, in which none of the players has an incentive to deviate. This must meet the following condition:

$$C \cdot \left(\delta^0 + \delta^1 + \delta^2 + \cdots + \delta^\infty\right) > D \cdot \delta^0 + N \cdot \left(\delta^1 + \delta^2 + \cdots + \delta^\infty\right). \tag{5.4}$$

It must be beneficial for every player to cooperate permanently, provided the other also plays a trigger strategy. The left side of (5.4) represents the payoff a player receives by choosing the cooperative strategy beginning in the current period ($t = 0$) and ending in "eternity" ($T = \infty$). This is the discounted sum (= present value) of all cooperative payoffs. This amount must be greater than what the player would receive by cheating (right side of the inequality). In the case of cheating, the player receives the high payoff D immediately ($t = 0$)—but only once: Since the player's opponent plays the trigger strategy, the latter switches to the cheating strategy for all subsequent periods, to which "cheating" is the best response. Thus, from period $t = 1$ to "eternity" ($T = \infty$) both play the cheating strategy (the Nash equilibrium) and henceforth receive the discounted sum of payoffs for mutual cheating, N.

Note and Tip In games with a stationary structure, many payoffs remain unchanged over several periods, which allows the discount factors to be multiplied. So, in the above example of the infinitely repeated prisoner's dilemma, the cooperative payoff C occurs in each period, which is why you can factor out the discount factors and sum them $(\delta^0 + \delta^1 + \delta^2 + \cdots + \delta^\infty) = \sum_{t=0}^{\infty} \delta^t$. This long chain of summed discount factors forms a *geometric series,* which for $0 < \delta < 1$ converges to a certain limit:

$$\lim_{T\to\infty} \sum_{t=0}^{T} \delta^t = \sum_{t=0}^{\infty} \delta^t = \frac{1}{1-\delta}.$$

If, for example, the discount factor is $\delta = 0.8$, then the sum of the discount factors converges to $\sum_{t=0}^{\infty} \delta^t = 1/(1-0.8) = 5$. In the case of a period payoff $C = 4$, the present value is thus $C \cdot \sum_{t=0}^{\infty} \delta^t = 4 \cdot 5 = 20$.

The term $\frac{1}{1-\delta}$ can also be derived without applying the geometric series, simply by determining the present value NPV of an infinitely repeated game:

$$NPV = C + \delta \cdot NPV \quad \Leftrightarrow \quad NPV = C \cdot \frac{1}{1-\delta}.$$

In the first period, you receive the period payoff C, in the next period you receive the discounted payoff of all subsequent periods. Since it does not matter which period you choose in an infinitely repeated game (the future always looks the same from any period), we can directly use the present value NPV for the payoff of all subsequent periods (the game essentially starts over again and again). This "short-circuiting" shortens a calculation that runs through an infinitely long loop (mathematically speaking, this is an *infinite regress*) and thus allows us to directly determine the limit.

With the knowledge of the limit of the infinite geometric series $\sum_{t=0}^{\infty} \delta^t$ it is also relatively easy to determine the limits of other time periods. For example, to determine the value of the cheating strategy in the prisoner's dilemma (right side of (5.4)), we still need the sum of the discount factors from period $t = 1$ to infinity, $\sum_{t=1}^{\infty} \delta^t$. Since the current period is not discounted or the discount factor of the current period is always $\delta^0 = 1$, we find the desired expression by subtracting only the current period from the already-known sum $\sum_{t=0}^{\infty} \delta^t$. In short, since the discounting starts tomorrow, the present value starting tomorrow ($t = 1$) needs to be discounted to today ($t = 0$):

$$\underbrace{\delta^0 + \delta^1 + \delta^2 + \ldots + \delta^\infty}_{\text{from today to eternity}} - \underbrace{\delta^0}_{\text{today}} = \underbrace{\delta^1 + \delta^2 + \ldots + \delta^\infty}_{\text{from tomorrow to eternity}}$$

$$\frac{1}{1-\delta} - 1 = \frac{\delta}{1-\delta}.$$

The limit for the infinite geometric series $\sum_{t=1}^{\infty} \delta^t$ is thus $\delta/(1-\delta)$. Following the same consideration, payoffs further in the future can also be discounted to today. In general,

$$\sum_{t=n}^{\infty} \delta^t = \frac{\delta^n}{1-\delta}.$$

Using the theorems for geometric series, we can thus simplify (5.4):

$$C \cdot \sum_{t=0}^{\infty} \delta^t > D \cdot \delta^0 + N \cdot \sum_{t=1}^{\infty} \delta^t \Rightarrow C \cdot \frac{1}{1-\delta} > D + N \cdot \frac{\delta}{1-\delta}.$$

This makes it easy to determine the critical discount factor for which the left side of the inequality is greater than the right. In other words, the critical discount factor for which the incentive to cheat disappears in the infinitely repeated game:

$$\delta > \tilde{\delta} \equiv \frac{D - C}{D - N}. \tag{5.5}$$

To enforce (or "support") cooperative behavior, the discount factor of the players must be greater than a critical threshold $0 < \tilde{\delta} < 1$, which depends on the payoffs. That is, if all players have at least the critical discount factor $\tilde{\delta}$, the cooperative solution can be supported as a subgame perfect Nash equilibrium through the described strategy. The lower the critical discount factor is, the easier it is to realize cooperation in the infinitely repeated game as even more impatient players are then able to cooperate. The fraction will be smaller the lower the difference between D and C is. $D - C$ corresponds to the players' incentive to deviate from the cooperative strategy (a player ultimately receives the highest payoff D for unilaterally cheating the other). The smaller the incentive to defect, the lower the discount factor can be so that cooperation is more feasible. The critical discount factor $\tilde{\delta}$ also becomes smaller when the difference in the denominator of the fraction, $D - N$, increases. This difference is the sum of the cheating incentive, $D - C$, and the net gain from cooperation, $C - N$. The only way to increase the denominator without simultaneously increasing the numerator is by having a higher (net) gain from cooperation for the players. Thus, the cheating incentive increases the critical discount factor, while the gain from cooperation decreases it.

Consequently, we were able to show that it is possible to use a trigger strategy to support cooperation in the infinitely repeated prisoner's dilemma. The specific trigger strategy considered above is just one possible variant, which due to the relentlessly long retaliation or punishment phase ("until eternity", hence sometimes referred to as "total retaliation") is also referred to as *grim trigger*. However, the punishment does not have to be that drastic to achieve the cooperative goal. The described mechanism basically always works when the base game has enough cooperative potential: The players must gain at least as much through bilateral cooperation that the once-achieved cheating gain does not compensate for the crash to the payoffs of the one-period Nash equilibrium. If this condition is met, the punishment mechanism can also be less drastic, which in some cases is even desirable as it allows for some flexibility and error tolerance.

An important prerequisite for an effective punishment mechanism, however, is its credibility, that is, the implemented strategy must be subgame perfect. But how should we generally proceed to check whether a strategy in the supergame is subgame perfect? The simplest case is when the same Nash equilibrium of the stage game is played in each period, regardless of the history of the repeated game. It is true that the repetition (even infinitely often) of a Nash equilibrium of the base game always represents a subgame-perfect Nash equilibrium. In all other cases, to check whether the equilibrium of a supergame is subgame perfect, the *one-shot deviation principle* is applied: A strategy profile s of an infinitely often repeated game satisfies the one-shot deviation principle if no player has an incentive to deviate from s in any period of the game—regardless of whether the players are on or off the equilibrium path in the considered period. To test whether a strategy profile has this property, it is sufficient to pick out any period and check the incentives to deviate for this period (hence the term "one-shot deviation").

With this consideration, we can show that the grim trigger strategy, in relation to the above prisoner's dilemma, represents a subgame perfect Nash equilibrium. Four initial cases ("four possible histories of the game") are conceivable: two on and two off the equilibrium path. *On the equilibrium path*, both players could have previously cooperated or cheated: For case (i) with mutual cooperation, (5.5) shows that there is no incentive to switch to cheating in any period. For case (ii) with mutual cheating, the grim trigger strategy prescribes to continue playing "cheating" consistently (the repetition of the Nash equilibrium). Given that the other player adheres to this rule, no player has an incentive to deviate from the grim trigger equilibrium strategy and unilaterally play "cooperation". Thus, on the equilibrium path, no player has an incentive to deviate, neither in cooperation nor in cheating. In the two cases off the equilibrium path, (iii) and (iv), the players each play different strategies: while one cooperates, the other cheats. Due to the symmetry of the game, these two cases can be combined into one. Since the grim trigger strategy prescribes to cheat permanently from the first observation of cheating *and* at the same time "cheating" is the best response to "cheating", it is always optimal for the players off the equilibrium path to cheat consistently. Thus, also *off the equilibrium path*, no player has an incentive to deviate from the equilibrium strategy in subsequent periods (or in other words: once deviated, the players have an incentive to return to grim trigger afterwards). Thus, Grim-Trigger is a subgame perfect Nash equilibrium.

Now let us consider *tit for tat* ("an eye for an eye"), another strategy that plays an important role in the context of supergames. Tit for tat includes the following instructions for the repeated prisoner's dilemma:

Start with "cooperation" in the first period and then always respond with the strategy that the opponent chose in the previous period.

This strategy embodies the principle of reciprocity, which plays a major role in the context of social cooperation. People tend to respond to the cooperative behavior of their fellow human beings: you cooperate if others do so, otherwise you also stop cooperating.

Tit for tat and the norm of reciprocity. In the late 1970s, political scientist Robert Axelrod invited scientists (mainly conflict researchers and game theorists) via a public call to submit strategies for the repeated prisoner's dilemma (Axelrod 1980). These strategies could be arbitrarily complex structured and should be programmed in BASIC or FORTRAN, to be able to compare them in the context of a computer tournament. A total of 14 algorithms were submitted, with the complexity of the submitted programs ranging from four to 77 lines of source code. The winning algorithm, which achieved the highest total payoff in this tournament, was also the simplest, as it corresponded to the tit for tat strategy. After the publication of these results, Axelrod invited a repetition of the competition on a larger scale, where computer enthusiasts and programmers were also called upon to participate. With a total of 63 submitted algorithms, the response to the second tournament was much larger than the first, and the complexity of the algorithms was considerable with up to 177 lines of source code. The tit for tat strategy also won in this larger, second tournament, which is particularly remarkable because all participants were informed about the successful performance of the tit for tat strategy in the first tournament.

Axelrod attributed the relative success of this strategy to four properties: being nice, retaliatory, forgiving, and clear. In the context of the prisoner's dilemma, a strategy is nice if it never chooses the cheating strategy first. Retaliatory means that niceness is immediately abandoned as soon as the counter-strategy cheats. Forgiveness implies the possibility, after a case of cheating, to return to the cooperative path at a later point in time. And clarity makes it understandable for everyone, so that cooperation is easier to implement. The impressive performance of the tit for tat strategy is an indication that the principle of reciprocity plays a central role in the context of interpersonal cooperation. However, it should be noted that based solely on two computer tournaments, no generalizing conclusions regarding the optimality or universality of tit for tat are permissible. Even in these competitions, tit for tat, for example, did not win any game in direct comparison, as it loses narrowly against offensive strategies, but can only draw level with cooperative strategies at a high level. In addition, tit for tat is not forgiving enough against a learning strategy that gradually approaches cooperation.[6]

It is easy to show that a cooperative Nash equilibrium in the prisoner's dilemma can be achieved with the tit for tat strategy: If the players consistently adhere to this strategy, they achieve the cooperative payoff C over all periods. If Bob deviates to "cheating" (while Alice sticks to tit for tat), then he receives the one-time deviation profit D. Since Bob knows that Alice will also respond with "cheating" according to tit for tat in the next period, it is better for him to play "cheating" again in the subsequent period for self-protection, whereby both players in the subsequent period each receive the payoff N. Since this logic also applies to the following period, etc., both players receive the payoff N from the second period into all eternity. Thus, the critical discount factor of the cooperative Nash equilibrium of the tit for tat strategy is also described by (5.4). If the players are sufficiently

[6] For a detailed and critical discussion of the results, see Milgrom (1984).

patient, $\delta > \tilde{\delta}$, cooperation can be sustained with this discount factor using both the grim trigger and the tit for tat strategy.

In contrast to grim trigger, the tit for tat strategy is not *subgame perfect* if the condition $2 \cdot C > D$ applies for the payoffs. While the players have no incentive according to (5.4) to deviate from the equilibrium path in case of mutual cooperation, they do have an incentive for $2 \cdot C > D$ to deviate from the tit for tat strategy as soon as one of the players has cheated. Suppose Bob had deviated and cheated once. Would he now have an incentive to stick to tit for tat from the second period onwards (if Alice also consistently plays tit for tat)? If he were to do this, he would have to choose the strategy that Alice has chosen in the previous period (i.e., "cooperation"), while Alice would respond to the strategy Bob has chosen in the previous period ("cheating"). Also in the following period, the players would reciprocate the previously played strategy of the co-player, that is, ("cooperation", "cheating") is followed by ("cheating", "cooperation"), etc. The players would therefore always choose different and permanently alternating strategies, whereby each receives the cheating gain D in one period and the cheating loss 0 in the next. Because of $2 \cdot C > D + 0$, it would be more attractive for each player to break the reciprocal pattern after a case of cheating off the equilibrium path to achieve permanent cooperation again.

Another concept often discussed in connection with repeated games is that of *renegotiation-proof equilibria.* The idea behind this is that despite the Grim Trigger strategy being subgame perfect, once the trigger has been activated and the players are on the punishment path for all eternity, they have an incentive to team up again to achieve a better solution. Why do the players possibly have such an incentive? One problem with the grim trigger strategy is that the punishment/retaliation affects not only the deviating player but also the punishing player, so both suffer payoff losses. Thus, both players have an incentive to renegotiate to somehow get back on the more advantageous cooperation path. Strategies that have this property—such as the grim trigger strategy—are therefore referred to as not renegotiation-proof. A problem that will naturally arise here is that a renegotiation or even just the potential possibility of a renegotiation undermines the credibility of the original threat and thus questions the feasibility of the initial equilibrium. What could a renegotiation-proof strategy look like? Such a strategy would be, for example: As soon as Bob deviates from "cooperation", Alice plays "deviation" in retaliation—but only as long as Bob shows no remorse. If Bob regrets the offense by choosing "cooperation" (thereby granting Alice a one-time advantage and punishing himself), Alice returns to "cooperation". The punishing player, Alice, has no incentive to renegotiate, as it is in the power of the punished player, Bob, to leave the punishment path at any time through remorse.

Collusion strategies in oligopoly. The mechanism of repeated games provides a plausible explanation for why cooperation problems of the prisoner's dilemma type are often much easier to overcome when the players interact repeatedly. A common application of this principle in industrial economics is the cooperation of companies within a *cartel.* Since price agreements (so-called *collusion*) are generally prohibited by law, companies cannot rely on legal means, such as formal contracts, to enforce these secretly made agreements. The following example illustrates the effect of the mechanism of repeated games on collusion strategies in an oligopoly. We derive this solution in general form, so that parameters effects on the equilibrium solution become more evident.

As a simplification, we assume that companies are symmetric and have only two strategies: They can either choose the (competitive) Cournot quantity or the collusive quantity. This results in the following payoff matrix:

A, B	Cournot Quantity	Collusion Quantity
Cournot Quantity	(π_{CC}, π_{CC})	(π_{CK}, π_{KC})
Collusion Quantity	(π_{KC}, π_{CK})	(π_{KK}, π_{KK})

To calculate the specific profits, we first determine the chosen quantities. For this, we assume the following general specification of linear (inverse) demand

$$p_A(x_A) = a - b_o \cdot x_A - b_c \cdot x_B \quad \text{and}$$
$$p_B(x_B) = a - b_o \cdot x_B - b_c \cdot x_A,$$

where b_o indicates the impact of the company's own quantity and b_c the impact of the competitor's quantity on the price of the own variant. For constant and identical marginal costs c the profit functions are

$$\pi_A(x_A, x_B) = (a - b_o \cdot x_A - b_c \cdot x_B) \cdot x_A - c \cdot x_A \quad \text{and}$$
$$\pi_B(x_A, x_B) = (a - b_o \cdot x_B - b_c \cdot x_A) \cdot x_B - c \cdot x_B.$$

To find the Cournot quantity, we partially derive each profit with respect to its own quantity, $\partial\pi_A/\partial x_A$ and $\partial\pi_B/\partial x_B$, and set the result to zero. This gives us the reaction functions

$$x_A(x_B) = \frac{a-c}{2 \cdot b_o} - \frac{b_c}{2 \cdot b_o} \cdot x_B$$
$$x_B(x_A) = \frac{a-c}{2 \cdot b_o} - \frac{b_c}{2 \cdot b_o} \cdot x_A.$$

If we insert these symmetrical functions into each other, we obtain both the symmetrical quantities and finally the symmetrical profits in the Nash equilibrium:

$$x_A^C = x_B^C = x^C = \frac{a-c}{2 \cdot b_o + b_c}$$
$$\pi_A = \pi_B = \pi_{CC} = \frac{b_c \cdot (a-c)^2}{(2 \cdot b_c + b_c)^2}.$$

The collusion quantity in turn results from the fact that the joint profit is maximized. This is

$$\pi(x_A, x_B) = (a - b_o \cdot x_A - b_c \cdot x_B) \cdot x_A - c \cdot x_A + (a - b_o \cdot x_A - b_c \cdot x_B) \cdot x_B - c \cdot x_B.$$

The profit is partially derived with respect to both quantities, $\partial\pi/\partial x_A$ and $\partial\pi/\partial x_B$, and the resulting two conditions are then set equal to zero, so that we obtain the following quantities:

$$x_A^K = x_B^K = x^K = \frac{a-c}{2 \cdot (b_o + b_c)},$$
$$\pi_A = \pi_B = \pi_{KK} = \frac{(a-c)^2}{4 \cdot (b_o + b_c)}.$$

We determine the asymmetric profits, π_{CK} andπ_{KC}, by inserting the respective quantities, x^C andx^K, into the profit function, e.g., in the case of company A, this is $\pi_{CK} = \pi_A\left(x_A^C, x_B^K\right)$ and$\pi_{KC} = \pi_A\left(x_A^K, x_B^C\right)$. This gives us the following

$$\pi_{KC} = \frac{b_o \cdot (a-c)^2 \cdot (2 \cdot b_o + 3 \cdot b_c)}{(2 \cdot b_o + b_c) \cdot 4 \cdot (b_o + b_c)^2},$$
$$\pi_{CK} = \frac{(a-c)^2 \cdot \left(2 \cdot b_o^2 + 2 \cdot b_o \cdot b_c + b_o^2\right)}{2 \cdot (b_o + b_c) \cdot (2 \cdot b_o + b_c)^2}.$$

Since $\pi_{CC} > \pi_{KC}$ and $\pi_{CK} > \pi_{KK}$, the choice of the Cournot quantity is a dominant strategy. In the infinitely repeated game, however, cooperation can be supported by choosing the grim trigger:

$$\sum_{t=0}^{\infty} \delta^t \cdot \pi_{KK} \geq \pi_{CK} + \sum_{t=1}^{\infty} \pi_{CC}$$
$$\frac{1}{1-\delta} \cdot \pi_{KK} \geq \pi_{CK} + \frac{\delta}{1-\delta} \cdot \pi_{CC}$$

which results in

$$\frac{1}{1-\delta} \cdot \frac{(a-c)^2}{4 \cdot (b_o + b_c)} \geq \frac{\left(2 \cdot b_o^2 + 2 \cdot b_o \cdot b_c + b_c^2\right) \cdot (a-c)^2}{2 \cdot (b_o + b_c) \cdot (2 \cdot b_o + b_c)^2} + \frac{\delta}{1-\delta} \cdot \frac{b_o \cdot (a-c)^2}{(2 \cdot b_o + b_c)^2}$$

and after many simplifications, we finally find $\delta \geq \tilde{\delta} \equiv 0.5$. This is quite an astonishing result, because it means that regardless of the concrete specification of linear demand, which corresponds to the standard formulation of oligopoly competition, the critical discount factor is always 0.5.

5.4 Comprehension Questions

1. Super games (infinitely repeated games with stationary structure) can be solved…
   A by applying the "one-shot deviation principle" due to the stationary structure.
   B by applying the principle of backward induction due to the stationary structure.
   C by only having to examine the first period because of the constant period payoffs and not having to discount the payoffs.
   D by easily calculating the discounted sum of all relevant payoffs based on the constant period payoffs using the theorems for geometric series.
2. In the prisoner's dilemma, cooperation can be achieved as a nash equilibrium if …
   A the game is *finitely* repeated.
   B the game is *finitely* repeated and both players have a sufficiently high discount factor.
   C the game is *infinitely* repeated and both players have a sufficiently high discount factor.
   D the game is not repeated.
3. The critical discount factor of a strategy for a super game is 0.6. In which cases can Alice and Bob implement this strategy?
   A $\delta_A = 0.4, \delta_B = 0.8$
   B $\delta_A = 0.8, \delta_B = 0.8$
   C $\delta_A = 0.4, \delta_B = 0.4$
   D $\delta_A = 0.6, \delta_B = 0.6$
4. If both players in the infinitely repeated prisoner's dilemma apply the following strategy, then cooperation can be realized as a *subgame perfect* Nash equilibrium.
   A Tit for tat
   B Grim trigger
   C Always choose "Cooperation"
   D Always choose "Defection"

1A,D; 2C; 3B, D (the discount factor must be above the threshold for all players involved); 4B,D (A is Nash equilibrium (NE) but not subgame perfect; C is no NE).

5.5 Problems

5.5.1 Sequential Game Versus Repeated Game

In Chap. 4, we did not consider repeated games, but sequential games in which the temporal structure also played an important role. The temporal structure of a sequential game becomes clear by dividing this game into stages. Isn't it therefore basically irrelevant whether the time periods of a game are called stages or periods? If not, what is the difference?

Solution in Sect. 5.6.1.

5.5.2 Perfect Tit for Tat

Consider the following prisoner's dilemma:

Alice, Bob	Cooperate	Betray
Cooperate	(4, 4)	(1, 5)
Betray	(5, 1)	(2, 2)

The trigger strategy "Perfect Tit for Tat" includes the following action instruction for the prisoner's dilemma:

Start with cooperate in the first period. If the result of the last period was either mutual cooperation (cooperate, cooperate) or mutual betrayal (betray, betray), choose to cooperate. Betray, if the result in the last period was either (betray, cooperate) or (cooperate, betray).

Show that the strategy "Perfect Tit for Tat" in the infinitely repeated game can support mutual cooperation as a subgame perfect Nash equilibrium, given the players' discount factor is sufficiently high!

Solution in Sect. 5.6.2.

5.5.3 Subgame Perfection and Repeated Games

The results of repeated games allow us to explain collusive behavior in a duopoly. The two companies, Alpha and Beta, can either set the low competition price p^C or the high collusion price p^M and thereby receive the payoffs:

Alpha, Beta	p_B^C	p_B^M
p_A^C	(400, 400)	(500, 300)
p_A^M	(300, 500)	(450, 450)

Assume that this game is repeated several times.

a. Can the collusion solution be sustained with finite repetition of the game?
b. Formulate the grim trigger strategy of the supergame! For which discount factors δ can collusion be realized as equilibrium?

Solution in Sect. 5.6.3.

5.6 Solutions

5.6.1 Sequential Game Versus Repeated Game

Problem Text in Sect. 5.5.1.

The temporal structure of a game indicates the order in which the players take their turn and at what point in the game which information is revealed. Figuratively speaking, the temporal structure serves as a script that tells the players when they have their turn and what has happened before or what is supposed to happen next. Stages thus organize the entire game in time steps. If one stage is omitted, it is no longer the same game, just as it would no longer be the same film if one scene were cut from the script. The repetition of a game, on the other hand, corresponds to the repetition of the entire film. If one period is omitted in the repeated game, then the repeated game changes from a game $\Gamma(T)$ to a game $\Gamma(T-1)$, but the underlying base game always remains unchanged.

5.6.2 Perfect Tit for Tat

Problem Text in Sect. 5.5.2.

In this case, a total of four histories are possible that could have taken place in the previous period. For each of these four histories, it must be checked whether the desired equilibrium (mutual choice of cooperation) can be supported. For reasons of symmetry, it is sufficient to consider only one player.

- Case 1: Both players have chosen to cooperate. The deviation to betray is worse than the repeated choice of cooperation, if

$$5+\delta \cdot 2+\delta^2 \cdot 4+\delta^3 \cdot 4+\ldots \leq 4+\delta \cdot 4+\delta^2 \cdot 4+\delta^3 \cdot 4+\cdots .$$

  Note that nothing changes from the third repetition onwards, so we can simply leave this out:

$$5+\delta \cdot 2 \leq 4+\delta \cdot 4,$$
$$\delta \geq \tilde{\delta} \equiv 0.5$$

- Case 2: Both players have chosen to betray. Maintaining to betray is worse than choosing to cooperate, if

$$5+\delta \cdot 2+\delta^2 \cdot 4+\delta^3 \cdot 4+\cdots \leq 4+\delta \cdot 4+\delta^2 \cdot 4+\delta^3 \cdot 4+\cdots$$
$$5+\delta \cdot 2 \leq 4+\delta \cdot 4$$
$$\delta \geq \tilde{\delta} \equiv 0.5$$

- Case 3: Alice has chosen to cooperate and Bob to betray. Should one of them choose cooperation instead of betrayal? No, this deviation is never beneficial, as it only leads to a poorer result for the player (from 2 to 1)!
- Case 4: Alice has chosen to betray and Bob to cooperate. Should one of them choose cooperation instead of betrayal? No, this deviation is never beneficial, as it only leads to a poorer result for the player (from 2 to 1)!

5.6.3 Subgame Perfection and Repeated Games

Problem Text in Sect. 5.5.2.

Part a)
The collusion solution $\left(p_A^M, p_B^M\right)$ is not a Nash equilibrium and therefore cannot be supported in finite repetition (see theorem on finitely repeated games).

Part b)
The grim trigger strategy would be: "Starting with the first period, choose the collusion price p_i^M in each period. If one company did not choose the collusion price p_i^M in the previous period, then choose the competition price p_i^C forever."

To determine the critical discount factor, the one-shot deviation principle must be applied. Basically, there is only one relevant history to check, namely the mutual choice of the collusion price. Furthermore, it is sufficient to check the deviation as early as the first period, as this is when the deviation gain is highest. With each subsequent deviation, this advantage will only decrease due to discounting:

$$\overbrace{\sum_{t=0}^{\infty} \delta^t \cdot 450}^{\text{NPV adherence}} \geq \overbrace{500 + \sum_{t=1}^{\infty} \delta^t \cdot 400}^{\text{NPV deviation}}$$

$$\frac{450}{1-\delta} \geq 500 + \frac{400 \cdot \delta}{1-\delta}$$

$$450 \geq 500 - 100 \cdot \delta$$

$$\Rightarrow \delta \geq \tilde{\delta} \equiv 0.5$$

Literature Recommendations

Andreoni J, Miller JH (1993) Rational cooperation in the finitely repeated prisoner's dilemma: experimental evidence. Econ J 103:570–585.

Axelrod R (1980) Effective choice in the prisoner's dilemma. J Conflict Resolut 24:3–25.

Mailath GJ, Samuelson L (2006) Repeated games and reputations: long-run relationships. Oxford University Press, Oxford/New York.

Milgrom PR (1984) Axelrod's the evolution of cooperation. Rand J Econ 15(2):305–309.

Incomplete Information

6

With complete information, the essential parameters of a game, especially the strategy space and the payoffs, are common knowledge for all players. Although this assumption is sufficient to describe many strategic situations, there are at least as many cases in which it is too restrictive. Often a decision-maker needs more information about the decision-making situation as well as certain characteristics of the other players, as there is *uncertainty ex ante*, i.e., at the time before the decision. We limit ourselves in the following to cases where the uncertainty only refers to the payoffs of the players (or at least one player),[1] which we simplified as the "type" of the player. For examplc, in a wage negotiation, the employer side may be uncertain about the willingness of the union leadership to take tough labor dispute. Employers must therefore make an assessment as to whether the type of union leadership is compromising or uncompromising. Or a company wants to source preliminary products from a new supplier but is uncertain whether this supplier can produce the preliminary products of the desired quality. In this case, the company would have incomplete information about whether the supplier is a type that can produce high quality or not. In both examples, the relevant players will make a probability assessment of the respective type of their opponent. For games with incomplete information, we therefore need all possible type manifestations for the "uncertainty-affected" players and a probability distribution over these types.

[1] Uncertainty about other parameters of the game can be described analogously.

F. Bartholomae and M. Wiens, *Game Theory and Applications*, Classroom Companion: Economics, https://doi.org/10.1007/978-3-658-44608-6_6

Table 6.1 Market entry game with incomplete information: high and low costs

A (at high costs), B	Enter (a_{B1})	Not enter (a_{B2})
Invest (a_{A1})	(0, −1)	(3, 0)*
Not invest (a_{A2})	(2, 1)*	(2, 0)

A (at low costs), B	Enter (a_{B1})	Not enter (a_{B2})
Invest (a_{A1})	(3, −1)	(5, 0)*
Not invest (a_{A2})	(2, 1)	(3, 0)

6.1 Bayes-Nash Equilibrium

The most important solution concept for games with incomplete information is the *Bayes-Nash equilibrium*, which is why these games are often also abbreviated as "Bayesian games". Before we formally describe the conditions for a Bayes-Nash equilibrium, let us first understand how incomplete information can be represented in a game.

Let us consider the following example[2]: A company (player B) plans to enter a market that has so far been supplied by an incumbent monopolist (player A). The monopolist can decide whether to invest in better technology that enables cheaper production of the good and thus offer it at a lower price. Player B can observe the investment of the monopolist but does not know how the investment will affect its production costs: Player B assumes that with probability ρ, Greek rho, the investment only slightly reduces the monopolist's costs, so its costs remain high despite the investment. With *high costs*, the investment would no longer be beneficial for the monopolist if he must compete against a new company. In this case, market entry would be profitable for player B. With the corresponding counter-probability $(1 - \rho)$, however, player B must fear that the investment causes such a significant cost reduction that the monopolist benefits in any case. If the monopolist achieves *low costs* through the investment, the strategy "invest" becomes strictly dominant. In this case, it would be better for player B to refrain from entering the market. The two matrices in Table 6.1 summarize the payoffs of the players.

In this example, player B (entrant) has incomplete information about the type of player A (incumbent), whereby the uncertainty refers to the incumbent's ability to make a sufficiently effective investment. The aspect of incomplete information is integrated into the conventional game representation using a "trick" that goes back to Harsanyi (1967, 1968a, 1968b): The random move associated with the uncertainty is carried out by the non-strategic or dummy player "nature"—we know her already from the lotteries in Chap. 1. Nature draws the types of players (in this example only that of player A) according to the given probability ρ. To show the opening move of nature, the rather unusual game tree representation for simultaneous games was chosen in Fig. 6.1.

[2] Cf. Fudenberg and Tirole (1991).

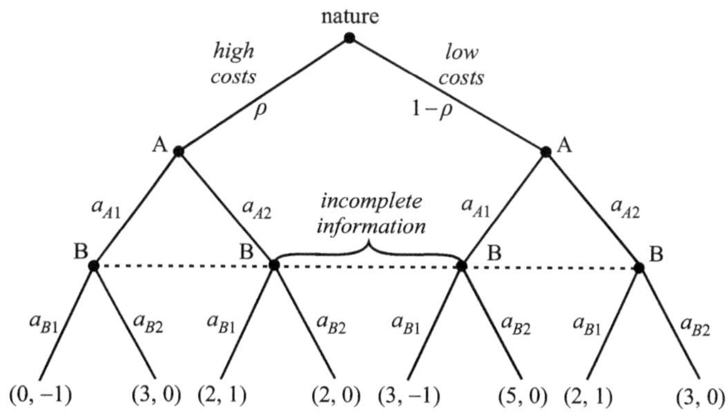

Fig. 6.1 Game tree representation for the simultaneous market entry game

After nature's move, the incumbent player A decides. The fact that player A's decision nodes are not connected to each other in Fig. 6.1 means that *player A knows his type*. This is an important assumption for Bayesian games, which we will discuss briefly below. After player A's investment decision, player B decides whether to enter the market or not. Since this is a simultaneous game, player A's preceding move is not observable to player B, which is expressed by the dashed lines between player B's decision nodes in the left and right subtree. The dashed connection line *between* the two subtrees, however, represents player B's incomplete information: Player B does not know whether he is in the left part of the game, where he would face a monopolist with high costs, or in the right part of the game, where the monopolist has low costs.

In a Bayesian game, the players always know the joint probability distribution over the player types, the *prior probabilities*, or "priors" for short. In relation to the example, the players know from the beginning the probability ρ. Thus, over time, the players' level of information changes according to the sequence shown in Fig. 6.2.

The variable t denotes the set of all player types, t_i stands for the type of player i and $\rho(\cdot)$ stands for the distribution of prior probabilities. *Ex ante*, i.e., *before* the game, all players only know the prior probabilities. *Interim*, i.e., after an intermediate step *during* the game, nature makes her random move, whereby each player learns (only) his own type. *Ex post*, i.e., *after* the game has ended, all players know the distribution of types.

You might now wonder what the point of the interim phase is, as the players are not playing against themselves, and it should therefore be irrelevant whether they

Fig. 6.2 Information revelation in incomplete information

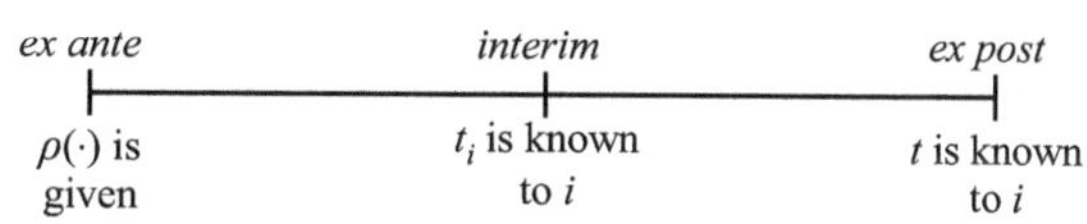

Table 6.2 Payoffs Common-Prior-Matrix

Alice, Bob	t_B^1	t_B^2
t_A^1	$\rho(t_A^1, t_B^1)$	$\rho(t_A^1, t_B^2)$
t_A^2	$\rho(t_A^2, t_B^1)$	$\rho(t_A^2, t_B^2)$

Alice, Bob	t_B^1	t_B^2	$\sum t_B$
t_A^1	0.5	0.125	0.625
t_A^2	0.25	0.125	0.375
$\sum t_A$	0.75	0.25	1

know their own type or not. This objection is valid in cases where there is one-sided uncertainty, i.e., only the type *of one* player is unknown. As soon as there is incomplete information about the types of two players, the information about one's own type is important as the respective player can draw conclusions about the type of the opponent from it. Poker is a good example to illustrate this point: The knowledge of one's own hand allows a Poker player to deduce which cards his fellow players must or cannot have. Or think of the children's game "Black Peter"—if you have the eponymous card, you know that none of the other players can have it. On the other hand, the player may also observe which action the other player chooses and draw conclusions from it, especially when, for example, the strategy spaces of the possible types are not identical, and the observed action is only available to a certain type, so that in extreme cases, his type is completely revealed.

The first matrix in Table 6.2 generally represents the distribution of prior probabilities for the types of two players. Each player (Alice and Bob) is either type 1 or type 2. As we already explained for the timeline in Fig. 6.2, the probabilities shown are known *ex ante* to both players. This is why the matrix is also referred to as the common prior matrix, i.e., there is common knowledge about this matrix. The second matrix in Table 6.2 shows a numerical example.

For example, the probability $\rho(t_A^1, t_B^1) = 0.5$ means that the probability that both players are of type 1 is 50%. As soon as each player learns his own type through the random drawing of nature, he can use this information to infer the type of the other player. Suppose Alice learns that she is of type 1. How likely is it now that Bob is also of type 1, if Alice already knows that she is type 1? Alice is thus interested in the conditional probability $\rho(t_B^1|t_A^1)$, i.e., the "probability for t_B^1 given t_A^1". This conditional probability can be easily calculated: If Alice knows that her type is 1, then she only needs to consider the top row of the common prior matrix: Either they are both of type 1, which occurs with a probability of 0.5, or only Alice is of type 1 while Bob is of type 2, which occurs with a probability of 0.125. Since only one of the two possibilities can occur, she divides the joint probability for the configuration of interest to her $\rho(t_A^1, t_B^1)$ by the sum of both probabilities $\rho(t_A^1)$ which is the probability that she is of type 1.

$$\rho(t_B^1|t_A^1) = \frac{\rho(t_A^1, t_B^1)}{\rho(t_A^1)} = \frac{0.5}{0.625} = 0.8$$

She has thus gained information: Before Alice knew her type, she assumed that Bob is of type 1 with probability $0.5 + 0.25 = 0.75$ and of type 2 with probability $0.125 + 0.125 = 0.25$, i.e., when she finds out that she is type 1 herself, it becomes more likely that Bob is also type 1. The procedure described here is Bayesian updating according to the Bayesian formula.

We now know the basic properties of games with incomplete information. A Bayesian game is consequently described by (N, S, T, ρ, u). Compared to a game with complete information (see Sect. 2.1), two elements have been added here: T and ρ, where $T = \{T_1, \ldots, T_n\}$ is the set of all player types $T_i = \{t_{i1}, \ldots, t_{iZ}\}$ with the different type manifestations of player i and ρ the probabilities of the individual types. Here, each type t_i is considered as an independent player with the payoff function $u_i(t_i) = \sum_{t_{-i}} \rho(t_{-i}|t_i) \cdot u_i[s_1(t_1), \ldots, s_n(t_n), t_1, \ldots, t_n]$ considered. As we have seen, such a game can be transformed into a game $\Gamma(T, S', u')$ with complete, but imperfect information (with S' as the set of $S_i(t_i)$ and u' as the set of $u_i(t_i)$).

With these basics, we can now define a *Bayes-Nash equilibrium*: In a Bayes-Nash equilibrium, each player chooses the best response to a given strategy of the other players for a given (prior) probability distribution over all player types and under the condition that each player i know their own type t_i.

$$\sum_{t \in T} u_i\left(s_i^*(t_i), s_{-i}^*(t_{-i}), t_i\right) \cdot \rho(t) \geq \sum_{t \in T} u_i\left(s_i(t_i), s_{-i}^*(t_{-i}), t_i\right) \cdot \rho(t)$$

This condition must be met for all players and strategies of the types. This is an extension of the already-known definition of a Nash equilibrium for games with incomplete information. Since each player i now must assume a probability distribution $\rho(\cdot)$ over all T possible types of his fellow players, the payoff of player i becomes an *expected payoff*. In the Bayes-Nash equilibrium, each player maximizes their expected utility given the (optimal) strategy of the opponent (s_{-i}^*) and knowing their own type (t_i). Thus, the above expression implies that the *beliefs* (i.e., the probability estimates) of the players along the equilibrium path are determined by the strategies of the players and by Bayesian updating.

After this very technical description, we now want to use our example, the simultaneous market entry game, to illustrate how we can determine the Bayes-Nash equilibrium. To do this, we first determine the best response of player B to the given strategies of all *type combinations of player A*. In the second step, we check whether, given the best answer from player B, each *type of player A* sticks to its strategy choice. Let us assume that both types of player A (i.e., both the high-cost incumbent and the low-cost incumbent) each choose "Invest" (a_{A1}). We see in the matrices in Table 6.1 that player B does not want to enter if player A invests—regardless of which type of A invests. The strategy combination "Invest" and "Not enter" considered here represents a Nash equilibrium in both matrices from Table 6.1, i.e., if player B stays away from the market, it is also optimal for both types of player A to invest. Thus, we have found the first Bayes-Nash equilibrium, which can be described by the strategy combination $\{(a_{A1}(t_A^1), a_{A1}(t_A^2)), a_{B2}\}^*$ and by the probability interval $\rho \in [0, 1]$. Any probability between 0 and 1 is part of this equilibrium, as here both types of player A choose the same strategy and it is therefore irrelevant for player B who he is dealing with.

Next, we consider the constellation $(a_{A1}(t_A^1), a_{A2}(t_A^2))$, where type 1 (high costs) chooses to "invest" (a_{A1}), but type 2 (low costs) decides for "not invest" (a_{A2}). If type 1 chooses to "invest", it is best for player B not to enter. Conversely, it remains optimal for type 1 to choose "Invest" if player B does not enter. But what about type 2? If player B does not enter, type 2 is better off if he also chooses to "invest". Thus, the strategy combination $\{(a_{A1}(t_A^1), a_{A2}(t_A^2)), a_{B2}\}$ cannot be a Bayes-Nash equilibrium, because here type 2 would deviate to "Invest" and we would end up in the first Bayes-Nash equilibrium again. Similarly, $\{(a_{A1}(t_A^1), a_{A2}(t_A^2)), a_{B1}\}$—player B chooses to "enter"—cannot be a Bayes-Nash equilibrium: If player B enters, it would be for type 1 better to choose to "not invest" and for type 2 better to choose to "invest". So, both types would have an incentive to change their strategy.

The third possible constellation is $(a_{A2}(t_A^1), a_{A1}(t_A^2))$. Here, type 1 chooses to "not invest" and type 2 chooses to "invest". If player B knew that he was facing type 1 (who chooses to "not invest"), then player B would enter. The best response to player B's market entry would again be to "not invest" from type 1 and to "invest" from type 2. Thus, the strategy combination $\{(a_{A2}(t_A^1), a_{A1}(t_A^2)), a_{B1}\}$ represents a Bayes-Nash equilibrium, if player B expects with a sufficiently high probability ρ to really face type 1. If, on the other hand, player B knew that he was facing type 2 (who chooses to "invest"), then it would be better not to enter. We can calculate the required, sufficiently high belief ρ as follows: If player B chooses to "not enter", then he always achieves a payoff of 0, regardless of the type of the opponent. If player B chooses to "enter", then he receives the expected payoff $\rho \cdot 1 + (1 - \rho) \cdot (-1)$. For the strategy combination $\{(a_{A2}(t_A^1), a_{A1}(t_A^2)), a_{B1}\}$ to be an equilibrium, $\rho \cdot 1 + (1 - \rho) \cdot (-1) > 0$ or $\rho > 1/2$ must therefore apply. Thus, we have found a second Bayes-Nash equilibrium with $\{(a_{A2}(t_A^1), a_{A1}(t_A^2)), a_{B1}\}$ for $\rho \in (1/2, 1]$. The remaining possibility, $\{(a_{A2}(t_A^1), a_{A1}(t_A^2)), a_{B2}\}$ where player B chooses to "not enter" when type 1 chooses "to invest" and type 2 to "not invest", is not a Bayes-Nash equilibrium: If player B does not enter, then it is optimal for both types to choose to "invest", i.e., player A would also invest, which brings us back to our first Bayes-Nash equilibrium $\{(a_{A1}(t_A^1), a_{A1}(t_A^2)), a_{B2}\}$.

Finally, the last constellation $(a_{A2}(t_A^1), a_{A2}(t_A^2))$ remains to be checked, where both types do "not invest". In this case, the best response from player B—regardless of the type of the opponent—would be to enter the market. But as soon as player B enters the market, it would be better for type 2 to switch to "invest". This

Table 6.3 Payoff matrix for different type combinations and probabilities

A, B	a_{B1}	a_{B2}
$a_{A1}(t_A^1), a_{A1}(t_A^2)$	$(0, 3), \rho \cdot (-1) + (1-\rho) \cdot (-1)$	$(3, 5), \rho \cdot 0 + (1-\rho) \cdot 0$
$a_{A1}(t_A^1), a_{A2}(t_A^2)$	$(0, 2), \rho \cdot (-1) + (1-\rho) \cdot 1$	$(3, 3), \rho \cdot 0 + (1-\rho) \cdot 0$
$a_{A2}(t_A^1), a_{A1}(t_A^2)$	$(2, 3), \rho \cdot 1 + (1-\rho) \cdot (-1)$	$(2, 5), \rho \cdot 0 + (1-\rho) \cdot 0$
$a_{A2}(t_A^1), a_{A2}(t_A^2)$	$(2, 2), \rho \cdot 1 + (1-\rho) \cdot 1$	$(2, 3), \rho \cdot 0 + (1-\rho) \cdot 0$

$\rho = 0.25$:

A, B	a_{B1}	a_{B2}
$a_{A1}(t_A^1), a_{A1}(t_A^2)$	$(0, \underline{3}), -1$	$(\underline{3}, \underline{5}), \underline{0}^*$
$a_{A1}(t_A^1), a_{A2}(t_A^2)$	$(0, 2), \underline{0.5}$	$(\underline{3}, 3), 0$
$a_{A2}(t_A^1), a_{A1}(t_A^2)$	$(\underline{2}, \underline{3}), -0.5$	$(2, \underline{5}), \underline{0}$
$a_{A2}(t_A^1), a_{A2}(t_A^2)$	$(\underline{2}, 2), \underline{1}$	$(2, 3), 0$

$\rho = 0.8$:

A, B	a_{B1}	a_{B2}
$a_{A1}(t_A^1), a_{A1}(t_A^2)$	$(0, \underline{3}), -1$	$(\underline{3}, \underline{5}), \underline{0}\,^*$
$a_{A1}(t_A^1), a_{A2}(t_A^2)$	$(0, 2), -0.6$	$(\underline{3}, 3), \underline{0}$
$a_{A2}(t_A^1), a_{A1}(t_A^2)$	$(\underline{2}, \underline{3}), \underline{0.6}^*$	$(2, \underline{5}), 0$
$a_{A2}(t_A^1), a_{A2}(t_A^2)$	$(\underline{2}, 2), \underline{1}$	$(2, 3), 0$

would bring us back to the third constellation discussed above, which led to the second Bayes-Nash equilibrium.

In summary, we have found two Bayes-Nash equilibria in this game: The first with the strategy combination $\{(a_{A1}(t_A^1), a_{A1}(t_A^2)), a_{B2}\}^*$ for $\rho \in [0, 1]$ and the second with $\{(a_{A2}(t_A^1), a_{A1}(t_A^2)), a_{B1}\}^*$ for $\rho \in (1/2, 1]$.

To better understand and follow this reasoning, let us look at a concrete numerical example. For the four constellations discussed above, we determine the expected payoffs by assuming the probabilities $\rho = 0.25$ and $\rho = 0.8$. Table 6.3 compares both payoff matrices.

The payoffs of both types of player A were recorded in a bracket (the first refers to type 1, i.e., high costs, and the second to type 2, i.e., low costs). Since player B has incomplete information about player A, we must calculate the expected payoff in each case.[3] We want to briefly illustrate this using the example of the first matrix for the strategy combination $\{(a_{A1}(t_A^1), a_{A2}(t_A^2)), a_{B1}\}$: If player B enters, type 1 of player A receives a payoff of 0 if he decides to "invest" ($a_{A1}(t_A^1)$) and type 2 receives a payoff of 2 if he chooses to "not invest" $a_{A2}(t_A^2)$. The payoffs of both

[3] Accordingly, if both players have incomplete information about the type of the opponent, then expected payoffs must be calculated for both players. There is a problem on this in Sect. 10.1.3.

types are thus summarized with the bracket (0, 2). Player B, on the other hand, receives a payoff of -1 if he encounters type 1 (this happens with ρ) and a payoff of 1 if he encounters type 2 (this happens with the counter probability $1-\rho$). The expected payoff for player B is therefore $\rho \cdot (-1) + (1-\rho) \cdot 1$.

In this way, we can now determine the payoffs of all players and player types. In addition to the general calculation, we also calculated the specific payoffs for a low ($\rho = 0.25$) and a high probability ($\rho = 0.8$) that player A is of type 1 (high cost). To find the equilibria, we can proceed in the same way as we know it from simultaneous games with complete information: We check for each cell the incentives of the players to deviate from it (in Table 6.3 the best responses were highlighted by underlining). However, it should be noted that an equilibrium strategy must always *be optimal for each player type* of player A. Therefore, only those cells represent a Bayes-Nash equilibrium where all payoffs, i.e., both those of player B as well as those of both player types of player A, are underlined. In conclusion, we find both equilibria again that we already identified with the general considerations made above: The first Bayes-Nash equilibrium $\{(a_{A1}(t_A^1), a_{A1}(t_A^2)), a_{B2}\}$ was independent of the level of probability ρ and thus it must appear in both examples (which is the case, it is always found in the top right corner). The second equilibrium $\{(a_{A2}(t_A^1), a_{A1}(t_A^2)), a_{B1}\}$ was only valid for $\rho > 1/2$ and accordingly we find it only in the second matrix. We can also check this using the general matrix: for $a_{A1}(t_A^1), a_{A1}(t_A^2)$, it is better for player B to choose a_{B1} than a_{B2} (and thus have a Nash equilibrium at a_{B1}) if $\rho \cdot 1 + (1-\rho) \cdot (-1) > \rho \cdot 0 + (1-\rho) \cdot 0$ or $\rho > 0.5$.

6.2 Bayes-Nash Equilibrium in Continuous Strategies

We can consider incomplete information in games with continuous strategies as well and determine the corresponding Bayes-Nash equilibrium. In this case, the player who is uncertain about the type of the opponent can no longer base the decision on a single and unique reaction function of the opponent; rather, they must—depending on the number of opponent types—also assume a corresponding number of reaction functions and will then derive a kind of "average reaction function" from this. In the literature, this is also occasionally referred to as an "expected reaction profile".

As an example, let us consider the competition between two companies. Both companies compete in quantity and face the market demand $p = 20 - X$, where p denotes the market price and $X \equiv x_A + x_B$ is the total quantity sold on the market (by company A and company B). There is incomplete information regarding the variable production costs $c_i \cdot x_i$: From the competitor's perspective, each of the two companies has a priori high ($c_i^h = 10$) or low ($c_i^l = 2$) production costs. Each of the two companies knows its own costs, but not those of the competitor. We assume that the costs of both companies are actually low. From the perspective of company A, let ρ_B be the probability that company B has high costs and correspondingly $1-\rho_B$ the probability of low costs. Accordingly, from the perspective of company B, let ρ_A be the probability of high costs for company A and $1-\rho_A$ for low costs.

For the following numerical example, we assume $\rho_A = 0.5$ and $\rho_B = 0.625$, i.e., company B expects on average medium costs from its competitor, and company A assumes rather high costs for its opponent. As usual, the probabilities are common knowledge of the players.

We now determine the Bayes-Nash equilibrium for simultaneous quantity competition of this duopoly with incomplete information. We proceed essentially in the same way as described in Sect. 3.4 for games with continuous strategies and complete information: Starting from the respective profit functions of the companies, the partial derivative with respect to the strategic variable is determined and set equal to zero, from which the reaction functions of the companies can be derived. In the duopoly game with complete information, we found the Nash equilibrium in continuous strategies from the intersection of the reaction functions.

In the case of incomplete information, however, there is a special feature to consider. Although each company knows its own costs and thus its own profit function does not contain any uncertainty *from its point of view*, it must consider that its competitor is deciding under uncertainty. Since the opponent is uncertain about the costs, each company must take the perspective of the competitor and base its own decision on this uncertainty. For example, it is important for company B to know that its competitor tends to consider it unproductive (although it knows itself that this is not the case): If its competitor considers it unproductive, it will expect a low output quantity, and will thus rather choose a higher quantity. Company B will therefore assume a higher production by company A and accordingly produce less itself. If company B, on the other hand, ignored its competitor's expectation, it could happen that the total quantity sold on the market is too high and both companies—including company B—make less profit. This optimal reaction of company B is thus a reaction to how its competitor estimates its costs and will decide accordingly.

Since, in our case, both companies have uncertainty about the cost level of the competitor, the expected reaction functions of both companies must be determined in a first step. The intersection of these expected reaction functions, however, does not yet represent the Bayes-Nash equilibrium, but first determines *the output quantities expected by both companies in equilibrium* $\overline{x}_A$ and $\overline{x}_B$. Each company then chooses based on the expected output quantity of the competitor in equilibrium $\overline{x}_j$ its optimal response $x_i^*(\overline{x}_j)$. For this, the expected strategy choice of the opponent is inserted into the own *deterministic reaction function* (i.e., into the actual reaction function without uncertainty).

To determine the expected reaction function of company A, we first set up the expected profit function. With this, we must consider that the revenue, $p \cdot x_A$, is certain, but the costs are uncertain.

$$\overline{\pi}_A(x_A, x_B) = (20 - x_A - x_B) \cdot x_A - \left[\rho_A \cdot c_A^h + (1 - \rho_A) \cdot c_A^l\right] \cdot x_A$$

The profit is maximized by setting the partial derivative with respect to the strategic variable x_A equal to zero, i.e., $\partial \overline{\pi}_A(x_A, x_B)/\partial x_A \overset{!}{=} 0$. From this, the expected reaction function of company A can be determined as

$$\overline{r}_A(x_B):\ \overline{x}_A(x_B) = 10 - \frac{1}{2}\cdot\left[x_B + \rho_A \cdot c_A^h + (1-\rho_A)\cdot c_A^l\right]$$

The bar over the variables indicates that this is the expected reaction of company A: company A forms an expectation about its own strategy choice and thus puts itself in the perspective of company B, which is characterized by uncertainty.

We obtain the expected reaction function of company B analogously:

$$\overline{r}_B(x_A):\ \overline{x}_B(x_A) = 10 - \frac{1}{2}\cdot\left[x_A + \rho_B \cdot c_B^h + (1-\rho_B)\cdot c_B^l\right]$$

In this case, company B puts itself in the perspective of company A and derives from this which reaction company A expects. If we now insert our numerical example values ($\rho_A = 0.5$, $\rho_B = 0.625/8$, $c_i^l = 2$, $c_i^h = 10$) into the reaction functions, we get:

$$\overline{r}_A(x_B):\overline{x}_A(x_B) = 7 - 0.5\cdot x_B \quad \text{or} \quad \overline{r}_A(x_B): x_B(\overline{x}_A) = 14 - 2\cdot\overline{x}_A$$
$$\overline{r}_B(x_A):\overline{x}_B(x_A) = 6.5 - 0.5\cdot x_A$$

The expected reaction function of company A was additionally solved for x_B so that both reaction functions can be plotted again in a common $x_A - x_B$-diagram. In Fig. 6.3, the expected reaction functions are shown as dashed lines.

The intersection of these functions yields $\overline{x}_A = 5$ and $\overline{x}_B = 4$ (point E in the graphic). These quantities are expected by both companies in equilibrium, and they will therefore base their decision on them. The expected quantities are different because the companies assess the costs of their competitor differently.

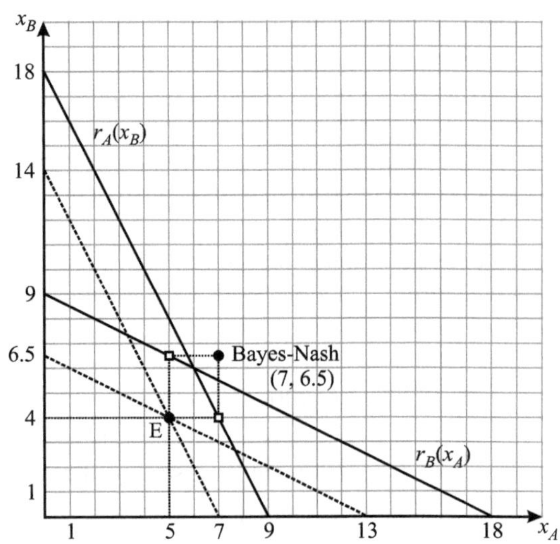

Fig. 6.3 Incomplete information in duopoly quantity competition

If, alternatively, both companies were aware of each other's low costs (situation with complete information), identical and higher quantities of $x_A^* = x_B^* = 6$ would result in equilibrium—you find them by setting the probabilities for both equal to zero, $\rho_A = \rho_A = 0$.[4] As we have seen, however, the different cost estimates mean that in equilibrium, company A is expected to produce a higher quantity and company B a lower quantity.

Based on these expectations, each company now finds its optimal production by determining the optimal response for the expected quantity of its competitor. We find the best response of each company on its (deterministic) reaction function, which results from the actual costs (in this case low for both companies). For example, in the figure, we find the best response of company B to the expected quantity $\overline{x}_A = 5$ of the competitor on its unbroken reaction function (marked with a small box). To calculate the optimal output quantity of company B we insert the value $\overline{x}_A = 5$ into the deterministic reaction function of company B and get $r_B(\overline{x}_A) : x_B^*(\overline{x}_A = 5) = 10 - 0.5 \cdot [5 + c_B^l] = 6.5$. Similarly, company A uses the expectation $\overline{x}_B = 4$ which results in $r_A(\overline{x}_B) : x_A^*(\overline{x}_B = 4) = 10 - 0.5 \cdot [5 + c_A^l] = 7$. The Bayes-Nash equilibrium is thus in the strategy combination $x_A^* = 7$ and $x_B^* = 6.5$ for given (and known) probabilities $\rho_A = 0.5$ and $\rho_B = 0.625$.

Obviously, both companies choose a higher quantity than they would produce if they had complete information about their low costs. The reason for this is simple: Since both companies—although with different intensities—overestimate the costs of the competitor, they expect too low quantities in equilibrium and react optimally (optimal for their given state of knowledge with incomplete information) with a higher production.

6.3 Signaling Games

Signaling games are sequential games with asymmetric, i.e., one-sided, incomplete information. The signal aspect of these games is because the uninformed player (receiver of the signal) can derive information about the type of the other player (sender of the signal) solely from the strategy choice of the co-player. The structure of a signaling game is as follows: First, nature draws the type of the sender from the set of all types according to a (known) probability distribution over all types (*ex ante*). We again assume the simplest case with just two types. The sender receives the information about his type (*interim*) and decides as a first mover about his strategy choice. The receiver observes the action of the first mover and tries to draw conclusions about the type of the sender based on this observation. A signaling game is therefore essentially about the *interpretation* of the first-mover strategy by the second mover.

[4] The case with complete information regarding low costs for both companies corresponds to the unbroken reaction functions in Fig. 6.3.

Note and Tip In a signaling game, the strategy choice of a player can provide clues about their type.

The following example is a modified version of the seminal signaling game by Spence (1973), an application of signaling theory in the case of personnel selection in the labor market. In this example, the employer (E, recipient of the signal) faces the problem of finding a productive employee/worker (W, potential sender of the signal) whose productivity is *ex ante* unobservable to the employer. With a probability of ρ she expects a productive employee, with whom she can achieve a high output or revenue R^h; with the counter-probability $1 - \rho$ she expects an unproductive employee, who only allows her a low revenue R^l. This probability distribution over the two types of employees is common knowledge of all players. Even if the employer cannot (directly) observe the productivity of the worker, she can pay attention to an observable characteristic of the worker, which she assumes is related to the productivity of the worker.

Let us assume for our example that the employer requests a work sample of the potential worker. This characteristic, which can be influenced by the sender (W) and observed by the recipient (E), represents the *signal* of the sender. The central prerequisite for the signal to be informative for the receiver, i.e., to serve as a *selection criterion*, is that it must be more difficult for the unproductive than for the productive type to generate this signal. In our example, the higher productivity of the desired worker type allows him to provide a high-quality work sample, while the unproductive type finds this rather difficult. We assume that the productive type incurs low effort costs of c^l for producing a work sample, while the unproductive type incurs high costs of c^h for doing so (with $c^l < c^h$). The recipient thus considers certain characteristics of the sender as a signal, which do not have to be directly related to the characteristic she is interested in. The important thing is only that the incentives of the sender types to generate the signal are correlated with the relevant characteristic.

In a signaling game, the sender's strategy choice thus consists in the decision to generate the signal (strategy S) or not (strategy N).[5] The recipient, in turn, faces the decision to use the sender's observed strategy choice as a distinguishing criterion or not. A signal only comes into question for him as a decision criterion if both sender types act differently, i.e., the productive type chooses strategy S (he provides a work sample), and the unproductive type chooses strategy N (he does not provide a work sample). If the recipient can distinguish between both sender types based on her observation, then she plays a *separating strategy* (strategy D). In our example, the separating strategy is synonymous with the employer paying different wages to both types, as they are distinguishable for her: She pays a high wage w^h

[5] From an information economic perspective, any activity of the sender, from which the receiver can draw conclusions about its type, can be considered a signal. Thus, even the non-sending of a signal can be interpreted as a signal. In the context of the example discussed in the text, however, for the sake of clarity, we only refer to a signal when the sender actively "generates" it, thus incurring costs.

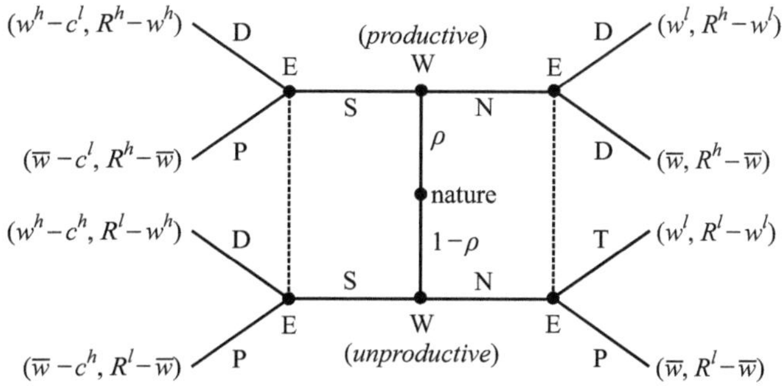

Fig. 6.4 Signaling game

to the productive type and a low wage w^l to the unproductive type. However, if the recipient cannot distinguish between the two sender types or she considers different sender behavior to be insignificant, i.e., she ignores every signal, then she plays a *pooling strategy* (strategy P). In this case, the recipient remains incompletely informed and pays each worker the average wage $\overline{w} = \rho \cdot w^h + (1 - \rho) \cdot w^l$.

Figure 6.4 represents the signaling game in extensive form.[6] We are now looking for the Bayes-Nash equilibria of this game. In signaling games two types of equilibria are always distinguished: the separating equilibrium on the one hand and the pooling equilibrium on the other. In the separating equilibrium both types play different strategies—for a given probability distribution over the sender types. In this case, it is optimal for the recipient to use the signal as a distinguishing criterion and play the separating strategy. In the pooling equilibrium, both types play the same strategy (either both N or both S). In this case, the types are indistinguishable for the recipient, so she can only choose the pooling strategy.

Let us first assume that both sender types play strategy N, i.e., they both do not deliver a work sample. This makes them indistinguishable for the recipient (employer), so she plays the pooling strategy and pays both the average wage $\overline{w}$. Could one of the two sender types now improve by deviating from his strategy and providing a work sample? Obviously, the productive type will first consider this, as he incurs lower signal costs c^l. The productive type will then deviate and choose strategy S if the higher wage he receives by identifying himself as a productive type, minus the signal costs, is higher than the average wage paid to him otherwise:

$$w^h - c^l \geq \rho \cdot w^h + (1 - \rho) \cdot w^l \Leftrightarrow \rho \leq 1 - \frac{c^l}{w^h - w^l} \tag{6.1}$$

[6] The game tree variant of signaling games differs slightly in appearance from the usual game tree representation, as here—in contrast to the conventional diagonal arrangement—the branches of the game tree are arranged horizontally and vertically. In this representation, the payoffs in the scenarios with or without a signal can be visually distinguished more easily due to the clear side arrangement (left and right). In addition, the payoffs are arranged one below the other on each side and are therefore easier to compare with each other.

This condition implies an upper limit for the probability ρ. The higher the probability of the productive type, the higher is the average wage $\overline{w}$ that the productive worker already gets without a signal in the pooling equilibrium. Ultimately, in the extreme case $\rho = 1$, every worker would be productive and thus the average wage would be equal to the wage of the productive type. This critical probability, above which it is rewarding to send the signal, is the smaller the higher the signal costs c^l and the larger the wage difference $(w^h - w^l)$ turns out to be. For example, if the wage differential is small and the signal is very expensive, the share/probability of productive workers must be extremely low so that they have an incentive to send the signal to receive the higher wage instead of the average wage.

However, condition (6.1) does not yet ensure that there will be a separation equilibrium. It must also be the case that the unproductive type has no incentive to generate the signal—since the productive type is now clearly identifiable to the employer, the unproductive type receives a lower wage. For the unproductive type not to send a signal, it must be the case:

$$\rho \cdot w^h + (1 - \rho) \cdot w^l - c^h \leq w^l \Leftrightarrow \rho \leq \frac{c^h}{w^h - w^l} \tag{6.2}$$

Condition (6.2) represents the critical upper limit for the probability. If the probability ρ is higher than this value, then the average wage paid by the employer increases so much that the unproductive type also wants to generate the signal. If ρ is below it, then the unproductive type has no incentive to generate the signal, when the productive one does. In addition, the employer cannot worsen by discriminating after observing the signal and playing a separation strategy (D). The joint conditions, (6.1) and (6.2), thus represent the conditions for the existence of a separation equilibrium of this game.

If at least one of the two conditions (6.1) and (6.2) is not met, then a pooling equilibrium exists, of which two variants need to be checked: A pooling equilibrium, in which both sender types do not generate the signal and one in which both generate the signal. Let us first assume that condition (6.1) is not met. In this case, the productive type would have no incentive anymore to distinguish itself from the unproductive type by generating a signal. Is it now possible that the unproductive type might have an interest in exploiting this situation, i.e., to generate a signal on his own if the productive one does not? If this were the case, then he would thereby force the productive type to continue generating the signal (this one otherwise faces the low wage w^l), even though it is no longer profitable for him. This incentive of the unproductive sender to put the productive one under pressure would exist if the following condition is met:

$$w^h - c^h \geq \rho \cdot w^h + (1 - \rho) \cdot w^l \Leftrightarrow \rho \leq 1 - \frac{c^h}{w^h - w^l} \tag{6.3}$$

Condition (6.3) again represents an upper limit for the critical probability and resembles condition (6.1). Specifically, we see that due to $c^l < c^h$ the critical limit (6.3) is always strictly smaller than the right-hand side of condition (6.1). So, if the critical probability from (6.1) is reached, at which the productive type loses interest, to distinguish himself from the unproductive type, then the critical probability from (6.3) has long since been exceeded. This implies that the unproductive sender type will never distinguish himself from the productive type by being the only one to generate the signal.

Now let us assume the other possibility, i.e., condition (6.2) is not met, but condition (6.1) is. In this case, the productive type would still have an incentive to distinguish himself from the unproductive type, but the unproductive one would now also want to generate the signal, to not be identifiable as unproductive. Since the productive type can then no longer distinguish himself, he would simply forego the signal and the unproductive type would immediately follow suit due to (6.3) and not generate a signal. We can conclude from both considerations that any pooling equilibrium can only be one in which neither of the two sender types generates the signal. The result is intuitive: Since a signal is always associated with costs, but pooling brings no benefit, it cannot be optimal to generate it.

Finally, we still must show that the described scenarios for a separation and a pooling equilibrium are also optimal for the employer as the receiver of the potential signal and that she has no incentive to deviate from them. In the above example, the employer is indifferent between a separation and a pooling strategy for *any probability* ρ. If the employer plays a *separation strategy*, then she pays the types of workers different wages (w^h or w^l), as she can distinguish them. Her expected profit from the recruitment strategy "separation" (D) is $\overline{\pi}(D) = \rho \cdot \left(R^h - w^h\right) + (1-\rho) \cdot \left(R^l - w^l\right)$, as she faces a productive worker with probability ρ and an unproductive one with probability $1-\rho$. On the other hand, if she plays a pooling strategy, then she pays to a representative worker, whose type is not observable for her, the average wage $\overline{w} = \rho \cdot w^h + (1-\rho) \cdot w^l$. Her expected profit with the recruitment strategy "pooling" (P) is $\overline{\pi}(P) = \rho \cdot R^h + (1-\rho) \cdot R^l - \overline{w}$. If we take a closer look at both expressions, we see that $\overline{\pi}(D) = \overline{\pi}(P)$ applies, i.e., the two expected profits are identical for each probability ρ, so the employer is always indifferent between both strategies. But this means nothing else than that both the separation and the pooling equilibrium are automatically also acceptable for the employer as a recipient ("incentive compatible") and she therefore has no incentive to deviate from it.

Finally, to illustrate the results, let us apply the following numerical example to the model described here: $c^l = 1$, $c^h = 3$, $w^l = 1$, and $w^h = 5$. The critical limit (6.1) is $\rho \leq 1 - \frac{1}{5-1} = 0.75$, for (6.2) $\rho \leq \frac{3}{5-1} = 0.75$, and for (6.3) $\rho \leq 1 - \frac{3}{5-1} = 0.25$. Thus, in the interval $\rho \in [0, 0.75]$ a separation equilibrium $\{(S, N), D\}^*$ exists and in the interval $\rho \in [0.75, 1]$ a pooling equilibrium $\{(N, N), P\}^*$—where the first expression in the round bracket always denotes the strategy of the productive worker and the second expression is the one of the unproductive type.

6.4 Comprehension Questions

1. Games with incomplete information can be represented as…
   A Lotteries, in which only nature decides the outcome of the game
   B Games with imperfect information about nature's moves
   C Games with imperfect information about the players' moves
   D Infinitely repeated games with uncertain future

2. In a simultaneous two-person game with mutual, incomplete information, the players have the following information *before* their strategy selection:
   A Each player knows the type of the opponent
   B Each player knows the probability distribution over their own type, but not the probability distribution over the type of the opponent
   C Each player knows their own type
   D Each player knows nothing about their opponent before the outcome of the game

3. In a game with incomplete information, …
   A Different types of players will always choose different strategies
   B A player's type can always be deduced from the observed behavior
   C All players consider the uncertainty of the other players regarding their type
   D There are always efficient outcomes

4. In a separating equilibrium of a signaling game…
   A The receiver can only identify the type of one sender, but not the type of the other
   B The types of the sender choose different strategies
   C The probability distribution over the sender types plays no role, as the receiver can clearly distinguish the types
   D The receiver has the incentive to make his strategy choice dependent on the observed behavior of the sender

1B; 2C; 3C (D is not correct, e.g., we see that too much is produced in Fig. 6.3); 4B, D (C is not correct, as the receiver can indeed distinguish the types clearly, but only based on the probability distribution, not without it.)

6.5 Problems

6.5.1 Information and Bayesian Equilibrium

Consider again the situation from problem 4.5.2: Two companies compete in a market with a time horizon of two periods. Company A can choose between the actions "price war" $\left(a_{A1}^{t}\right)$ or "market division" $\left(a_{A2}^{t}\right)$. Company B observes this decision in $t = 0$ and decides at the beginning of period $t = 1$ on the actions "market exit" $[a_{B1}$ with $\pi_{B}^{t}\left(a_{A2}^{t}\right) = 0]$ and "remain in the market" $[a_{B2}^{t}$ with $\pi_{B}^{t}\left(a_{A1}^{t}\right) = -1, \pi_{B}^{t}\left(a_{A2}^{t}\right) = 2]$. This decision can be observed by company A. The period profit of company A is $\pi_{A,D}^{t}\left(a_{A1}^{t}\right) = -1$ and $\pi_{A,D}^{t} = 2$ if both companies are active in the market (duopoly) and $\pi^{t}_{A,M} = 6$ if company B has exited the market (monopoly of A).

Now assume that company B is uncertain whether company A is a "normal" company with the payoffs indicated above or an "aggressive" company that always prefers the action "price war"—a priori, company B assumes that company A is "normal" with probability ρ and "aggressive" with $1 - \rho$.

a. How can this game with incomplete information be translated into a formally equivalent game with imperfect information? Draw the game tree of this game with imperfect information! Explain, depending on the probability estimate about the type of company A, whether it is attractive for company B to be in the market in period 1!
b. Company B adjusts its probability estimate for the second period based on the observed behavior in period 1. Which action does company B choose if it observes the action "market division"? Is it plausible that company B concludes from the behavior "price war" that company A is "aggressive" for sure? Does your statement change if the payoff in the monopoly case for a "normal" company is $\pi^M_A = 4$ instead of $\pi^M_A = 6$?

Solution in Sect. 6.6.1.

6.5.2 Bayesian Equilibrium in Continuous Strategies

Consider the following game, which is related to a socio-political context. At the beginning of a year, the government and a (representative) worker face a decision problem: The worker has a job offer and must commit to how many months L in the year he wants to pursue this activity with a monthly wage of w. In the months when he is not working, he receives a government transfer of t per month as an unemployed person. The number of working months L is continuous and lies in the interval $L \in [0, 12]$—e.g., the worker can also work for about 7.26 months. His work income consists of wage and transfer income and amounts to $L \cdot w + (12 - L) \cdot t$. In addition to his annual income, the unemployed person also considers the opportunity costs of taking up work (especially giving up leisure time), which are assumed to be $\beta \cdot L^2$. Here, β is a measure of the motivation and discipline of the unemployed person. If $\beta = 0.5$, the unemployed person is "motivated", while if $\beta = 1$, the unemployed person is "unmotivated". The utility of the government depends linearly on the annual income (=wage + transfer) of the worker—the happier the worker, the higher the probability of re-election—however, the transfer payments t are a burden on the government budget and thus cause costs of t^2 for the government.

a. Assume a one-year time horizon and formulate the payoff function of both players in an appropriate manner! What is the strategic decision variable for the worker and the government, respectively?

b. Now assume that the wage rate is $w = 12$. Determine the reaction functions for the simultaneous game (continuous strategies) of the government with a motivated worker! Determine graphically and numerically the Nash equilibrium of the simultaneous game including the payoffs achieved!
c. Assume that the government does not know whether the employee is "motivated" (type 1) or "unmotivated" (type 2). A priori, both cases are equally likely. In reality, the worker is motivated. Determine the government's reaction function in case of incomplete information (based on the given probability) as well as for the motivated type of worker! Determine graphically and numerically the Bayes-Nash equilibrium! Compare this result with the one that would occur with complete information about the worker's motivation! Would the motivated worker have an incentive to reveal his true type? What would be better from the government's perspective?
d. Repeat the analysis from (c) for the case that the worker is actually unmotivated. Show whether an unmotivated worker has an incentive to reveal his true type! What would be better from the government's perspective?

Solution in Sect. 6.6.2.

6.5.3 Signaling Games

Consider the signaling game described in the text between the employer and the applicant. Now assume that the employer would choose a freely determined, medium wage w^m with $w^l < w^m < w^h$ instead of the probability-weighted average wage $\overline{w}$ in a pooling strategy. Show how this change affects the incentives of the receiver as well as the two sender types and thus the resulting equilibria of the signaling game!

Solution in Sect. 6.6.3.

6.6 Solutions

6.6.1 Information and Bayesian Equilibrium

Problem text in Sect. 6.5.1.

Part a)
This game with incomplete information about the type of company A can be converted into a game with imperfect information. To do this, we introduce the additional non-strategic player "nature" who determines the type of company A based on the probability distribution.

From the solution to problem 1b) in Sect. 4.6.2, we already know that part of the game tree that can be observed with a "normal" company A. Since an "aggressive" company A can only play the action "price war", we find the payoffs for company

B by only considering the path payoffs of the game tree in which company A chooses "price war". This yields to following game tree, in which company B is only uncertain about the type of company A when it chooses "price war"—the "aggressive" company A cannot choose "division", which is why the observation of this action reveals a "normal" company A:

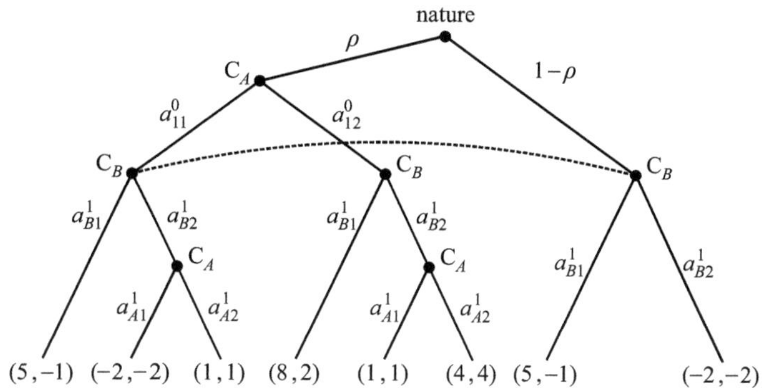

Next, we want to check whether action a_{B2}^1 ("remain in the market") is attractive for company B at all. This depends on what it observes in $t = 0$: If it observes that company A is dividing the market, it can be certain that A is a "normal" company and will therefore stay in the market. But what if company A has chosen "price war" (a_{A1}^0)? Let us consider the expected payoffs of company B depending on the choice of its actions—a "normal" company will choose "division" in $t = 1$ in any case:

$$E\left(a_{B1}^1\right) = -1$$

$$E\left(a_{B2}^1\right) = \underbrace{\rho \cdot 1}_{\text{"normal"}} + \underbrace{(1-\rho)\cdot(-2)}_{\text{"aggressive"}} = 3 \cdot \rho - 2$$

Thus, company B will decide to "remain" if $E\left(a_{B2}^1\right) > E\left(a_{B1}^1\right)$ or $3 \cdot \rho - 2 > -1$ applies. This is the case for $\rho > 1/3$.

Part b)

Formally, the a priori probability can be adjusted using the Bayesian formula. Nevertheless, we can also address the situation intuitively. We can see from the game tree in a) that a "normal" company can be identified with certainty from the action "division", as this action is not available to an "aggressive" company. "Division" thus clearly reveals the type of company A to which "remain" is the best response for company B.

Should company B, conversely, conclude with certainty from "price war" that A is an "aggressive" company? If company A can assume that company B always concludes from "price war" that it is an "aggressive" company and therefore leaves the market (which is the best response), it is optimal for a "normal" company

A to choose "price war" in $t = 0$, thereby increasing its payoff from 4 (path: $\rho - a^0_{A2} - a^1_{B2} - a^1_{A2}$) to 5 (path: $\rho - a^0_{A1} - a^1_{B1}$). So, it cannot be concluded with certainty from "price war" that A is an "aggressive" company.

The changed monopoly profit changes the payoff structure of the game. The payoff relevant for the analysis (path: $\rho - a^0_{A1} - a^1_{B1}$) drops from 5 to 3. In this situation, it is no longer advantageous for a "normal" company A to choose "price war" (a^0_{A1}), as it would be worse off, even if company B considers it "aggressive" and leaves the market (3, path: $\rho - a^0_{A1} - a^1_{B1}$) instead of 4 (path: $\rho - a^0_{A2} - a^1_{B2} - a^1_{A2}$)). In this case, the decision of company A reveals its type.

6.6.2 Bayesian Equilibrium in Continuous Strategies

Problem text in Sect. 6.5.2.

Part a)
The utility functions of the two players, worker (W) and government (G), are as follows:

$$U_W(L,t) = w \cdot L + (12 - L) \cdot t - \beta \cdot L^2$$
$$U_G(L,t) = w \cdot L + (12 - L) \cdot t - t^2$$

The strategic decision variable of the representative worker is the annual working time L; the strategy of the government consists in choosing the optimal transfer rate t.

Part b)
To find the reaction functions of the representative worker, we calculate the partial derivation of U_W with respect to L and set it equal to zero, $\partial U_W(L,t)/\partial(L) \stackrel{!}{=} 0$. Solving for L give:

$$\bar{r}_W(t): \ \bar{L}(t) = \frac{1}{2 \cdot \beta} \cdot (w - t)$$

The bar above the variables indicates that this is the expected reaction of the worker, as the government does not know the motivation of the worker.

The reaction function of the government (G) is accordingly (partial derivation of U_G with respect to t, setting to zero, and solving for t):

$$r_G(L): \ t(L) = \frac{1}{2} \cdot (w - L)$$

For the graphical representation, we solve one of the reaction functions for the independent variable. E.g., we obtain for the worker $(\bar{L}) = w - 2 \cdot \beta \cdot \bar{L}$. For the given wage rate $w = 12$, we get $t(\overline{L}) = 12 - 2 \cdot \beta \cdot \overline{L}$ for the worker and $t(L) = 6 - 0.5 \cdot L$ for the government.

The following figure represents the reaction functions of the worker and the government graphically, where the nonbroken outer reaction function of the worker is that of the motivated type: $(\overline{L}, \beta = 0.5) = 12 - \overline{L}$.

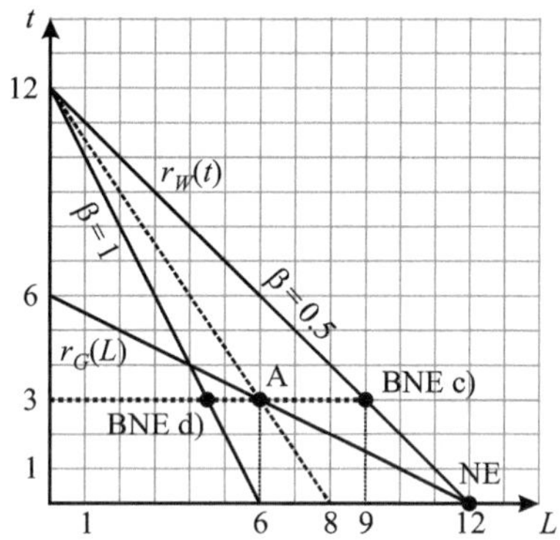

Thus, we obtain $L^* = 12$ and $t^* = 0$ as Nash equilibrium (NE in the figure). The utility of the worker is $U_W(L = 12, t = 0) = 72$ and that of the government $U_G(L = 12, t = 0) = 144$.

Part c)

Since the government has incomplete information about the type of worker, the reaction function of the worker is the expected reaction function from the perspective of the government. The expected reaction function of the worker results from his expected motivation, $\overline{\beta} = 0.5 \cdot \beta^L + 0.5 \cdot \beta^H = 0.75$, and is $t(\overline{L}, \overline{\beta} = 0.75) = 12 - 1.5 \cdot \overline{L}$. This function is shown in the graph in a) as a dashed line.

From the intersection of this expected reaction function of the worker with the reaction function of the government, we find $L = 6$ and $t = 3$, which both players expect from each other in equilibrium (point A). Since the worker is actually motivated ($\beta = 0.5$), he will respond to an expected transfer of $t = 3$ according to his true reaction function $L(t) = 12 - t$ (see a) with an optimal work effort of $L^* = 9$. In the Bayes-Nash equilibrium, BNE c), the worker's utility is therefore $U_W(L = 9, t = 3) = 76.5$ and the government's utility is $U_G(L = 9, t = 3) = 108$.

If the motivated worker were to reveal his type, as shown in a), he would receive no transfer and have a utility of 72. Thus, he clearly has no incentive to reveal his true type. The government, on the other hand, worsens from 144 to 108, meaning it would be better for it to know the true type.

Part d)

The reaction function of an unmotivated worker is already plotted in the graph. This is the inner solid line, which is given by $t(\overline{L}, \beta = 1) = 12 - 2 \cdot \overline{L}$.

As we can see from the graph, an unmotivated worker indeed has an incentive to reveal himself: With a transfer of $t = 3$, according to his reaction function $L(t) = 6 - 0.5 \cdot t$, he works $L = 4.5$, BNE d), thus achieving a utility of $U_W(L = 4.5, t = 3) = 56.25$. If he reveals himself, the transfer increases to $t = 4$, L reduces to 4, so his utility with $U_W(L = 4, t = 4) = 64$ is higher.

In the Bayes-Nash equilibrium, as shown in (c), the government has a utility of 108. If it knows the type of the worker, it adapts optimally and receives a utility of $U_G(L = 4, t = 4) = 64$, that is, it worsens through knowledge of the worker's true type.

6.6.3 Signaling Games

Problem text in Sect. 6.5.3.

First, we need to re-establish the conditions for the separation and pooling equilibrium. We always evaluate the derived conditions for the numerical example chosen in the text $c^l = 1$, $c^h = 3$, $w^l = 1$, and $w^h = 5$ as well. If the employer can freely choose the medium wage w^m, then the following conditions for a separation equilibrium must be met:

(I) $w^h - c^l \geq w^m$ or $4 \geq w^m$
(II) $w^l \geq w^m - c^h$ or $4 \geq w^m$

Under condition (I), the productive type has an incentive to distinguish himself from the unproductive type and send the signal. Under condition (II) the unproductive type has no incentive to do the same and will refrain from generating the signal. For the chosen numerical example, both inequalities consistently demand that the medium wage w^m must not be higher than 4. If it is higher, the pooling equilibrium is more attractive for both types, in which they both do not send out a signal. We now consider the incentives of the employer in a pooling and a separation strategy. The expected profit in a separation strategy of the employer remains unchanged and is still $\overline{\pi}(D) = \rho \cdot (R^h - w^h) + (1 - \rho) \cdot (R^l - w^l)$. Her expected profit in the pooling strategy is now $\overline{\pi}(P) = \rho \cdot R^h + (1 - \rho) \cdot R^l - w^m$. The employer is better off with a separation strategy if $\overline{\pi}(D) \geq \overline{\pi}(P)$ holds. This leads to the condition

(III) $w^m \geq \rho \cdot w^h + (1 - \rho) \cdot w^l$ or $w^m \geq 1 + 4 \cdot \rho$

Thus, in the separation equilibrium, the employer will pay a medium wage w^m of $1 + 4 \cdot \rho = w^m \leq 4$. The absolute amount of 4 will not be exceeded as long as the probability $\rho \leq 0.75$ holds. For a probability $\rho > 0.75$, the separation equilibrium cannot be maintained. In the pooling equilibrium, the employer pays both types of employees the medium wage $w^m = 4$, because this is the lowest wage that the productive type just accepts in the pooling equilibrium (for the unproductive type, the switch to the pooling equilibrium is always optimal).

Literature Recommendations

Fudenberg, D., & Tirole, J. (1991). *Game theory* (p. 209 ff.). MIT.

Harsanyi, J. (1967). Games with incomplete information played by 'Baysian' players, Part I: The basic model. *Management Science, 14*, 159–182.

Harsanyi, J. (1968a). Games with incomplete information played by 'Baysian' players, Part II: The basic model. *Management Science, 14*, 320–334.

Harsanyi, J. (1968b). Games with incomplete information played by 'Baysian' players, Part III: The basic model. *Management Science, 14*, 486–502.

Spence, M. (1973). Job market signaling. *The Quarterly Journal of Economics, 87*(3), 355–374.

Part II

Problems

Methodology 7

7.1 Problems

7.1.1 Decision under Uncertainty (to Sect. 1.3)

The Winstone family is a traditional entrepreneurial family in Bedrock Island. The utility functions of the family members are

$$u(y) = 2 \cdot \frac{y^z}{z} \text{ with } z > 0$$

Specifically, dad Fred Winstone is described by $z > 1$, mom Wilma Winstone by $z = 1$, and daughter Pebbles by $z < 1$.

Daughter Pebbles runs a small ice cream shop on Bedrock Lake as a hobby and is weighing two alternatives: She can either sell ice cream or help her father in the quarry. However, she does not know what the weather will be like. According to the weather report, there is a probability of $\rho = 0.75$ for sunny and warm weather and a probability of $(1 - \rho) = 0.25$ for rainy and cold weather. In case of rainy and cold weather, Pebbles can only earn $y_1 = 4$ dollars at her ice cream stand, but $y_2 = 64$ dollars if the weather is sunny and warm. Her father pays her 36 dollars regardless of the weather if she helps in the quarry. The trip to and from Bedrock Lake takes so long that Pebbles cannot change her decision once made.

a. Describe the curvature of the individual utility function for all family members! Based on this, characterize the risk aversion of the family members!
b. Let $z = 0.5$ apply for Pebbles. Draw the expected utility function for Pebbles' lottery a! Determine the expected payoff $E(a)$, the expected utility of the lottery $u(a)$, and the utility of the expected payoff $u[E(a)]$ mathematically and graphically! What can be concluded from the comparison of expected utility and expected payoff for Pebbles' risk aversion?

F. Bartholomae and M. Wiens, *Game Theory and Applications*, Classroom Companion: Economics, https://doi.org/10.1007/978-3-658-44608-6_7

c. Explain the term "certainty equivalent"! Determine Pebbles' ($z = 0.5$) certainty equivalent y_C as well as her risk premium numerically and graphically! How will Pebbles decide between selling ice cream and working in the quarry? How much would her father have to pay her at least to get her to help in the quarry?

Solution in Sect. 7.2.1.

7.1.2 Arrow-Pratt Measure (to Sect. 1.3)

In the wake of a global financial crisis, the well-established investment bank Silverwoman Sachs has run into slight liquidity problems. To settle the short-term liabilities, the bank needs to quickly get fresh money. For this, it can either rely on a (safe) government rescue package (a_1) of 4 billion euros or invest in a risky project (a_2), which is successful with a probability of $\rho = 0.2$ resulting in a payoff of 36 billion euros but generates a loss of 4 billion euros with the counter-probability. The decision on the form of financing is made based on the expected utility function $u(y) = k \cdot y^m$, where $k = 5$ is assumed.

a. So far, the company policy was $m = 1$. Start by generally defining the Arrow Pratt measure! Which risk attitude does this utility function express? Determine which form of financing Silverwoman Sachs will choose! Explain whether the company would decide differently for other positive values of k!
b. The two CEOs Sid Scary and Gigi Gambler disagree on which risk policy would be better considering the financial crisis: Sid thinks $m_S = 0.5$ is appropriate, while Gigi opts for $m_G = 2$. Calculate the Arrow-Pratt measure for both and show which form of financing each prefers! What is the underlying attitude to risk in each case? Reason which attitude you think is more justified!
c. Sketch Sid's and Gigi's utility functions and draw the expected value, utility of the expected value, expected utility, certainty equivalent, and risk premium for the risky project! Explain formally and verbally the relationship between the four concepts!

Solution in Sect. 7.2.2.

7.1.3 Nash Equilibrium with Three Players (to Sect. 2.4)

Consider the following simultaneous three-person game, in which Alice chooses the row, Bob the column, and Charlie the matrix. Alice receives the first payoff listed, Bob the second, and Charlie the third. Determine all Nash equilibria in pure strategies!

s_{C1}		
	s_{B1}	s_{B2}
s_{A1}	(0, 0, 0)	(3, 2, 3)
s_{A2}	(2, 3, 3)	(2, 2, 2)

s_{C2}		
	s_{B1}	s_{B2}
s_{A1}	(1, 1, 3)	(2, 2, 2)
s_{A2}	(3, 2, 1)	(0, 2, 0)

s_{C3}		
	s_{B1}	s_{B2}
s_{A1}	(2, 1, 2)	(3, 1, 1)
s_{A2}	(3, 3, 3)	(2, 2, 2)

Solution in Sect. 7.2.3.

7.1.4 Iterated Elimination of Strictly Dominated Strategies and Nash Equilibrium (to Sect. 3.1)

Determine the equilibrium in the matrix by elimination of strictly dominated strategies! Is this a Nash equilibrium? Is this a "coincidence" or is there a fundamental connection between the two solution concepts?

Alice, Bob	s_{B1}	s_{B2}	s_{B3}	s_{B4}
s_{A1}	(5, 2)	(1, 4)	(−1, 1)	(2, −1)
s_{A2}	(4, 0)	(2, 1)	(3, 0)	(3, −1)
s_{A3}	(−2, −2)	(1, −1)	(2, 3)	(2, 2)
s_{A4}	(1, 1)	(1, 3)	(1, 2)	(4, 1)

Solution in Sect. 7.2.4.

7.1.5 Mixed Strategies (to Sect. 3.3)

Determine all Nash equilibria in the following game! Represent your results graphically (diagram with reaction correspondences)!

Alice, Bob	s_{B1}	s_{B2}
s_{A1}	(0, 0)	(0, 50)
s_{A2}	(50, 0)	(−10, −10)

Solution in Sect. 7.2.5.

7.1.6 Mixed Strategies with Three Strategies (to Sect. 3.3)

Determine the reaction correspondences of the two players and all Nash equilibria for the following game!

Alice, Bob	s_{B1}	s_{B2}
s_{A1}	(5, 6)	(0, 4)
s_{A2}	(0, 4)	(4, 8)
s_{A3}	(3, 4)	(3, 0)

Solution in Sect. 7.2.6.

7.1.7 Mixed Strategies, Dominated Strategy, and Outside Option (to Sects. 3.1 and 3.3)

The following game is given in matrix form:

Alice, Bob	s_{B1}	s_{B2}	s_{B3}
s_{A1}	(2, 8)	(3, 3)	(0, 0)
s_{A2}	(1, 0)	(2, 1)	(2, 2)
s_{A3}	(4, 1)	(5, 2)	(1, 1)

a. Eliminate all strictly dominated strategies and then determine all Nash equilibria!
b. Determine the reaction correspondences of the players!
c. Suppose Alice had a fourth strategy s_{A4}, which always generates a payoff of x for her. What is the minimum value of x so that she chooses s_{A4} regardless of Bob's decision? If so, what kind of strategy is this? Also, discuss the rationality requirements for both players!

Solution in Sect. 7.2.7.

7.1.8 Mixed Strategies and Dominance (to Sects. 3.1 and 3.3)

Systematically examine the following game for Nash equilibria and apply the procedure for determining mixed strategies as well!

Alice, Bob	s_{B1}	s_{B2}	s_{B3}
s_{A1}	(1, 0)	(1, 2)	(0, 1)
s_{A2}	(0, 3)	(1, 1)	(2, 0)

Solution in Sect. 7.2.8.

7.1.9 Subgame Perfection and Imperfect Information (to Sect. 4.2)

Determine all subgame perfect Nash equilibria for the following game given in extensive form!

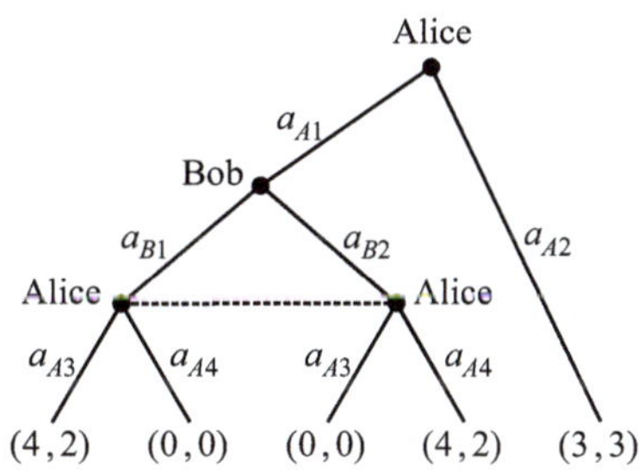

Solution in Sect. 7.2.9.

7.1.10 Bargaining Game (to Sect. 4.2)

The car salesman Rob Racketeer and his potential customer Babsi Bargain are negotiating the price of an exclusive sports car. It is generally known that at the beginning of the negotiations, Babsi's maximum willingness to pay is 100,000 euros and the reservation price for Rob is 90,000 euros. Rob proposes a price p, which Babsi can accept or reject. If she accepts it, she purchases the car at this price; if she rejects it, Babsi makes a final offer at a price q.

Mr. Racketeer is a hands-on entrepreneur with a flourishing business. He is not necessarily dependent on the deal with Ms. Bargain, so he negotiates a little more confidently: If Ms. Bargain rejects the offer and makes her own price offer in the second stage Mr. Racketeer increases his reservation price to $90 + \delta$, with $\delta < 10$, in this second negotiation round.

a. Specify the actions, strategies, and payoffs of the two players!
b. Find the price $p \in [90, 100]$ for which a "take-it-or-leave-it" offer from the seller that is accepted by the buyer is a Nash equilibrium! Explain why the only subgame-perfect equilibrium is reached when the seller proposes his second-stage reservation price, and the buyer accepts this offer!

Solution in Sect. 7.2.10.

7.1.11 Rubinstein Bargaining Game (to Sect. 5.2)

In a Rubinstein bargaining game, two players negotiate the division of a fixed amount of money (share of the "cake") against a potentially infinite time horizon. The size of the cake is usually normalized to 1. The first player, Alice, makes an offer to the second, Bob, in the form that she proposes a share $s_A \in [0, 1]$ for herself and thus the remaining share $(1 - s_A)$ for Bob. Then Bob decides whether to accept or reject. If he accepts, the game ends. If he rejects, the game continues in the next round and he makes a counterproposal, in which he grants Alice a share of s_B and claims $(1 - s_B)$ for himself. If the proposal is accepted, the game ends, if not, the game continues, and Alice makes another offer. The negotiation will continue until the offer is accepted. The discount factor of Alice is δ_A and of Bob is δ_B.

Determine the subgame perfect Nash equilibrium of this bargaining game! Has one of the two players an advantage? Also generally discuss the equilibrium division depending on the discount factors!

Solution in Sect. 7.2.11.

7.1.12 Signaling Game (to Sect. 6.3)

Assume the following situation in a labor market: The productivity of a worker is either low, $m^l = 1$, or high, $m^h = 3$, where the proportion of the two types in the total population is ρ and $1 - \rho$, respectively. The employer can only observe the education level e, but not the productivity m. The workers must incur costs of $c^l = 4 \cdot e$ or $c^h = e$ to achieve an education level e (the education level has no influence on productivity, it only serves as a signal). For simplification, we assume that only two education levels are possible: 0 (unskilled) and e^* (with vocational training).

The employer can pay attention to the signal and pay the workers according to the assumed productivity, i.e., if he observes an education level of e^*, he assumes high productivity and pays the wage $w^h = 3$ accordingly (separation). If he does not pay attention to the signal, he pays all workers according to the expected productivity $\overline{w} = E(m) = \rho + 3 \cdot (1 - \rho)$ (pooling).

a. Assume that there is only one employer and one worker. Draw the game tree! What are the conditions for a separating equilibrium or a pooling equilibrium?
b. Now consider the situation with two employers and several workers. Discuss whether a pooling equilibrium is plausible if the corresponding parameter constellation allows a separation equilibrium!

Solution in Sect. 7.2.12.

7.2 Solutions

7.2.1 Decision Under Uncertainty

Problem text in Sect. 7.1.1.

Part a)
To determine the curvature of the utility function, the second derivative (derivative of the first derivative) must be calculated:

$$\frac{du(y)}{dy} = 2 \cdot y^{z-1}$$
$$\frac{d^2u(y)}{dy^2} = 2 \cdot (z-1) \cdot y^{z-2}$$

The second derivative takes different signs depending on the value of $(z-1)$. This implies different curvatures of the utility function and, thus, different risk attitudes of the decision-makers. The following statements can be made:

- For $z > 1$ (dad), the second derivative is greater than zero and the function is left-curved (convex). A convex utility function characterizes a risk-seeking decision-maker.
- For $z = 1$ (mom), the second derivative is equal to zero and the function has no curvature—it is linear. A linear utility function characterizes a risk-neutral decision-maker.
- For $z < 1$ (daughter), the second derivative is less than zero and the function is right-curved (concave). A concave utility function characterizes a risk-averse decision-maker.

Part b)
We can calculate the following values:

- Expected value:

$$E(a) = 0.25 \cdot 4 + 0.75 \cdot 64 = 49$$

- Expected utility:

$$E[u(a)] = 0.25 \cdot 2 \cdot \frac{4^{0.5}}{0.5} + 0.75 \cdot \frac{64^{0.5}}{0.5} = 26$$

- Utility of the expected value:

$$u[E(a)] = 2 \cdot \frac{49^{0.5}}{0.5} = 28$$

Since the utility of the expected value is greater than the expected utility, Pebbles can be characterized as risk-averse: Pebbles derives greater utility from the certain payment of the expected value of the lottery than from the lottery itself. This implies that the risk from the lottery leads to a loss of utility for her, i.e., she is risk averse.

Graphical representation:

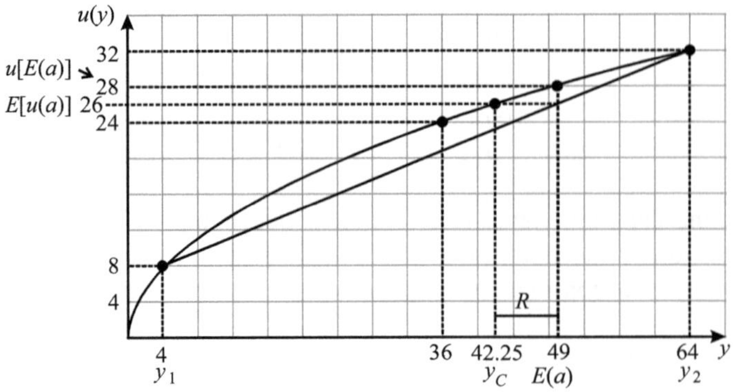

Part c)
The certainty equivalent is defined as the certain payment that provides the decision-maker with the same utility as the lottery. Thus, the calculation is as follows:

$$u(y_C) = E[u(a)]$$

$$2 \cdot \frac{(y_C)^{0.5}}{0.5} = 26$$

$$\Rightarrow y_C = 42.25.$$

The risk premium is finally calculated from the difference between the expected value and the certainty equivalent,

$$R = 49 - 42.25 = 6.75.$$

Her father would have to pay Pebbles at least the certainty equivalent of 42.25 dollars for her to be willing to work in the quarry. For 36 dollars, she tolerates the weather risk associated with selling ice cream.

Both the certainty equivalent and the risk premium are already marked in the figure in (b).

7.2.2 Arrow-Pratt Measure

Problem text in Sect. 7.1.2.

Part a)

In general, the Arrow-Pratt measure is defined as

$$r_A(y) = -\frac{d^2u(y)/dy^2}{du(y)/dy}.$$

Specifically, for $m = 1$ the utility function is $u(y) = 5 \cdot y^1$ and the following value for $r_A(y)$ can be calculated:

$$r_A(y) = -\frac{0}{5} = 0.$$

Thus, it is a risk-neutral utility function.

Silverwoman Sachs can choose between two lotteries with the following payoffs:

	$w_1(p_1 = 0.2)$	$w_2(p_2 = 0.8)$
Rescue package (a_1)	4	4
Project (a_2)	36	−4

The expected utility from both lotteries is

$$E[u(a_1)] = 0.2 \cdot 5 \cdot 4 + 0.8 \cdot 5 \cdot 4 = 20$$
$$E[u(a_2)] = 0.2 \cdot 5 \cdot 36 + 0.8 \cdot 5 \cdot (-4) = 36 - 16 = 20$$

Thus, a risk-neutral decision-maker is indifferent between both lotteries.

For other positive values of k, the decision would not change, as it would only be a linear positive transformation of the utility function.

$$E[u(a_1)] = 0.2 \cdot k \cdot 4 + 0.8 \cdot k \cdot 4 = 4 \cdot k$$
$$E[u(a_2)] = 0.2 \cdot k \cdot 36 + 0.8 \cdot k \cdot (-4) = 4 \cdot k.$$

Therefore, it still holds that $E[u(a_1)] = E[u(a_2)]$. The Arrow-Pratt measure remains unchanged:

$$r_A(y) = -\frac{0}{k} = 0$$

Part b)
The Arrow-Pratt measures of the two CEOs with $u_S(y) = 5 \cdot y^{0.5}$ and $u_G(y) = 5 \cdot y^2$ are

$$r_A^S(y) = -\frac{5 \cdot 0.5 \cdot (-0.5) \cdot y^{-1.5}}{5 \cdot 0.5 \cdot y^{-0.5}} = -\frac{1}{2 \cdot y} < 0$$
$$r_A^G(y) = -\frac{5 \cdot 2}{5 \cdot 2 \cdot y} = \frac{1}{y} > 0$$

Thus, Sid exhibits a risk-averse and Gigi exhibits a risk-seeking risk attitude.

Accordingly, their expected utilities are calculated differently. Sid decides based on the comparison between the utility values

$$E[u_S(a_1)] = 0.2 \cdot 5 \cdot 4^{0.5} + 0.8 \cdot 5 \cdot 4^{0.5} \cdot 4 = 10$$
$$E[u_S(a_2)] = 0.2 \cdot 5 \cdot 36^{0.5} + 0.8 \cdot 5 \cdot \left[-\left(4^{0.5}\right)\right] = 6 - 8 = -2$$

for the rescue package a_1, while Gigi opts due to

$$E[u_G(a_1)] = 0.2 \cdot 5 \cdot 4^2 + 0.8 \cdot 5 \cdot 4^2 \cdot 4 = 80$$
$$E[u_G(a_2)] = 0.2 \cdot 5 \cdot 36^2 + 0.8 \cdot 5 \cdot \left[-\left(4^2\right)\right] = 1296 - 64 = 1232$$

clearly for the risky project a_2.

The argument for Sid's risk-averse attitude is that the company can be saved with certainty and thus continues to exist. The argument for Gigi's behavior is the "gambling for resurrection" argument, which means that in a relatively hopeless situation, a project is undertaken that, if successful, leads to a revival of the company. The associated high risks are consciously accepted, as the company would otherwise no longer exist anyway (and possibly the liability is also limited downwards).

Part c)
According to their risk attitude, Sid's risk-averse attitude is a concave utility function and Gigi's risk-seeking attitude is a convex utility function:

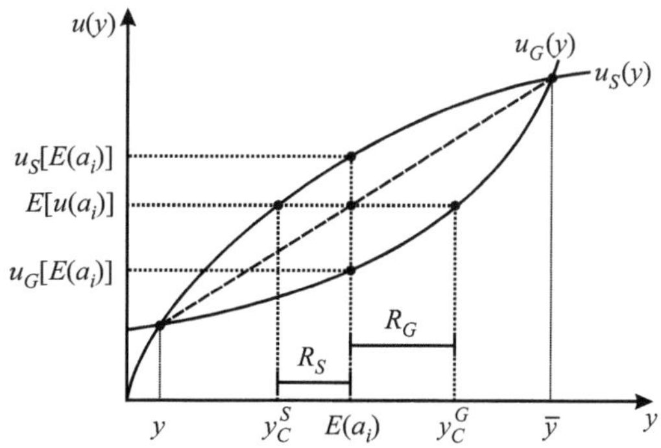

The expected value of the project is

$$E(a_2) = 0.2 \cdot 36 + 0.8 \cdot (-4) = 4.$$

Accordingly, the utilities of the expected value for the two are

$$u_S[E(a_2)] = 5 \cdot 4^{0.5} = 10$$
$$u_G[E(a_2)] = 5 \cdot 4^2 = 80$$

We know from (b) the expected utilities with $E[u_S(a_2)] = -2$ and $E[u_G(a_2)] = 1232$. While the expected value weighs the monetary payoffs with their respective probability of occurrence, the expected utility weighs the payoffs evaluated with the utility function with their probability of occurrence. In the figure, the expected value is therefore read off the horizontal axis (payoffs) and the expected utility on the vertical axis (utility). The certainty equivalent is determined by

$$u_S\left(y_C^S\right) = 5 \cdot \left(y_C^S\right)^{0.5} = -2 \Rightarrow y_C^S = -0.16$$
$$u_G\left(y_C^G\right) = 5 \cdot \left(y_C^G\right)^2 = 1232 \Rightarrow y_C^G \approx 15.70$$

The comparison with the expected value of the lottery shows that the risk-averse Sid would forego the risky project for a lower certain amount than the expected value—this would even be the case with a negative amount, although the expected value is positive! On the other hand, the risk-seeking Gigi would have to be paid a significantly higher positive amount for her to be willing to forego the project. Accordingly, the risk premiums are

$$R_S = 4 - (-0.16) = 4.16$$
$$R_G = 4 - 15.70 = -11.70.$$

Likewise, these show that the risk-averse Sid would be willing to pay to avoid the risk (positive risk premium) while the risk-seeking Gigi would have to be paid (negative risk premium) to forego the risk of the project.

7.2.3 Nash Equilibrium with Three Players

Problem text in Sect. 7.1.3.

To determine the Nash equilibrium, we first solve each matrix for Alice and Bob as usual. I.e., to find Alice's best response, we look for Alice's highest payoff in each column (Bob's strategy) and underline it. Similarly, we underline Bob's highest payoff in each row.

s_{C1}		
	s_{B1}	s_{B2}
s_{A1}	$(0, 0, 0)$	$(\underline{3}, \underline{2}, \underline{\mathbf{3}})$
s_{A2}	$(\underline{2}, \underline{3}, \underline{\mathbf{3}})$	$(2, 2, \underline{\mathbf{2}})$

s_{C2}		
	s_{B1}	s_{B2}
s_{A1}	$(1, 1, \underline{\mathbf{3}})$	$(\underline{2}, \underline{2}, 2)$
s_{A2}	$(\underline{3}, \underline{2}, 1)$	$(0, \underline{2}, 0)$

s_{C3}		
	s_{B1}	s_{B2}
s_{A1}	$(2, \underline{1}, 2)$	$(\underline{3}, \underline{1}, 1)$
s_{A2}	$(\underline{3}, \underline{3}, \underline{\mathbf{3}})$	$(2, 2, \underline{\mathbf{2}})$

To find Charlie's best response, we must compare the individual cells: Let us start with cell (s_{A1}, s_{B1}). We see that Charlie has a payoff of 0 in matrix s_{C1}, 3 in matrix s_{C2}, and 2 in matrix s_{C3}. Thus, they have the highest payoff at s_{C2}. For clarity, this payoff has been highlighted in bold. We proceed analogously for the remaining cells: For cell (s_{A1}, s_{B2}) s_{C1} provides the highest payoff of 3, for cell (s_{A2}, s_{B1}) s_{C1} and s_{C3} provide the highest payoff, and for cell (s_{A2}, s_{B2}), this is also the case for s_{C1} and s_{C3}. Thus, we can identify three strategy combinations as Nash equilibria in pure strategies: $\{s_{A1}, s_{B2}, s_{C1}\}$, $\{s_{A2}, s_{B1}, s_{C1}\}$, and $\{s_{A2}, s_{B1}, s_{C3}\}$.

7.2.4 Iterated Elimination of Strictly Dominated Strategies and Nash Equilibrium

Problem text in Sect. 7.1.4.

In the elimination of strictly dominated strategies, those strategies are successively eliminated that the player will never choose. This is the case if another strategy is strictly better, i.e., it provides a higher payoff for any given strategy of the other player.

Let us take a closer look at the matrix. We see, for example, that strategy s_{A3} always provides lower payoffs for Alice than strategy s_{A2}:

- $s_{A3}(s_{B1}) = -2 < 4 = s_{A2}(s_{B1})$,
- $s_{A3}(s_{B2}) = 1 < 2 = s_{A2}(s_{B2})$,
- $s_{A3}(s_{B3}) = 2 < 3 = s_{A2}(s_{B3})$, and
- $s_{A3}(s_{B4}) = 2 < 3 = s_{A2}(s_{B4})$.

We can therefore eliminate the strategy s_{A3}, as no rational player will choose this strategy, i.e., this strategy is *not rationalizable*. Bob, of course, knows that Alice will never choose s_{A3} and will also not consider this strategy in his strategy selection. Thus, we can reduce the matrix to

Alice, Bob	s_{B1}	s_{B2}	s_{B3}	s_{B4}
s_{A1}	(5, 2)	(1, 4)	(−1, 1)	(2, −1)
s_{A2}	(4, 0)	(2, 1)	(3, 0)	(3, −1)
s_{A4}	(1, 1)	(1, 3)	(1, 2)	(4, 1)

In this matrix, we see that Bob will never choose s_{B3}, as his second strategy s_{B2} always provides a higher payoff. The only strategy of Alice for which this would not apply was the eliminated strategy s_{A3}. However, Alice will never choose this strategy and thus Bob does not consider outcomes related to this strategy either. Consequently, Bob will never choose s_{B3} and we can eliminate this strategy. Hence, the complexity of the 4×3 matrix can further be reduced to a 3×3 matrix:

Alice, Bob	s_{B1}	s_{B2}	s_{B4}
s_{A1}	(5, 2)	(1, 4)	(2, −1)
s_{A2}	(4, 0)	(2, 1)	(3, −1)
s_{A4}	(1, 1)	(1, 3)	(4, 1)

In the same way, we can eliminate further strategies: s_{B4} is strictly dominated by s_{B2}, s_{A4} by s_{A2}, s_{B1} by s_{B2}, and s_{A1} by s_{A2}. Thus, the only rationalizable strategies that remain are s_{A2} and s_{B2}.

All strategies that survive the iterated elimination of strictly dominated strategies are rationalizable. The order of elimination does not matter. For example, we could have first eliminated s_{B1} (strictly dominated by s_{B2}) and then proceeded with the elimination of $s_{A1}[s_{A2}]$, $s_{A3}[s_{A2}]$, $s_{B3}[s_{B2}]$, $s_{B4}[s_{B2}]$, and $s_{A4}[s_{A2}]$—in square brackets, we indicated the dominating strategy. Finally, (s_{A2}, s_{B2}) would result again.

But is this also a Nash equilibrium? We can check this using the "underline" method:

Alice, Bob	s_{B1}	s_{B2}	s_{B3}	s_{B4}
s_{A1}	$(\underline{5}, 2)$	$(1, \underline{4})$	$(-1, 1)$	$(2, -1)$
s_{A2}	$(4, 0)$	$(\underline{2}, \underline{1})$	$(\underline{3}, 0)$	$(3, -1)$
s_{A3}	$(-2, -2)$	$(1, -1)$	$(2, \underline{3})$	$(2, 2)$
s_{A4}	$(1, 1)$	$(1, \underline{3})$	$(1, 2)$	$(\underline{4}, 1)$

We see that only at (s_{A2}, s_{B2}) both payoffs are underlined, which means that we have found the (unique) Nash equilibrium. Since the iterated elimination of strictly dominated strategies only leaves rationalizable strategies and only these strategies are considered as best responses at all, it is not surprising that equilibria found using this method are always Nash equilibria.

7.2.5 Mixed Strategies

Problem text in Sect. 7.1.5.

We can identify (s_{A1}, s_{B2}) and (s_{A2}, s_{B1}) as Nash equilibria. Since the number of Nash equilibria is always odd, finding only two means that there must be another equilibrium in mixed strategies. To find it, let us assign probabilities to the pure strategies of the players as follows:

- Alice plays strategy s_{A1} with probability p_A and s_{A2} with probability $1 - p_A$.
- Bob plays strategy s_{B1} with probability p_B and s_{B2} with probability $1 - p_B$.

We can calculate the probabilities of the equilibrium mixed strategies in two ways—using the optimization approach or the expected utility comparison.

Optimization Approach
First, we set up the payoff function for Alice

$$\begin{aligned} E[u_A(p_A, p_B)] &= p_A \cdot p_B \cdot 0 + p_A \cdot (1 - p_B) \cdot 0 + (1 - p_A) \cdot p_B \cdot 50 \\ &\quad + (1 - p_A) \cdot (1 - p_B) \cdot (-10) \\ &= 60 \cdot p_B + 10 \cdot p_A - 60 \cdot p_A \cdot p_B - 10 \end{aligned}$$

Second, we determine the probability p_A that maximizes Alice's expected payoff. For this purpose, we determine the partial derivative of the function with respect to p_A (Alice's strategic variable) and set it equal to zero:

$$\frac{\partial E[u_A(p_A, p_B)]}{\partial p_A} = 10 - 60 \cdot p_B \overset{!}{=} 0$$

This condition is therefore fulfilled if $p_B = \frac{1}{6}$.

Third, we perform the same procedure for Bob, i.e., we set up the payoff function and optimize it with respect to Bob's strategic variable p_B:

$$\begin{aligned}
E[u_B(p_A, p_B)] &= p_A \cdot p_B \cdot 0 + p_A \cdot (1 - p_B)50 + (1 - p_A) \cdot p_B \cdot 0 \\
&\quad + (1 - p_A) \cdot (1 - p_B) \cdot (-10) \\
&= 60 \cdot p_A + 10 \cdot p_B - 60 \cdot p_A \cdot p_B - 10 \\
\frac{\partial E[u_A(p_A, p_B)]}{\partial p_B} &= 10 - 60 \cdot p_A \stackrel{!}{=} 0 \\
&\Rightarrow p_A^* = \frac{1}{6}
\end{aligned}$$

Comparison of Expected Utility

First, we determine the probability p_B, at which it is better (or equally good) for Alice to play s_{A1}, i.e., choose $p_A = 1$. For this purpose, we compare the expected utility of strategy s_{A1} and strategy s_{A2}:

$$\begin{aligned}
E[u_A(p_A = 1, p_B)] &\geq E[u_A(p_A = 0, p_B)] \\
p_B \cdot 0 + (1 - p_B) \cdot 0 &\geq p_B \cdot 50 + (1 - p_B) \cdot (-10) \\
p_B &\leq 1/6
\end{aligned}$$

Second, we perform the same procedure for Bob to determine the probability p_A that Bob prefers s_{B1} (i.e., choose $p_B = 1$) to s_{22} (or is indifferent):

$$\begin{aligned}
E[u_B(p_A, p_B = 1)] &\geq E[u_B(p_A, p_B = 0)] \\
p_A \cdot 0 + (1 - p_A) \cdot 0 &\geq p_A \cdot 50 + (1 - p_A) \cdot (-10) \\
p_A &\leq 1/6
\end{aligned}$$

The expected utility comparison is the easiest way to formally determine the reaction correspondences: Alice will choose strategy s_{A1} for sure (i.e., $p_A = 1$) if $p_B < 1/6$; similarly, she chooses strategy s_{A2} for sure (i.e., $1 - p_A = 1$ or $p_A = 0$) if $p_B > 1/6$. Since the same reasoning applies to Bob's decision, we find

$$p_A^*(p_B) = \begin{cases} 1 & p_B \in \left[0, \frac{1}{6}\right) \\ [0, 1] & p_B = \frac{1}{6} \\ 0 & p_B \in \left(\frac{1}{6}, 1\right] \end{cases} \qquad p_B^*(p_A) = \begin{cases} 1 & p_A \in \left[0, \frac{1}{6}\right) \\ [0, 1] & p_A = \frac{1}{6} \\ 0 & p_A \in \left(\frac{1}{6}, 1\right] \end{cases}$$

Thus, we can draw the reaction correspondences as follows:

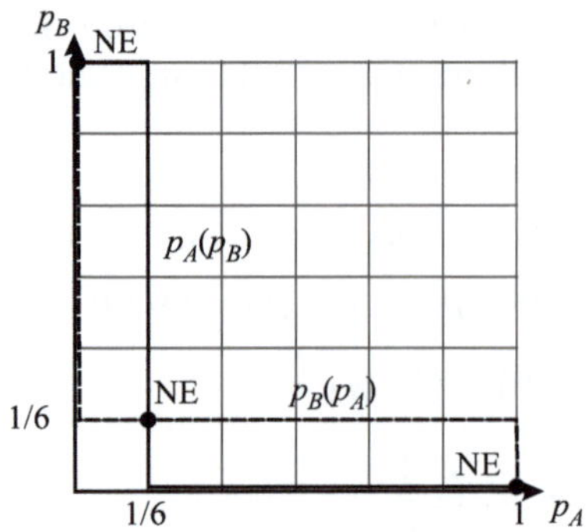

There are three Nash equilibria (NE), two in pure and one in mixed strategies: (s_{A1}, s_{B2}) with the payoff combination (0, 50), (s_{A2}, s_{B1}) with the payoff combination (50, 0), and $(p_A = 1/6, p_B = 1/6)$ with the (expected) payoff combination (0, 0)—the latter is calculated by inserting the equilibrium probabilities in the expected utility.

7.2.6 Mixed Strategies with Three Strategies

Problem text in Sect. 7.1.6.

We find two Nash equilibria in pure strategies, (s_{A1}, s_{B1}) and (s_{A2}, s_{B2}):

Alice, Bob	s_{B1}	s_{B2}
s_{A1}	$(\underline{5}, \underline{6})$	(0, 4)
s_{A2}	(0, 4)	$(\underline{4}, \underline{8})$
s_{A3}	$(3, \underline{4})$	(3, 0)

As the number of equilibria is even, there must be another equilibrium in mixed strategies. Let p_{A1} be the probability that Alice decides for s_{A1}, p_{A2} for s_{A2}, and thus $1 - p_{A1} - p_{A2}$ for s_{A3}. Similarly, let p_B be the probability that Bob opts for s_{B1} and with $1 - p_B$ for s_{B2}.

Depending on p_B, Alice has different expected payoffs for the individual strategies:

- $s_{A1} : E[u_A(p_{A1} = 1, p_{A2} = 0, p_B)] = 5 \cdot p_B + 0 \cdot (1 - p_B) = 5 \cdot p_B$
- $s_{A2} : E[u_A(p_{A1} = 0, p_{A2} = 1, p_B)] = 0 \cdot p_B + 4 \cdot (1 - p_B) = 4 - 4 \cdot p_B$
- $s_{A3} : E[u_A(p_{A1} = 0, p_{A2} = 0, p_B)] = 3 \cdot p_B + 3 \cdot (1 - p_B) = 3$

To better understand at which values of p_B Alice prefers which strategy, we graphically represent the expected payoffs depending on p_B:

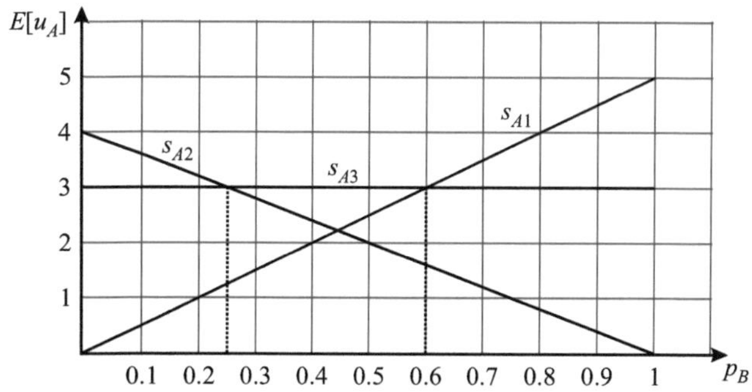

We see that there are two intersections where two strategies are exactly equally good and otherwise one strategy is always clearly better. We will therefore now look at these intersections. Let us first compare s_{A2} and s_{A3}, i.e., s_{A2} is better if

$$4 - 4 \cdot p_B \geq 3 \quad \Rightarrow p_B \leq 0.25$$

Next, we first compare s_{A3} and s_{A1}, i.e., s_{A3} is better if

$$5 \cdot p_B \leq 3 \quad \Rightarrow p_B \leq 0.6$$

This allows us now to determine Alice's reaction correspondence:

$$r_A(p_B) = \begin{cases} p_{A1} = 0, \ p_{A2} = 1 & p_B \in [0, 0.25) \\ p_{A1} = 0, \ p_{A2} \in [0, 1] & p_B = 0.25 \\ p_{A1} = 0, \ p_{A2} = 0 & p_B \in (0.25, 0.6) \\ p_{A1} \in [0, 1], \ p_{A2} = 0 & p_B = 0.6 \\ p_{A1} = 1, \ p_{A2} = 0 & p_B \in (0.6, 1] \end{cases}$$

To understand the reaction correspondence and to be able to read it better, you may find it helpful to have a look at the figure: If $p_B < 0.25$, Alice chooses $p_{A1} = 0$ and $p_{A2} = 1$, i.e., s_{A2} is played with certainty. The figure clearly shows that Alice has the highest expected payoff with this strategy. At $p_B = 0.25$, as $pA_1 = 0$, s_{A1} is not chosen. For p_{A2}, all values between 0 and 1 are possible, which means that Alice is indifferent between s_{A2} and s_{A3}—in the figure, both deliver the same payoff. We would proceed analogously for the other cases.

Let us now turn to Bob. Depending on his strategy choice, Bob has the following expected payoffs:

$$E[u_B(p_{A1}, p_{A2}, p_B = 1)] = 6 \cdot p_{A1} + 4 \cdot p_{A2} + 4 \cdot (1 - p_{A1} - p_{A2}) = 2 \cdot p_{A1} + 4$$
$$E[u_B(p_{A1}, p_{A2}, p_B = 0)] = 4 \cdot p_{A1} + 8 \cdot p_{A2} + 0 \cdot (1 - p_{A1} - p_{A2}) = 4 \cdot p_{A1} + 8 \cdot p_{A2}$$

Bob prefers s_{B1} (i.e., $p_B = 1$) over s_{B2} (i.e., $p_B = 0$) if:

$$2 \cdot p_{A1} + 4 \geq 4 \cdot p_{A1} + 8 \cdot p_{A2} \text{ or } 4 - 2 \cdot p_{A1} - 8 \cdot p_{A2} \geq 0$$

Thus, Bob's reaction correspondence is:

$$r_B(p_{A1}, p_{A2}) = \begin{cases} p_B = 1 & 4 - 2 \cdot p_{A1} - 8 \cdot p_{A2} > 0 \\ p_B \in [0, 1] & 4 - 2 \cdot p_{A1} - 8 \cdot p_{A2} = 0 \\ p_B = 0 & 4 - 2 \cdot p_{A1} - 8 \cdot p_{A2} < 0 \end{cases}$$

In the Nash equilibrium in mixed strategies, both players must be indifferent between their pure strategies, which means that

(i) $4 - 2 \cdot p_{A1} - 8 \cdot p_{A2} = 0$ and $p_{A1} = 0$ (Alice is indifferent between s_{A2} and s_{A3}) or
(ii) $4 - 2 \cdot p_{A1} - 8 \cdot p_{A2} = 0$ and $p_{A2} = 0$ (Alice is indifferent between s_{A1} and s_{A3})

must be fulfilled. We can exclude case (ii): If Alice does not play s_{A2} with positive probability, then Bob will always choose s_{B1} since s_{B2} is the only strategy where Bob responds with s_{B2}. In addition, the two conditions in (ii) cannot be fulfilled together: $4 - 2 \cdot p_{A1} - 8 \cdot 0 = 0$ is only fulfilled for $p_{A1} = 2$—but a probability that is larger than 1 is not possible!

However, case (i) is possible, and we find the solution:

$$4 - 2 \cdot 0 - 8 \cdot p_{A2} = 0 \Leftrightarrow p_{A2} = 0.5$$

Thus, the Nash equilibrium in mixed strategies is $\{(p_{A1} = 0, p_{A2} = 0.5), p_B = 0.25\}$.

7.2.7 Mixed Strategies, Dominated Strategy, and Outside Option

Problem text in Sect. 7.1.7.

Part a)
The elimination of strictly dominated strategies shows: s_{A1} is strictly dominated by s_{A3} and s_{B1} is strictly dominated by s_{B2}. An elimination of further strategies is not possible, so we can reduce the game to a 2×2 matrix:

Alice, Bob	s_{B2}	s_{B3}
s_{A2}	(2, 1)	($\underline{2}$, $\underline{2}$)
s_{A3}	($\underline{5}$, $\underline{2}$)	(1, 1)

Thus, we can find two Nash equilibria in pure strategies, (s_{A2}, s_{B3}) and (s_{A3}, s_{B2}). Since this number is even, there must be another equilibrium in mixed strategies. We want to draw the reaction correspondences in (b), thus, we recommend the method of expected utility comparison.

Remember that a mixed strategy assigns a probability to each of the player's pure strategies. Let p_{A1} be the probability that Alice decides for s_{A1}, p_{A2} for s_{A2}, and thus $1 - p_{A1} - p_{A2}$ for s_{A3}. Accordingly, let p_{B1} be the probability that Bob decides for s_{B1}, p_{B2} for s_{B2}, and thus $1 - p_{B1} - p_{B2}$ for s_{B3}. However, we know that each player has a non-rationalizable strategy, which was proven by the iterated elimination. We can therefore assign a probability of 0 to this strategy, i.e., $p_{A1} = 0$ and $p_{B1} = 0$.

- Alice prefers to play s_{A2} instead of s_{A3} if

$$E[u_A(p_{A1} = 0, p_{A2} = 1, p_{B1}, p_{B2})] > E[u_A(p_{A1} = 0, p_{A2} = 0, p_{B1}, p_{B2})]$$
$$2 \cdot p_{B2} + 2 \cdot (1 - p_{B1} - p_{B2}) > 5 \cdot p_{B2} + 1 \cdot (1 - p_{B1} - p_{B2})$$

Because Bob will never play s_{B1}, it follows that $p_{B1} = 0$ and the condition simplifies to

$$2 \cdot p_{B2} + 2 \cdot (1 - p_{B2}) > 5 \cdot p_{B2} + (1 - p_{B2})$$
$$p_{B2} > 1/4$$

- Bob prefers to play s_{B2} instead of s_{B3} if

$$E[u_B(p_{A1}, p_{A2}, p_{B1} = 0, p_{B2} = 1)] > [u_B(p_{A1}, p_{A2}, p_{B1} = 0, p_{B2} = 0)]$$
$$1 \cdot p_{A2} + 2 \cdot (1 - p_{A1} - p_{A2}) > 2 \cdot p_{A2} + 1 \cdot (1 - p_{A1} - p_{A2})$$

Because Alice will never play s_{A1}, it follows that $p_{A1} = 0$ and the condition simplifies to

$$p_{A2} + 2 \cdot (1 - p_{A2}) > 2 \cdot p_{A2} + (1 - p_{A2})$$
$$p_{A2} < 1/2$$

Thus, the Nash equilibrium in mixed strategies is: $(\{p_{A1} = 0, p_{A2} = 1/2\}, \{p_{B1} = 0, p_{B2} = 1/4\})$.

Part b)
Based on the results from (a), the following reaction correspondences are obtained:

$$r_A(p_{B1}, p_{B2}) = \begin{cases} p_{A1} = 0, p_{A2} = 0 & p_{B2} \in \left[0, \frac{1}{4}\right) \text{ and } p_{B1} \in [0, 1] \\ p_{A1} = 0, p_{A2} \in [0, 1] & p_{B2} = 1/4 \quad \text{ and } p_{B1} \in [0, 1] \\ p_{A1} = 0, p_{A2} = 1 & p_{B2} \in \left(\frac{1}{4}, 1\right] \text{ and } p_{B1} \in [0, 1] \end{cases}$$

$$r_B(p_{A1}, p_{A2}) = \begin{cases} p_{B1} = 0, p_{B2} = 0 & p_{A2} \in \left[0, \frac{1}{2}\right) \text{ and } p_{A1} \in [0, 1] \\ p_{B1} = 0, p_{B2} \in [0, 1] & p_{A2} = 1/2 \quad \text{and } p_{A1} \in [0, 1] \\ p_{B1} = 0, p_{B2} = 1 & p_{A2} \in \left(\frac{1}{2}, 1\right] \text{ and } p_{A1} \in [0, 1] \end{cases}$$

For the sake of completeness, the probability of the strategies not being played. i.e., s_{A1} and s_{B1}, are also always given here. As we see, it does not matter how likely it is that this strategy will be played—but if the players are rational, they will never choose this strategy.

Part c)
We need to determine the payoff x for strategy s_{A4} so that Alice chooses it for each of Bob's strategies. It must therefore always deliver a higher payoff than the best of the other strategies: For s_{B1} Alice will choose s_{A3} with payoff 4, for s_{B2} s_{A3} with 5, and for s_{B3} s_{A2} with 2. Therefore, x must be at least 5 for Alice to always choose s_{A4}. This would result in a dominant strategy. This concept only requires Alice to be rational, while Bob does not have to fulfill this assumption.

7.2.8 Mixed Strategies and Dominance

Problem text in Sect. 7.1.8.

A closer examination of the strategies reveals that Bob has one dominated strategy: Strategy s_{B3} is strictly dominated by s_{B2} and can be eliminated. In the remaining 2×2 matrix, Alice has the weakly dominant strategy s_{A1}. Bob's best response to this is s_{B2}. This results in the (not necessarily unique) Nash equilibrium (s_{A1}, s_{B2}). However, no further Nash equilibria can be found:

Alice, Bob	s_{B1}	s_{B2}	s_{B3}
s_{A1}	$(\underline{1}, 0)$	$(\underline{1}, \underline{2})$	$(0, 1)$
s_{A2}	$(0, \underline{3})$	$(\underline{1}, 1)$	$(\underline{2}, 0)$

The search for an equilibrium in mixed strategies is unnecessary, as the number of Nash equilibria in pure strategies is already odd.

7.2.9 Subgame Perfection and Imperfect Information

Problem text in Sect. 7.1.9.

To determine the subgame perfect Nash equilibria, we use backward induction and therefore start by determining the equilibrium of the last subgame. This begins when Alice has chosen a_{A1} and is characterized by an imperfect information structure. It is therefore advisable to represent and solve the game in this stage in matrix form:

Alice, Bob	a_{B1}	a_{B2}
a_{A3}	(4, 2)	(0, 0)
a_{A4}	(0, 0)	(4, 2)

This game has two Nash equilibria in pure strategies: (a_{A3}, a_{B1}) and (a_{A4}, a_{B2}). Consequently, there must be another one in mixed strategies. Let p_A be the probability that Alice chooses a_{A3} and p_B be the probability that Bob chooses a_{B1}. Alice prefers to choose a_{A3} (i.e., $p_A = 1$) instead of a_{A4} if

$$E[u_A(p_A = 1, p_B)] > E[u_A(p_A = 0, p_B)]$$
$$4 \cdot p_B + 0 \cdot (1 - p_B) > 0 \cdot p_B + 4 \cdot (1 - p_B)$$
$$p_2 > 0.5$$

Bob prefers to choose a_{B3} (i.e., $p_B = 1$) instead of a_{B2} if

$$E[u_B(p_A, p_B = 1)] > E[u_B(p_A, p_B = 0)]$$
$$2 \cdot p_A + 0 \cdot (1 - p_A) > 0 \cdot p_A + 2 \cdot (1 - p_A)$$
$$p_A > 0.5$$

Thus, the Nash equilibrium in mixed strategies is $(p_A = 0.5, p_B = 0.5)$.

Since we know how the players react at the last stage, we can represent the game tree in reduced form. However, we must consider that there are three possible outcomes of the game at the last stage due to the imperfect information, which affect the resulting payoffs differently:

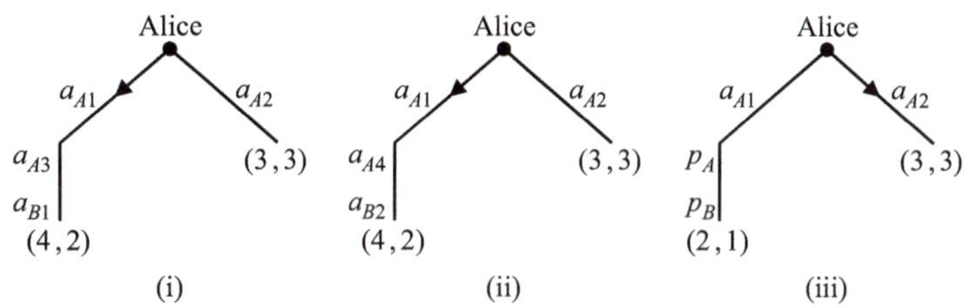

- Case (i) describes the relevant payoffs if (a_{A3}, a_{B1}) has resulted in Nash equilibrium at the last stage. In this case, Alice will choose a_{A1} in the first stage yielding the subgame perfect Nash equilibrium $\{(a_{A1}, a_{A3}), a_{B1}\}$.
- In case (ii) (a_{A4}, a_{B2}) has resulted in Nash equilibrium at the last stage, here too, Alice will choose a_{A1} in the first stage. The subgame perfect Nash equilibrium is $\{(a_{A1}, a_{A4}), a_{B2}\}$.
- In the third case, (iii), the mixed strategies $(p_A = 0.5, p_B = 0.5)$ were chosen in the last stage, resulting in expected payoffs of $0.5 \cdot 4 + 0.5 \cdot 0 = 2$ for Alice and $0.5 \cdot 2 + 0.5 \cdot 0 = 1$ for Bob. Compared to (a) and (b), this changes the decision of Alice in the first stage to a_{A2}. The subgame perfect Nash equilibrium is $\{(a_{A2}, p_A = 0.5), p_B = 0.5\}$.

7.2.10 Bargaining Game

Problem text in Sect. 7.1.10.

Part a)
The game consists of a maximum of two stages, in which the two players, seller (Rob) and buyer (Babsi), can determine their actions. Rob can propose a price, p, in the first stage and "accept" or "reject" Babsi's price proposal in the second stage. Babsi, in turn, can "accept" or "reject" Rob's price proposal in the first stage, and, in the second stage, offer her own price proposal q if necessary. Thus, the strategies for the two players are $s_R = \{p, \text{(accept, reject)}\}$ and $s_{\text{B}} = \{\text{(accept, reject)}, q\}$. The resulting payoffs correspond to the surpluses of the players: Upon conclusion of the deal, Rob receives the sales price less his reservation price, and zero otherwise. Babsi receives the difference between her willingness to pay and the purchase price when the deal is concluded, and zero otherwise as well.

The sequence of the game along with the resulting payoffs are shown in the following figure:

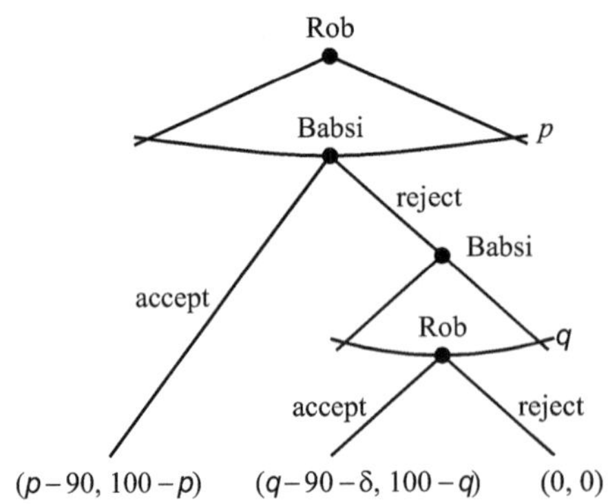

Part b)
Rob makes an offer, which can either be accepted by Babsi or not—a counterproposal from Babsi is no longer possible ("take-it-or-leave-it-offer"). The strategies of the players are thus $s_R = \{p\}$ and $s_B = \{\text{accept, reject}\}$. Rob will propose a price from the interval $p \in [90, 100]$ which a rational buyer will also accept: Even if Rob would claim 99.99 for himself, Babsi would still have a positive payoff of $0.01 > 0$.

It remains to check whether the proposed offer is subgame-perfect. In accordance with the principle of backward induction, we start the considerations in the second round of negotiations (period 1). In period 1, Babsi will submit exactly Rob's reservation price as a counterproposal, as Rob will not accept a lower price in period 1. Hence, if the negotiations come to a second round, Rob's reservation price $p = 90 + \delta$ will be proposed by Babsi and accepted by Rob.

What does this realization mean for the first round (period 0)? In period 0, Babsi will never accept a price that is higher than Rob's reservation price. For Bob, on the other hand, it makes no sense to submit a starting proposal in period 0 that (i) is higher than his reservation price (Babsi will reject this) or (ii) is lower than his reservation price (Bob can improve here through the second round of negotiations). Consequently, Bob will propose exactly the reservation price in period 0 and Babsi will accept it because she cannot improve with the second round of negotiations. Only for $p = 90 + \delta$, she is indifferent between today's payoff upon acceptance and tomorrow's payoff upon rejection.

7.2.11 Rubinstein Bargaining Game

Problem text in Sect. 7.1.11.

To simplify the analysis, we are looking for an equilibrium with stationary strategies, i.e., strategies that are independent of the period under consideration. Since the discount factors $0 < \delta_A < 1$ and $0 < \delta_B < 1$ apply, the negotiation becomes increasingly costly for the parties over time. Conceptually, the "cake" shrinks like ice in the sunshine from period to period, thus creating a pressure on the parties to reach an agreement rather sooner than later.

Alice makes Bob an offer, which he can accept or reject:

(1) $1 - s_A = \delta_B \cdot (1 - s_B)$

The left side indicates Bob's share that he would receive upon acceptance. If Bob were to reject, then the game would continue with the next round and he would demand for himself $(1 - s_B)$, which due to the discounting would only be worth $\delta_B \cdot (1 - s_B)$. As the cake shrinks, each portion becomes smaller in absolute terms—e.g., if the cake shrinks by 50%, the whole cake tomorrow will be as big as half the cake today.

Accordingly, Bob makes an offer, which Alice can just about accept:

(2) $s_B = \delta_A \cdot s_A$

Both conditions (1) and (2) must be fulfilled simultaneously. If we substitute (2) into (1), we get $1 - s_A = \delta_B \cdot (1 - \delta_A \cdot s_A)$ or $s_A = 1 - \delta_B + \delta_A \cdot \delta_B \cdot s_A$ and, consequently,

$$s_A = \frac{1 - \delta_B}{1 - \delta_A \cdot \delta_B}.$$

Substituting into (2) yields

$$s_B = \delta_A \cdot \frac{1 - \delta_B}{1 - \delta_A \cdot \delta_B}.$$

As a subgame perfect Nash equilibrium, we thus obtain the following distribution:

$$\left\{\frac{1-\delta_B}{1-\delta_A\cdot\delta_B}, \frac{(1-\delta_B)\cdot\delta_A}{1-\delta_A\cdot\delta_B}\right\}$$

This division is already realized in the first round as any delay would only reduce the cake.

As we can see from the equilibrium, the player who decides first is clearly at an advantage (first-mover advantage), as her share will always be higher than that of the second mover.

It also shows that the higher a player's relative discount factor is, the higher his share generally will be. In other words, a low preference for present consumption or a high patience is manifested here in the form of a more advantageous distribution. It is also noticeable that the higher and more similar the discount factors are, the more evenly the distribution occurs. For instance, if $\delta_A = \delta_B = 0.99$, Alice receives 53% and Bob receives 47%, while at $\delta_A = \delta_B = 0$, it is very unequal with 100% for Alice and 0% for Bob.

7.2.12 Signaling Game

Problem text in Sect. 7.1.12.

Part a)

The game tree can be represented as in Sect. 6.3, Fig. 6.4. Our usual tree representation is also possible (but unusual for signal games), nevertheless we would like to use it here—to also show that it is a little more chaotic. The game starts with the random assignment of productivity by nature. In the second stage, the worker (W) can choose between the two education levels 0 and e^*. Subsequently, the employer (E) can decide whether to pay attention to the signal (a) or ignore it (i):

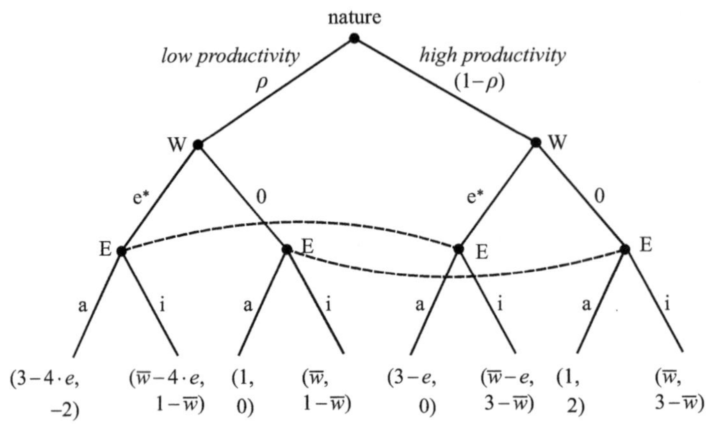

Here, $\overline{w}$ denotes the expected wage which results from the expected productivity, $\overline{w} = E(m) = \rho + 3 \cdot (1 - \rho)$.

For a separating equilibrium to occur, the productive worker must have an incentive to invest in education, while the unproductive worker refrains from doing so. Thus, education, to have a signaling effect, must not only be too expensive, so that no worker takes it but also not too cheap, so that every worker invests in it. The specification of the education in this example can meet these conditions, as it is more costly for the unproductive worker than for the productive type (the unproductive type must spend four times as much as the productive type, $c^l = 4 \cdot e = 4 \cdot c^h$). However, this requires the employer to pay attention to the signal, i.e., if he observes a training ($e = e^*$), he should assume high productivity and pay the wage $w^h = m^h = 3$ to the worker; and with no training ($e = 0$), he should assume low productivity and accordingly pay a wage of $w^l = m^l = 1$. A credible signal thus requires:

- A productive worker ($m^h = 3$) will invest in training if her payoff with training, $3 - e$, is higher than without training, 1. The signal may therefore cost at most $e_h^* \leq 2$.
- An unproductive worker ($m^l = 1$) will not invest in training if her payoff with training, $3 - 4 \cdot e$, is lower than without training, 1. The signal must therefore cost at least $e_l^* \geq 0.5$.

As long as $e^* \in [0.5, 2]$ applies, a separating equilibrium exists in which the productive type sends the signal, and the unproductive type does not.

However, the condition just derived is necessary, but not sufficient for a separating equilibrium, because the productive type might still prefer not to invest in training if she is better off with the average wage. To make this easier to understand, let us assume that there are only productive types. In this situation, none of the productive workers will invest in training: The average wage corresponds to the expected productivity $E(m) = 0 + 3 \cdot (1 - 0) = 3$. At the same time, however, education causes costs without increasing wages, so education is simply not worthwhile. Whether education is worthwhile, therefore, crucially depends on the expected productivity and the proportion of productive workers. These will only invest in training if their achievable payoff, $3 - e$, is higher than the average wage, $\overline{w} = \rho + 3 \cdot (1 - \rho)$. Thus, the maximum cost of the signal is $e \leq 2 \cdot \rho$. This means, if the education costs exceed $2 \cdot \rho$, no worker has an incentive to invest in education and a pooling equilibrium occurs.

Part b)

If a separation equilibrium is possible, it is not plausible for two employers to have a pooling equilibrium where employers do not pay attention to the signal. This is intuitively understandable: If an employer pays attention to the signal, all productive workers will come to him, as they receive a higher net pay there. Conversely, all unproductive workers will go to the employer who does not pay attention to the signal, as the average wage, which considers the entire population, is higher than the wage of the unproductive workers. As a result, the employer who does not pay attention to the signal incurs losses—he pays too high wages in comparison to his workers' productivity. Therefore, every employer will pay attention to the signal and a separating equilibrium will occur.

Industrial Economics Application Examples

8

8.1 Problems

8.1.1 Format War

Two companies, Queenston (player A) and Eastern Digital (player B), would like to establish a new format for digital memory cards on the market. The options are either the newly developed technology "UltraStore" by Queenston or the technology "HyperStash" already tested in field trials by Eastern Digital.

Essentially, a company has a higher payoff than its competitor if "its" technology becomes the standard. However, Queenston's specific payoffs also depend on whether the management makes a specific investment of V in the further development of its own technology "UltraStore" prior to the companies' technology choice. In the case of success, i.e., its technology becomes the standard, Queenston achieves a payoff of $2+V$. In the case of failure, i.e., Eastern Digital's technology prevails, the investment must be written off and the payoff is $4-V$. To simplify the further analysis, it is assumed on the one hand that the profits of Eastern Digital are not directly influenced by the investment, and on the other hand, the payoffs in the case of a "format war" (i.e., the choice of different technologies) are normalized to zero. The following matrix then gives the payoffs depending on the technology choice, where s_{i1} stands for the strategy "UltraStore" and s_{i2} for the strategy "HyperStash" of player $i = A, B$:

A, B	s_{B1}	s_{B2}
s_{A1}	(2 + V, 1)	(0, 0)
s_{A2}	(0, 0)	(4−V, 5)

In the following, assume positive investments in the interval $4 > V \geq 0$ and assume that the technology choice of the companies is simultaneous and without agreement.

F. Bartholomae and M. Wiens, *Game Theory and Applications*, Classroom Companion: Economics, https://doi.org/10.1007/978-3-658-44608-6_8

a. Find all Nash equilibria in pure strategies by determining the best responses $r_i(s_j)$ (reaction correspondences)! Is the present game situation a coordination or a cooperation problem? Which focal points can be derived from the payoffs depending on V? Determine the value of V for which the game has the structure of the game "battle of the sexes"!
b. Is there a further Nash equilibrium in mixed strategies in this game? If yes, determine this equilibrium depending on V! Draw the reaction correspondences for $V = 3$ and mark all Nash equilibria! Which Nash equilibrium do you think is a plausible solution? Does your statement depend on V?
c. Now assume that the technology choice is sequential. Draw the resulting two-stage game with perfect information in extensive form for $V = 3$ if Eastern Digital (player B) moves first! Determine the subgame perfect equilibrium of this game! What is a "first-mover advantage"? Is there one here? Which two properties must be fulfilled in the technology choice under real conditions so that the situation is accurately described by the two-stage game with perfect information?
d. Now assume that V can take any positive value. How could Queenston (player A) achieve that there is no longer a first-mover advantage for Eastern Digital? Why is the result independent of who moves first? Provide all solutions of the sequential and simultaneous game depending on V and the order of moves!

Solution in Sect. 8.2.1.

8.1.2 Price Versus Quantity Competition

Two companies compete in a duopoly market with differentiated products. The demand functions for the two companies i, j are each given by $x_i(p_i, p_j) = 30 - p_i + 0.5 \cdot p_j$. To simplify the calculations, it is assumed that average costs are constant at zero.

a. Determine graphically and numerically the reaction curves and the Nash equilibrium of the simultaneous game in case of price competition!
b. Determine graphically and numerically the reaction curves and the Nash equilibrium of the simultaneous game in case of quantity competition!
c. Compare your results from (a) and (b)!
d. Use the reaction functions to explain how the Nash equilibrium can result from a dynamic adjustment process!
e. Is the Nash equilibrium Pareto-optimal? Prove your answer!

Solution in Sect. 8.2.2.

8.1.3 Location Choice (Hotelling Problem)

Two hot dog vendors are competing on a beach. There are a total of 60 consumers, who are evenly distributed over the beach with a length of one. Each consumer wants to buy exactly one hot dog. Since price is legally prohibited and quality competition is not worthwhile for the two vendors, the hot dog vendors only compete for location. The consumers will always buy at the nearest hot dog stand. The utility of a vendor is given by the number of the customers. If both stands are exactly in the same place, the customers share equally.

a. First consider the following discrete distribution of consumers along the beach:

West	Center	East
20 people	20 people	20 people

Represent this game (players, strategies, payoffs) in strategic form (matrix) and solve it! Is the solution unique?

b. Show that the positioning of both vendors in the middle (median[1] of the distribution) represents the unique Nash equilibrium of the game!
c. Illustrate your result from (b) graphically using the reaction functions!

Solution in Sect. 8.2.3.

8.1.4 Strategic Manager Compensation

On a homogeneous market, two companies with identical costs $C(x_i) = 5 + 6 \cdot x_i$ face each other in quantity competition. The market demand is described by the price-response function $p = 24 - X$, where X denotes the total sales in the market.

a. Determine the reaction functions and calculate the Nash equilibrium of the simultaneous game and the payoffs achieved!
b. Determine the subgame perfect equilibrium and the resulting payoffs if company A sets the quantity in stage 1 bindingly and observably for company B, which chooses its quantity in stage 2!
c. Company A is run by a manager who receives the wage $w = s \cdot \pi_A + (1 - s) \cdot R_A + w^0(s)$ from the owner, where π_A denotes the profit and R_A the revenue of the company. $s \in [0, 1]$ represents a weighting factor between these two

[1] The median refers to the value that halves a distribution, meaning, 50% of the values in the distribution are smaller and 50% are larger than this value. It is also referred to as the central value or 0.5-quantile.

values. The residual wage $w^0(s)$ is chosen in a way that the manager always receives her reservation wage, that is $w(x_A, x_B, s) = w^R$ (this ensures that the manager always wants to work for the owner). A two-stage game is considered, in which in the first stage the weighting factor s is set by the owner in a binding and observable way, and in the second stage the manager and company B simultaneously determine the sale quantities, with the manager orienting herself on her payoff function $w(x_A, x_B, s)$. Determine the subgame perfect equilibrium of this two-stage game and compare it with (b)!

Solution in Sect. 8.2.4.

8.1.5 Strategic Investment

Two companies are competing in a homogeneous market. The price-response function is $p(X) = 100 - X$. For each unit produced, company i incurs costs of c_i. Both companies can reduce the variable production costs through investment of r_i in research and development (R&D): $c_i(r_i, r_j) = 50 - r_i - s \cdot r_j$ with $i \neq j$, $i = 1, 2$.

The parameter s measures the (positive) external effect for company i when company j invests in (R&D). Such a positive external effect can occur, e.g., when research results cannot be kept secret or a company invests in public research institutions (universities), from which other companies also benefit. However, research and development also cause costs for the company. It is assumed that a certain level of R&D causes disproportionately high costs, i.e., the research laboratories are operated with decreasing returns to scale: $0.5 \cdot r_i^2$.

The game runs in two stages: In the first stage, both companies choose their R&D investments λ_i simultaneously. In the second stage, they compete in terms of quantity and simultaneously decide on their optimal output quantity x_i (Cournot competition).

a. Determine the subgame perfect equilibrium of this two-stage game for $s = 0.5$!
b. How does the spillover s influence the result in general?

Solution in Sect. 8.2.5.

8.1.6 Company and Competition Authority

In most countries there are antitrust laws that prohibit certain actions by dominant companies as anti-competitive (e.g., exclusivity obligations). Against this background, we analyze the following stylized situation: Let us assume that there are two players, a company and the competition authority. The company may behave in an anti-competitive manner, thereby generating profit gain of $g > 0$ (otherwise the profit is 0). The competition authority can control the behavior of the company (incurring costs of $c > 0$). If the control reveals anti-competitive behavior, the company must pay a fine of $f > g$, and the utility of the competition authority is $u - c > 0$.

a. Assume that the competition authority knows before the decision about the control whether the company has behaved anti-competitively. Draw the game tree! What is the subgame perfect equilibrium? Would your analysis change if the authority would first commit to its decision (observable for the company)?
b. Now assume that both players make their decision simultaneously. Explain why no equilibrium in pure strategies exists! Determine the equilibrium in mixed strategies and explain how this depends on the parameters! Graphically represent the reaction functions and identify the Nash equilibrium! How does a change in the fine affect this equilibrium?

Solution in Sect. 8.2.6.

8.1.7 Collusion Among Many Companies

There are n companies in a market that produce a homogeneous good with identical and constant marginal costs c and compete in price. With collusion, they could achieve a total profit of π^K.

a. Show that $p = c$ represents the Nash equilibrium in price competition (Bertrand paradox)! For what discount factors δ is it possible to realize collusion as equilibrium in the infinitely repeated game with the grim trigger strategy? How does the critical discount factor depend on the number of companies in the market?
b. Assume that the market grows at a rate g each period, where $g \cdot \delta < 1$. What statement does this model make about the question of whether collusion is more likely in growing or shrinking markets?

Solution in Sect. 8.2.7.

8.1.8 Cartels and Coordinated Behavior Among Many Companies

In most cases, a mere cartel prohibition, i.e., a prohibition of contractually agreed collusion, is not sufficient. Companies manage to behave collusively even without contractual arrangements. For this reason, a ban on coordinated behavior must also be established. Let us consider why this is the case.

In a market with demand $X = 24 - p$, n companies compete in Cournot competition. These produce at identical and constant average costs $AC = 0$.

a. Determine the optimal quantity and profit for n companies! Complete for $n = 2$ the profits for the following payoff matrix:

A, B	Collusion	Cournot
Collusion	(π_A^{MM}, π_B^{MM})	(π_A^{MC}, π_B^{MC})
Cournot	(π_A^{CM}, π_B^{CM})	(π_A^{CC}, π_B^{CC})

Which strategy combination will be established in the static game, if (i) the companies can conclude a binding contract, (ii) this is not possible due to the cartel prohibition? Is a ban on coordinated behavior necessary?

b. Alternatively, assume that the $n = 2$ companies make their sales decisions anew in each period and can observe the decision of the other company in the previous period. Formulate the grim trigger strategy for this! For which discount factors δ can collusion in the infinitely repeated game be realized as equilibrium? Let the companies' discount factor be (i) $\delta^l = 0.25$ or (ii) $\delta^h = 0.75$. Is a ban on coordinated behavior in both cases necessary? Can collusive behavior be realized by such a strategy if the companies are only active in the market for a finite number of periods?
c. Determine the optimal deviation of a company if the remaining $n - 1$ companies behave collusively! How does the willingness of companies to behave collusively change when the number n of companies increases? Assume additionally that the market grows each period by a factor of g, where $g \cdot \delta < 1$. Is collusion thus more likely in growing ($g > 1$) or shrinking ($g < 1$) markets?

Solution in Sect. 8.2.8.

8.1.9 Advertising

The two companies Pear and Sumsang are competing in the tablet market. The demand for Sumsang's product is less price elastic, but the customers' maximum willingness to pay is lower in the initial situation. Specifically, the price-response functions are given by $p_S(x_S, x_P) = 100 - x_S - x_P$ and $p_P(x_S, x_P) = 230 - 2 \cdot x_P - x_S$. Both companies produce with identical costs $C_i = 400 + 10 \cdot x_i$. Sumsang now can invest in advertising measures a, observable by Pear, at a cost of $0.5 \cdot a^2$, which increases the maximum willingness to pay of the customers in $p_S(x_S, x_P)$ from 100 to $100 + a$. Subsequently, the companies simultaneously determine their quantities.

a. Formulate this situation as a game: Who are the players? What strategies are available to them? What are the payoff functions depending on the chosen actions? Characterize the temporal structure and the information structure of the game!

b. Now determine the reaction functions of the two players in competition in the sales market depending on a!
c. Calculate the equilibrium quantities depending on a! Graphically represent the equilibrium for $a = 0$! Use the graph to explain why the requirements formulated in the equilibrium concept are fulfilled!
d. There are three possible advertising packages: $a_l = 0$, $a_m = 70$, and $a_h = 105$. Which package should Sumsang choose? What impact does the choice of package have on Pear's profits? Show in your graph in (c) how the reaction functions and thus the equilibrium changes due to the advertising investment! What would happen if Pear could not observe the advertising investment and therefore continued to produce the original sales quantity?
e. Suppose the success of the advertising measure is unclear, i.e., it can increase the willingness to pay as stated or it can simply fizzle out. Since Sumsang carries out the advertising measures, it knows about the effectiveness of the advertising. Briefly explain why the refinement concept used for sequential games with complete information no longer works with incomplete information and what characterizes the equilibrium concept used for dynamic games with incomplete information! Suppose Pear observes that Sumsang is running the advertising. Discuss whether it can automatically conclude from this that the advertising measures are effective!

Solution in Sect. 8.2.9.

8.1.10 Lobbying

To support an industry, a country's government plans to implement subsidy measures. However, it does not know how much money is really needed. It therefore relies on additional information from the industry. To avoid a "wrong" decision by the government, the industry's two dominant corporations A and B decided to provide the government with "advisors". The granted subsidies depend on how much consulting the two corporations conduct, $90 \cdot (h_A + h_B) + h_A \cdot h_B$, where h denotes the number of advisory sessions. This subsidy is evenly divided between the two corporations. However, the on-site advisors also incur costs for the corporation. Since the advisors always must compile new information, the costs increase exponentially. In total, costs of $C_i(h_i) = h_i^2$ with $i = 1, 2$ are incurred.

a. Formulate this situation as a game: Who are the players? What strategies can they choose? Consider a simultaneous game with discrete strategies. Assume that both corporations can choose between thirty sessions ($h_i = 30$) and ninety sessions ($h_i = 90$). Determine the payoffs for all strategy combinations and represent the resulting game in matrix form! Also determine all Nash equilibria of the game! Is the equilibrium Pareto-optimal?
b. Consider now continuous strategies. Determine the reaction functions! Find the Nash equilibrium of the simultaneous game including the payoffs achieved!

c. The companies decide to abandon their competing information policy and instead jointly advise the government. Determine the total number of advisory sessions the government must endure! What payoffs do the companies now receive? What are external effects? Based on your results, argue what kind of external effects must have existed before the joint advice! What statement can you make about the old Nash equilibrium?

Solution in Sect. 8.2.10.

8.1.11 Holdup Problem

The start-up Blubberama plans to build an online platform for management seminars on the Internet, where young executives can access the experience of qualified business experts in online courses and private chat rooms. Since special software is needed to build the platform, the software company Nerdsoft has been commissioned to create such a program.

Nerdsoft is considering whether it should develop a corresponding software for Blubberama or better refrain from it. The costs for the development of the software would be $c = 200{,}000$ euros; if development is waived, Nerdsoft incurs no costs, but also no value is created for Blubberama. Blubberama estimates the value of the software for itself at $v = 800{,}000$ euros. However, Blubberama is not willing to conclude a preliminary contract for the creation of the software, as it does not know whether Nerdsoft will then actually deliver a suitable software. It only makes the promise that it will make a purchase offer after the development of the software. The offer can either be high, so that Blubberama pays half of the value ($v/2$) to Nerdsoft, or it is low and Blubberama only pays 100,000 euros. The offer is binding and Nerdsoft can therefore only decide whether to accept the offer or not. If Nerdsoft accepts, it receives the offered purchase price, if it declines, it receives no money, as the platform is company-specific and cannot be sold to anyone else.

a. Formulate this situation as a game: Who are the players? What possible actions are available to them? Characterize the information structure and represent the payoffs depending on the chosen strategy combinations in a suitable form! Solve the game and state the equilibrium strategies completely!
b. Suppose Blubberama could commit before the start of the game to pay a penalty of f into a fund to support exploited programmers if it makes a low offer. Represent the modified game! How high must the penalty be so that the equilibrium from (a) does not result? Does Blubberama have an incentive to make such a voluntary commitment?
c. Now assume that contrary to the previous assumption, the software could also be used by another company. This company offers Nerdsoft a payment of 300,000 euros for exclusive use, regardless of Blubberama's bid. What implications arise for the game? Explain in a nutshell if the competition is harmful

to Blubberama in this situation! What conclusion do you come to if the other company offers 600,000 euros?

Solution in Sect. 8.2.11.

8.1.12 Competition with Incomplete Information

In a Cournot duopoly with linear price-response function $p(X) = 12 - X$, where $X \equiv x_A + x_B$, the costs of company A are common knowledge, while the costs of company B are only known to this company. Specifically, $C_A(x_A) = 3 \cdot x_A$ (i.e., there are constant unit costs of 3). With probability ρ the marginal costs of company B are 100% higher, i.e., $C_B^h(x_B) = 6 \cdot x_B$, and with probability $(1 - \rho)$ they are 100% lower, i.e., $C_B^l(x_B) = 0$.

a. Determine the reaction function of company A (based on the probability assessment regarding company (B) and for each of the two types of company B!
b. Since company A does not know the actual type of company B, it does not base its own quantity on the reaction functions of one of the two types of company B, but on a fictitious "expected" reaction of company B (the quantities of the two types are each weighted with the corresponding probability). When determining their sales quantity, the corresponding types of company B must consider this. Determine graphically and numerically the Bayes-Nash equilibrium, i.e., the equilibrium sales quantities of company A and the two types of company B, for $\rho = 0.5$!
c. Compare your result with incomplete information from (b) with the one that would result with complete information about the costs of company B! What can be said about efficiency in each case?

Solution in Sect. 8.2.12.

8.2 Solutions

8.2.1 Format War

Problem Text in Sect. 8.1.1.

Part a)
For $0 \leq V < 4$ the reaction correspondences are

$$r_A(s_{B1}) = \{s_{A1}\} \qquad r_A(s_{B2}) = \{s_{A2}\}$$
$$r_B(s_{A1}) = \{s_{B1}\} \qquad r_B(s_{A2}) = \{s_{B2}\}$$

Thus, there are two Nash equilibria in pure strategies, (s_{A1}, s_{B1}) and (s_{A2}, s_{B2}). This is a coordination problem, as it involves aligning common interests. Depending on V, the individual equilibria are preferred to varying degrees by the players.

As an exogenous focal point of the game, (s_{A2}, s_{B2}) is a possibility, as this would involve the use of Eastern Digital's already tested technology. For $V < 1$, this equilibrium can also be recommended as an endogenous focal point of the game.

In the "battle of the sexes", there is a symmetrical payoff structure for both players. For $V = 3$, such a structure can be achieved:

A, B	s_{B1}	s_{B2}
s_{A1}	(5, 1)	(0, 0)
s_{A2}	(0, 0)	(1, 5)

Part b)

Since the number of Nash equilibria is always odd and we found only two in pure strategies, a third equilibrium must exist in mixed strategies. Let p_A be the probability that player A chooses s_{A1} and p_B the probability that player B chooses s_{B1}. Player A prefers to choose s_{A1} (i.e., $p_A = 1$) instead of s_{A2} if

$$\begin{aligned} E[u_A(p_A = 1, p_B)] &> E[u_A(p_A = 0, p_B)] \\ (2 + V) \cdot p_B + 0 \cdot (1 - p_B) &> 0 \cdot p_B + (4 - V) \cdot (1 - p_B) \\ 2 \cdot p_B + V \cdot p_B &> 4 - 4 \cdot p_B - V + V \cdot p_B \\ p_B &> (4 - V)/6 \end{aligned}$$

Correspondingly, player B prefers to choose s_{B1} (i.e., $p_B = 1$) instead of s_{B2} if

$$\begin{aligned} E[u_B(p_A, p_B = 1)] &> E[u_B(p_A, p_B = 0)] \\ 1 \cdot p_A + 0 \cdot (1 - p_A) &> 0 \cdot p_A + 5 \cdot (1 - p_A) \\ p_A &> 5/6. \end{aligned}$$

Therefore, the reaction correspondences of the two players are as follows:

$$p_A(p_B) = \begin{cases} 1 & p_B \in \left(\frac{(4-V)}{6}, 1\right] \\ [0, 1] & p_B = \frac{(4-V)}{6} \\ 0 & p_B \in \left[0, \frac{(4-V)}{6}\right) \end{cases}, \quad p_B(p_A) = \begin{cases} 1 & p_A \in \left(\frac{5}{6}, 1\right] \\ [0, 1] & p_A = \frac{5}{6} \\ 0 & p_A \in \left[0, \frac{5}{6}\right) \end{cases}$$

Thus, $\left(p_A = \frac{5}{6}, p_B = \frac{4-V}{6}\right)$ results as a Nash equilibrium in mixed strategies.

For $V = 3$, we obtain the following graphical representation of the reaction correspondences:

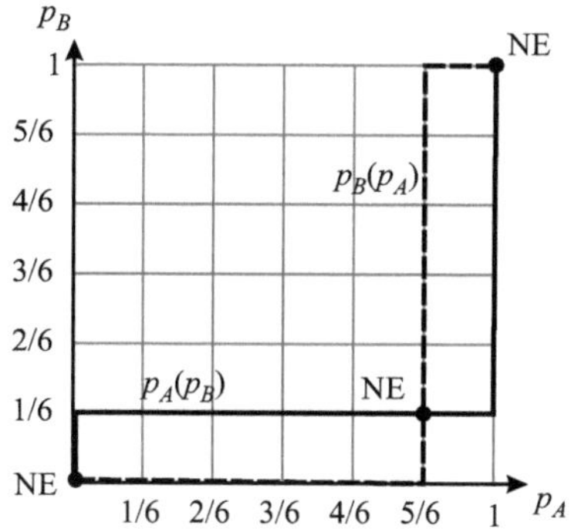

In this case, the Nash equilibrium in mixed strategies seems not plausible, as it can happen that both choose a different technology—which is precisely not their intention. This is also not an anti-coordination game (inspection game) where such an equilibrium would be more plausible. The statement about the plausibility of one of the equilibria is dependent on V, as it was shown in (a) that for $V < 1$ an endogenous focus point exists, which suggests (s_{A2}, s_{B2}) as an equilibrium.

Part c)
For $V = 3$, we obtain the following game tree:

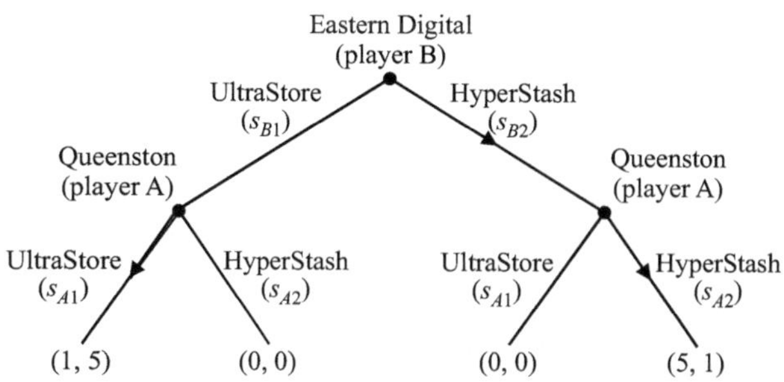

By backward induction, we find the strategy combination $\{\{s_{A1}(s_{B1}), s_{A2}(s_{B2})\}, \{s_{B2}\}\}$ as the subgame perfect Nash equilibrium of the game.

The first-mover advantage states that the player who moves first has an advantage, as he can restrict the strategy space of the following player and thus bring about an outcome that is advantageous for him. In the present case, there is such an advantage, as Eastern Digital can use its first move to set the technology standard that is more advantageous for it.

The technology choice must be irreversible, i.e., the company cannot change its choice afterwards, e.g., by recouping investments made through sales. Furthermore, the technology choice must be observed by the other company, i.e., Queenston must receive information about Eastern Digital's choice.

Part d)
If $V > 4$, s_{A1} would be a strictly dominant strategy for Queenston, i.e., Queenston would always choose s_{A1} regardless of Eastern Digital's decision. It also does not matter whether the decisions are made simultaneously or sequentially. Thus, Eastern Digital's first-mover advantage no longer exists:

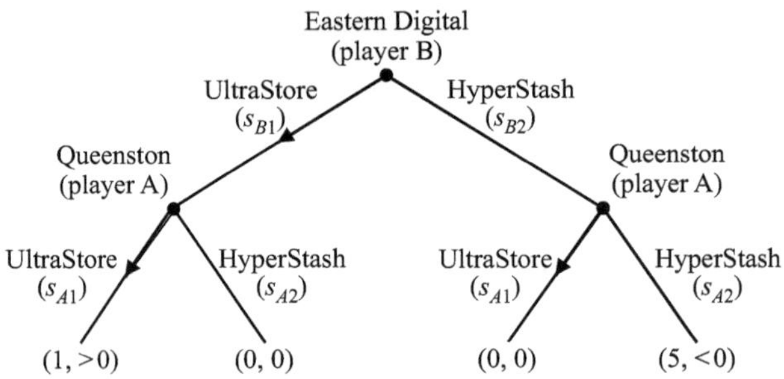

This results in the following solutions depending on V:

- $0 \leq V \leq 4$:
  - Simultaneous decision:(s_{A1}, s_{B1}),(s_{A2}, s_{B2}), $\left(p_A = \frac{5}{6}, p_B = \frac{4-V}{6}\right)$
  - Eastern Digital moves first: $\{\{s_{A1}(s_{B1}), s_{A2}(s_{B2})\}, \{s_{B2}\}\}$
  - Queenston moves first: $\{\{s_{A1}\}, \{s_{B1}(s_{A1}), s_{B2}(s_{A2})\}\}$
- $V > 4$:
  - Simultaneous decision: (s_{A1}, s_{B1})
  - Eastern Digital moves first: $\{\{s_{A1}(s_{B1}), s_{A1}(s_{B2})\}, \{s_{B1}\}\}$
  - Queenston moves first: $\{\{s_{A1}\}, \{s_{B1}(s_{A1}), s_{B2}(s_{A2})\}\}$

8.2.2 Price Versus Quantity Competition

Problem Text in Sect. 8.1.2.

Part a)
In price competition, the price is the strategic variable chosen by the companies. Therefore, we must set up the profit function (payoff function) in such a way that the profit depends on the price. This is already fulfilled by the given demand function. The information about the cost function is completely given by the assumption of constant average costs of zero: average costs of zero imply both fixed costs and marginal cost of zero: $MC = (FC + x \cdot VC)/x = FC/x + VC = 0$—this condition can only be met if both the fixed and the marginal costs are zero.

The relevant profit function for company i is therefore

$$\pi_i = p_i \cdot x_i(p_i, p_j) = p_i \cdot (30 - p_i + 0.5 \cdot p_j).$$

Due to symmetry, the same profit function applies to both companies, which means that it is sufficient to consider only one company in the further analysis. To determine the profit maximum, we calculate the first-order condition with respect to p_i,

$$\frac{\partial \pi_i}{\partial p_i} = 30 - 2 \cdot p_i + 0.5 \cdot p_j \stackrel{!}{=} 0.$$

The second derivate is $-2 < 0$, which ensures a maximum. To determine the reaction function, it is now necessary to solve the condition for the strategic variable of player i, i.e., p_i,

$$p_i(p_j) = 15 + 0.25 \cdot p_j$$

As we see, the resulting reaction function depends positively on the strategic variable of the other company, p_j. Due to the symmetry, we can conclude the reaction function of company j as $p_j(p_i) = 15 + 0.25 \cdot p_i$. To determine the Nash equilibrium, we substitute the reaction functions of the two companies into each other:

$$p_i = 15 + 0.25 \cdot (15 + 0.25 \cdot p_i) = 18.75 + 0.0625 \cdot p_i$$
$$\Rightarrow p_i^* = 20\left(= p_j^*\right)$$

The Nash equilibrium is therefore $(p_i = 20, p_j = 20)$. Substituting this into the demand function results in the quantities,

$$x_i(p_i = 20, p_j = 20) = 30 - 20 + 0.5 \cdot 20 = 20 = x_j,$$

and profits of the companies,

$$\pi_i = 20 \cdot (30 - 20 + 0,5 \cdot 20) = 400 = \pi_j.$$

In the following figure, we see the reaction functions of the two companies and the resulting Nash equilibrium.

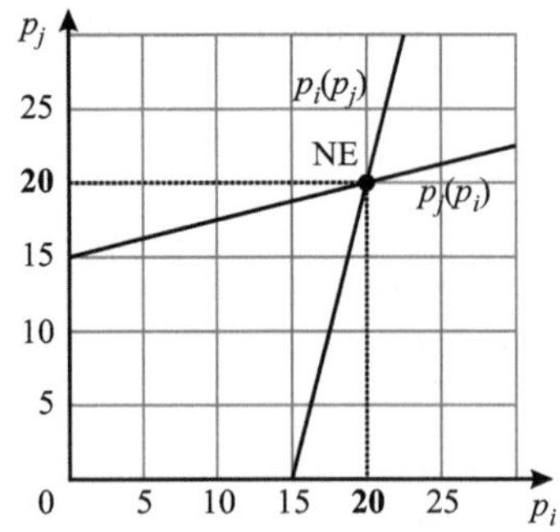

Part b)

In quantity competition, the quantity is the strategic variable chosen by the companies. Therefore, the profit function must be set up in such a way that the profit depends on the quantity. Since the given demand functions depend on the prices, we first must calculate the price-response function. To do this, we solve the system of equations

$$x_i(p_i, p_j) = 30 - p_i + 0.5 \cdot p_j$$
$$x_j(p_j, p_i) = 30 - p_j + 0.5 \cdot p_i$$

with respect to the quantities. Such a system of equations can be solved in various ways. One possibility is to solve $x_j(p_j, p_i)$ for p_j and substitute it into $x_i(p_i, p_j)$:

$$p_j = 30 - x_j + 0.5 \cdot p_i$$
$$x_i = 30 - p_i + 0.5 \cdot (30 - x_j + 0.5 \cdot p_i) = 45 - 0.5 \cdot x_j - 0.75 \cdot p_i$$
$$\Rightarrow p_i = 60 - \frac{4}{3} \cdot x_i - \frac{2}{3} \cdot x_j$$

Similarly, $p_j(x_j, x_i)$ can be determined. Due to the symmetry, $p_j = 60 - 4/3 \cdot x_j - 2/3 \cdot x_i$.

Since we now know the relevant price-response functions, we can determine the profit function:

$$\pi_i = p_i(x_i, x_j) \cdot x_i = \left(60 - \frac{4}{3} \cdot x_i - \frac{2}{3} \cdot x_j\right) \cdot x_i$$

To determine the profit maximum, we calculate the first-order condition with respect to x_i,

$$\frac{\partial \pi_i}{\partial x_i} = 60 - \frac{8}{3} \cdot x_i - \frac{2}{3} \cdot x_j = 0. \quad (8.1)$$

The second derivative is $-8/3 < 0$, which ensures a maximum. To determine the reaction function, it is necessary to solve the condition for the strategic variable of player i, x_i:

$$x_i(x_j) = 22.5 - 0.25 \cdot x_j$$

As we see, the resulting reaction function depends negatively on the strategic variable of the other company, x_j. Due to symmetry, we can conclude the reaction function of company j as $x_j(x_i) = 22.5 - 0.25\cdot_i$. To determine the Nash equilibrium, we substitute the reaction functions of the two companies into each other:

$$\begin{aligned} x_i &= 22.5 - 0.25 \cdot (22.5 - 0.25 \cdot x_i) = 16.875 + 0.0625 \cdot x_i \\ &\Rightarrow x_i^* = 18\left(= x_j^*\right) \end{aligned}$$

The Nash equilibrium is therefore $(x_i = 18, x_j = 18)$. The equilibrium prices are obtained by substituting the quantities into the price-response function:

$$p_i(x_i = 18, x_j = 18) = 60 - \frac{4}{3} \cdot 18 - \frac{2}{3} \cdot 18 = 24 = p_j$$

The profits of the companies are determined by substituting the quantities into the profit function:

$$\pi_i = \left(60 - \frac{4}{3} \cdot 18 - \frac{2}{3} \cdot 18\right) \cdot 18 = 432 = \pi_j.$$

Finally, we plot the reaction functions of the two companies and the resulting Nash equilibrium in the following figure:

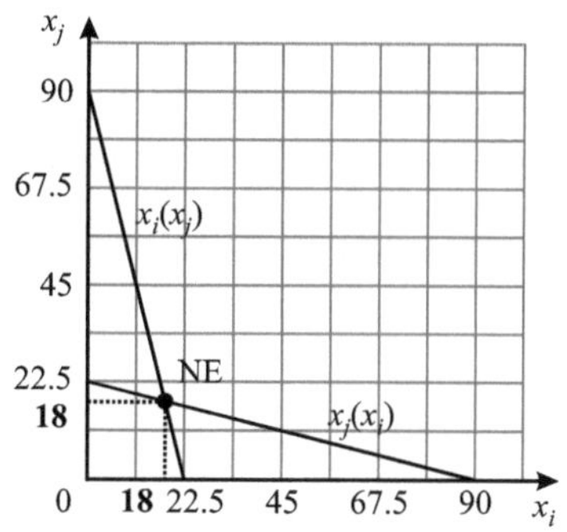

Part c)

When we compare the results from a) and b), we can identify two differences: On the one hand, the reaction functions run differently: While they have a positive slope in price competition, they run downwards in quantity competition. On the other hand, despite identical demand, the companies show a lower profit in price competition than in quantity competition.

The different slope of the reaction functions results from the different relationships of the strategic variables to each other. We can already see in the demand function that, in the case of price competition, an increase in the competitor's price has a positive effect on the quantity sold. This is also reflected later in the reaction function: The reaction to a competitor's higher price is also a price increase (for which there is potential). In this case, we speak of *strategic complements*—the strategic variables behave in the same direction.

The opposite is the case with quantity competition: here, the price reaction function shows that an increase in the competitor's quantity lowers the own price. As a result, according to the reaction function, the company reacts to an increase in the competitor's quantity by reducing its own quantity (to stabilize the price). This is referred to as *strategic substitutes*—the strategic variables run in opposite directions.

The different profits in both forms of competition can be explained by the varying intensity of competition: In price competition, the competition is significantly more intense than in quantity competition. We see this clearly in the equilibrium prices: In price competition, they are at 20, whereas in quantity competition, they are at 24.

Part d)

We want to show now how the Nash equilibrium results from a dynamic adjustment process. Let us start with the extreme situation where company i produces nothing and company j therefore offers the monopoly quantity of 22.5 (this point is on the reaction function of company j when company i offers zero). However, the point $\left(x_i = 0, x_j = 22.5\right)$ does not lie on the reaction function of company i. According to its reaction function, given the quantity $x_j = 22.5$, company i will offer $x_i = 22.5 - 0.25 \cdot 22.5 = 16.875$ which results in the quantity combination $\left(x_i = 16.875, x_j = 22.5\right)$. However, company j will react to a quantity of $x_i = 16.875$ with the quantity $x_j = 22.5 - 0.25 \cdot 16.875 \approx 18.281$, so that in the second step the quantity combination $\left(x_i = 16.875, x_j = 18.281\right)$ is established. The following table illustrates the adjustment process:

x_i	x_j	Adjustment by
0	22.500	–
16.875	22.500	i
16.875	18.281	j
17.930	18.281	i
17.930	18.018	j

(continued)

(continued)

x_i	x_j	Adjustment by
17.996	18.018	i
17.996	18.001	j
18.000	18.001	i
18.000	18.000	j

As we see, after several adjustments by the players based on their reaction functions, a stable equilibrium at $(x_i = 18, x_j = 18)$, the Nash equilibrium, is established. The adjustment illustrates the idea of the Nash equilibrium: It is the pair of mutually best responses, where no player has an incentive to deviate.

Part e)

In general, a Nash equilibrium is Pareto-optimal when there are no external effects. Thus, we need to show that a payoff combination is possible in which both players are better off than in the Nash equilibrium. For this, we do not maximize the individual payoffs, but the total payoff. In practice, this would correspond to a cartel agreement or a merger between companies. We want to calculate this exemplarily for the case of quantity competition.

The total profit in quantity competition is

$$\pi_{\text{total}} = \pi_i + \pi_i = \left(60 - \frac{4}{3} \cdot x_i - \frac{2}{3} \cdot x_j\right) \cdot x_i + \left(60 - \frac{4}{3} \cdot x_i - \frac{2}{3} \cdot x_j\right) \cdot x_j$$

If we optimize over both output quantities, i.e., partially derive with respect to x_i and x_j, we obtain the following first-order conditions:

$$\frac{\partial \pi_{\text{total}}}{\partial x_i} = 60 - \frac{8}{3} \cdot x_i - \frac{2}{3} \cdot x_j - \frac{2}{3} \cdot x_j \stackrel{!}{=} 0$$
$$\frac{\partial \pi_{\text{total}}}{\partial x_j} = 60 - \frac{8}{3} \cdot x_j - \frac{2}{3} \cdot x_i - \frac{2}{3} \cdot x_i \stackrel{!}{=} 0.$$

We can already deduce from this that there must be a negative external effect. Compared to the first-order condition when considering only one company, (8.1), the quantity decision of the other company is taken more into account—as can be seen from the additional term $-2/3 \cdot x_j$. Since this is negative, the original output quantity was too high. The more intensive consideration of the quantity of the other company is also reflected in the reaction functions:

$$x_i(x_j) = 22.5 - 0.5 \cdot x_j$$
$$x_j(x_i) = 22.5 - 0.5 \cdot x_i$$

whereas the competitor's quantity was previously only considered with a factor of 0.25 in the own quantity decision, this has now increased to 0.5. The solution results in a production of $x_i = x_j = 15$, a sales price of $p_i = p_j = 30$ and thus profits for both companies $\pi_i = \pi_j = 15 \cdot 30 = 450$. Hence, the profit for both companies is higher than in the Nash equilibrium. Thus, the Nash equilibrium cannot be Pareto-optimal.

8.2.3 Location Choice (Hotelling Problem)

Problem Text in Sect. 8.1.3.

Part a)
The players are the two hot dog vendors (Alice and Bob), their strategies are each the location choices "West", "Center", and "East". The payoffs result from the distribution of consumers. From the allocation rule described in the problem, it follows that consumers will divide equally between the two sellers if they choose the same location, i.e., (West, West), (Center, Center), or (East, East). The payoffs of the two vendors are thus half of all 60 consumers on the beach, (30, 30). If both choose a location at opposite ends, i.e., West and East, they will attract the consumers localized there as customers, whereas the consumers from the middle will be evenly distributed between the two, as they are equally distant. The payoff here is again (30, 30). If a hot dog vendor chooses the center while the other is at one end of the beach (Center and West or East), the vendor in the center will receive both the consumers from the center and those positioned at the opposite end from the other vendor's location. Thus, the vendor in the center has a payoff of 40, while the other vendor can only attract 20 consumers. The described payoffs are summarized in the following matrix:

A, B	West	Center	East
West	(30, 30)	(20, 40)	(30, 30)
Center	(40, 20)	(30, 30)	(40, 20)
East	(30, 30)	(20, 40)	(30, 30)

The matrix shows that "Center" is a strictly dominant strategy for each player. This results in the unique Nash equilibrium (Center, Center).

Part b)
Now, the two players have a continuous strategy choice, meaning they can choose any position between 0 (West of the beach) and 1 (East of the beach). Since the consumers are uniformly distributed, the median of the distribution is at 0.5. Thus, we must show that the strategy combination $\{s_A^* = 0.5, s_B^* = 0.5\}$ represents the Nash equilibrium of the game. To do this, we check whether a unilateral deviation from this strategy combination is profitable. Due to symmetry, it is sufficient to consider only one player.

Let us assume that Bob is located at the center, $s_2^* = 0.5$.

- The following figure shows the case when Alice chooses a value $s_A < 0.5$, thus locating herself west of Bob.

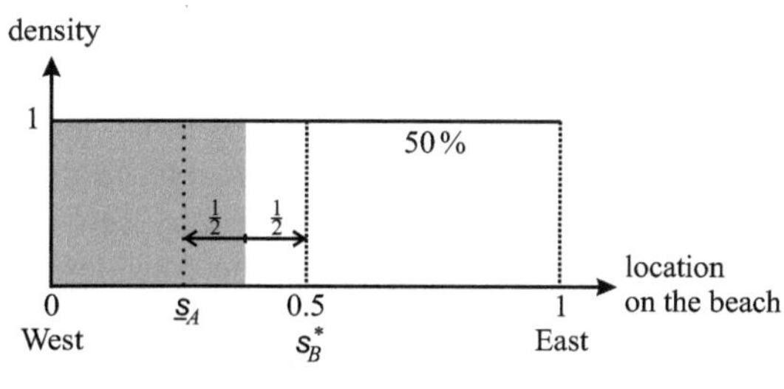

Alice's payoff is represented by the gray area. On the one hand, she receives all customers who are west of her, i.e., in the range $\left[0, \underline{s}_A\right]$, and all customers who are east of her but closer to her location than to Bob's. This applies to all consumers who are west of the midpoint of the distance between her and Bob, i.e., in the range $\left[\underline{s}_A, \frac{1}{2} \cdot \left(s_B^* - \underline{s}_B\right)\right]$. Her payoff is therefore:

$$\left[\underline{s}_A + \frac{1}{2} \cdot \left(s_B^* - \underline{s}_A\right)\right] \cdot 60 = \left[\underline{s}_A + \frac{1}{2} \cdot \left(0.5 - \underline{s}_A\right)\right] \cdot 60 = 15 + 30 \cdot \underline{s}_A$$

We see that the payoff increases in $\underline{s}_A$. Furthermore, in the range $\underline{s}_A \in [0; 0.5)$ it is always less than 30. With $s_A^* = 0.5$, however, her payoff would be 30 and thus higher than at position $s_A < 0.5$.

- If Alice chooses a value $\overline{s}_A > 0.5$ and thus locating herself east of Bob, she reaches all customers who are east of her, i.e., in the range $[\overline{s}_A, 1]$, and all customers who are west of her but closer to her location than to Bob's, i.e., in the range $\left[\frac{1}{2} \cdot \left(\overline{s}_A - s_B^*\right), \overline{s}_A\right]$. Her payoff is therefore:

$$\left[(1 - \overline{s}_A) + \frac{1}{2} \cdot \left(\overline{s}_A - s_B^*\right)\right] \cdot 60 = \left[(1 - \overline{s}_A) + \frac{1}{2} \cdot (\overline{s}_A - 0.5)\right] \cdot 60 = 45 - 30 \cdot \overline{s}_A$$

We see here that the payoff decreases in $\overline{s}_1$. Furthermore, in the range $\overline{s}_A \in (0.5; 1]$ it is always less than 30. Thus, $s_A^* = 0.5$ always yields a higher payoff than $s_A > 0.5$.

In summary, we see that with Bob's location choice of $s_B^* = 0.5$ a deviation from $s_A^* = 0.5$ (center) neither to $s_A < 0.5$ (east) nor to $s_A > 0.5$ (west) is profitable. The strategy combination $\left\{s_A^* = 0.5, s_B^* = 0.5\right\}$ therefore represents a Nash equilibrium.

Part c)
To graphically represent the equilibrium, we need the reaction correspondences, i.e., we need to determine what the optimal reaction looks like given a location choice of the opponent. Basically, there are three options: positioning to the west, east, or exactly at the location of the other vendor. As we have seen, depending on

where the location of the other vendor is, different payoffs result. If Bob is located west of the center/median of the beach, Alice should not position herself west of him, as she can only reach the few consumers west of Bob, for whom the journey to Bob would be too far. Bob, in turn, would serve all consumers to the east of him in addition to the remaining, closer buyers in the west. Alice's best response here is therefore to choose a location to the east of him—and thus closer to the center -, but at the same time as close to him as possible. In this way, she receives all consumers east of her location and loses only very few consumers who are west of her location and east of Bob's location. The consideration is analogous if Bob has positioned himself east of the median. In this case, Alice should position herself to the west but as close as possible to Bob's location. The rule is thus to always choose a location closer to the center than the other, but still very close to the location of the competitor. Only when the other vendor is located in the center, the same location should be chosen because any deviation would make the vendor worse off, as the considerations from (b) have shown.

Formally, the reaction functions of the players are

$$s_A(s_B) = \begin{cases} s_B + \varepsilon & s_B \in [0, 0.5) \\ 0.5 & s_B = 0.5 \\ s_B - \varepsilon & s_B \in (0.5, 1] \end{cases}, \quad s_B(s_A) = \begin{cases} s_A + \varepsilon & s_A \in [0, 0.5) \\ 0.5 & s_A = 0.5 \\ s_A - \varepsilon & s_A \in (0.5, 1] \end{cases}$$

where ε, epsilon, represents a very small value and means that the location choice should be very close to that of the competitor but not directly on their location. The figure graphically represents the reaction functions. These are very close to each other and intersect only at the point ($s_A = 0.5, s_B = 0.5$), the Nash equilibrium.

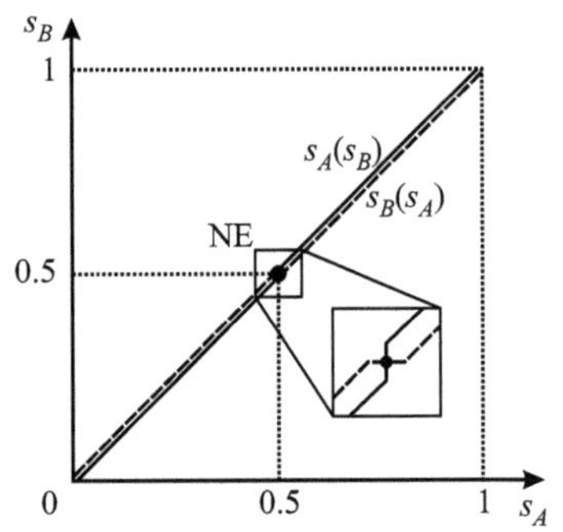

8.2.4 Strategic Manager Compensation

Problem Text in Sect. 8.1.4.

Part a)
Since the companies are in quantity competition with each other, the own quantity x_i is the strategic variable. In addition, the companies are symmetric, which is why the profit for both companies is represented the same depending on the quantities:

$$\pi_i(x_i) = \left(24 - x_i - x_j\right) \cdot x_i - (5 + 6 \cdot x_i).$$

Optimization with respect to the own quantity x_i results in

$$\frac{\partial \pi_i}{\partial x_i} = 24 - 2 \cdot x_i - x_j - 6 \stackrel{!}{=} 0.$$

We solve this maximization condition for the own quantity to obtain the respective reaction function,

$$x_i\left(x_j\right) = 9 - 0.5 \cdot x_j.$$

If we insert the two symmetrical reaction functions into each other,

$$x_i = 9 - 0.5 \cdot (9 - 0.5 \cdot x_i),$$

we get the profit-maximizing quantities, which allows us to determine the Nash equilibrium with $x_i^* = 6$ or $\{x_A^* = 6, x_B^* = 6\}$. This results in an identical profit for both companies of $\pi_A = \pi_B = 31$.

Part b)
We use backward induction to determine the subgame perfect equilibrium (or, as it is also called in this context, the Stackelberg equilibrium). The order of moves is shown in the following figure.

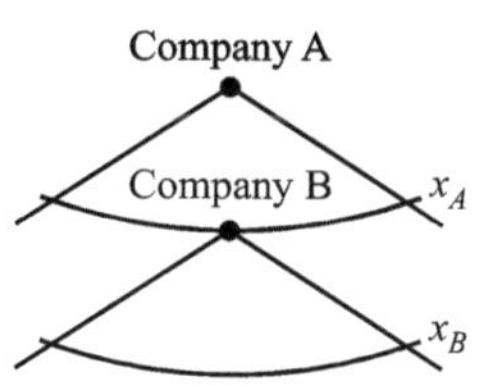

At the last stage, the follower (company B) chooses its quantity with the quantity decision of the leader (company A) already known at this stage, $\overline{x}_A$,

$$\pi_B(x_B, \overline{x}_A) = (24 - x_B - \overline{x}_A) \cdot x_B - (5 + 6 \cdot x_B).$$

Company B maximizes its profit with respect to its quantity,

$$\frac{\partial \pi_B(x_B, \overline{x}_A)}{\partial x_B} = 18 - 2 \cdot x_B - \overline{x}_A \stackrel{!}{=} 0.$$

From this, the reaction function of company B can be determined, which corresponds to the reaction function from the simultaneous game in a). There is a difference, however, in that the quantity decision of company A is already known and can therefore be reacted to

$$x_B = 9 - \frac{1}{2} \cdot \overline{x}_A.$$

In the first stage, company A anticipates this reaction of company B. As a result, the profit of company A is only dependent on its own quantity,

$$\pi_A(x_A) = [24 - x_A - (9 - 0.5 \cdot x_A)] \cdot x_A - (5 + 6 \cdot x_A) = 9 \cdot x_A - 0.5 \cdot x_A^2 - 5.$$

From the maximization

$$\frac{\partial \pi_A(x_A)}{\partial x_A} = 9 - x_A \stackrel{!}{=} 0$$

the quantities $x_A^* = 9$ and thus $x_B^* = 4.5$ result. We see that company A, compared to the simultaneous game, expands its quantity (to the monopoly quantity!) and company B restricts its quantity. Accordingly, the profits $\pi_A = 35.5$ and $\pi_B = 15.25$ emerge, which shows that company A increases its profit while company B suffers a loss of profit. Thus, there is a first-mover advantage.

Part c)
We are now considering the strategic manager compensation. This extension adds not only new players, but also new actions to the known oligopoly model. In the present game, there are now three players: the owner of company A, the manager of company A, and company B (e.g., this one is owner-managed, so there are no relevant conflicts of interest between owner and management as in company A).

The players have the following strategies:

- Owner of company A: $s_{AO} = \{a_{AO}^0 = s\}$, i.e., choice of the weighting factor.
- Manager of company A: $s_{AM} = \{a_{AM}^1(h_{AO}) = x_A(s)\}$, i.e., choice of the quantity depending on the chosen weighting factor (= history of the game).
- Company B: $s_B = \{a_B^1(h_{AO}) = x_B(s)\}$, i.e., choice of the quantity depending on the chosen weighting factor (= history of the game).

The manager's payoff is given by her compensation. If we specify and simplify the compensation function, we see that the weighting factor determines the extent to which the manager considers the production costs in her decision:

$$w(x_A, x_B, s) = s \cdot \pi_A + (1 - s) \cdot R_A + w^0(s)$$

$$= s \cdot [p(X) \cdot x_A - C(x_A)] + (1 - s) \cdot [p(X) \cdot x_A] + w^0(s)$$
$$= p(X) \cdot x_A - s \cdot C(x_A) + w^0(s)$$

Since $w^0(s)$ is only determined after the game in such a way that $w = w^R$ applies, it does not matter *during* the game and we do not need to consider it further, i.e., the expression simplifies to

$$w(x_A, x_B, s) = p(X) \cdot x_A - s \cdot C(x_A)$$

The higher s, the more the costs are considered. For $s = 1$, the costs are fully considered, and the manager behaves like a "normal" profit-maximizing company. For $s = 0$, however, the manager focuses exclusively on revenue, which tends to lead to her aiming for a higher output quantity (more aggressive behavior to increase the market share).

Since the game consists of two stages, we apply backward induction. We therefore first consider the decision in the second stage. Here s has already been chosen and the manager is in a simultaneous quantity competition with company B, so we must determine $x_A(s)$ and $x_B(s)$. The manager's reaction function is as follows:

$$w(x_A, x_B, s) = x_A \cdot (24 - x_A - x_B) - s \cdot (5 + 6 \cdot x_A)$$
$$\frac{\partial w(x_A, x_B, s)}{\partial x_A} = 24 - 2 \cdot x_A - x_B - 6 \cdot s \overset{!}{=} 0$$
$$\Rightarrow x_A = 12 - 3 \cdot s - 0.5 \cdot x_B$$

The reaction function of company B remains unchanged from (b). From this, the profit-maximizing quantities x_A^* and x_B^* can be calculated:

$$x_A = 12 - 3 \cdot s - 0.5 \cdot (9 - 0.5 \cdot x_A)$$
$$x_A^* = 10 - 4 \cdot s$$
$$\Rightarrow x_B^* = 9 - 0.5 \cdot (10 - 4 \cdot s) = 4 + 2 \cdot s$$

Now we turn to the first stage. Here, the owner of company A sets the weighting factor s to maximize the company's profit:

$$\pi_A(s) = \left(24 - x_A^* - x_B^*\right) \cdot x_A^* - C\left(x_A^*\right)$$
$$= [24 - (10 - 4 \cdot s) - (4 + 2 \cdot s)] \cdot (10 - 4 \cdot s) - [5 + 6 \cdot (10 - 4 \cdot s)]$$
$$= [10 + 2 \cdot s] \cdot (10 - 4 \cdot s) - [65 - 4 \cdot s] = 35 + 4 \cdot s - 8 \cdot s^2$$
$$\frac{\partial \pi_A(s)}{\partial s} = 4 - 16 \cdot s \overset{!}{=} 0$$
$$\Rightarrow s^* = 0.25$$

If we insert s^* into the quantity choice of the manager and of company B, the results are $x_A^* = 9$ and $x_B^* = 4.5$, i.e., the same result as in the Stackelberg case is obtained. Through strategic compensation, company A can secure the advantage of a first-mover despite the simultaneous competition.

After the game, the owner will finally set $w^0(s) < 0$ to achieve a positive profit—but this has no effect on the manager's behavior, as a deviation from her optimal choice would only result in her no longer being able to receive her reservation wage.

8.2.5 Strategic Investment

Problem Text in Sect. 8.1.5.

Part a)

This game is solved by backward induction, i.e., we start with the simultaneous Cournot competition on the second stage. Since the exact specification of the cost function $c_i(r_i, r_j)$ is not important here, we ignore it for the sake of simplicity. Similarly, the R&D costs are not yet relevant.

The profit function can thus be simplified as follows:

$$\pi_i(x_i, x_j, r_i, r_j) = (100 - x_i - x_j) \cdot x_i - c_i(r_i, r_j) \cdot x_i. \tag{8.2}$$

The partial derivative with respect to the strategic variable, the company's produced quantity x_i, is

$$\frac{\pi_i(x_i, x_j, r_i, r_j)}{\partial x_i} = 100 - 2 \cdot x_i - x_j - c_i(r_i, r_j)$$

From this, we determine the reaction function of company i by setting the derivative equal to zero and solving for the strategic variable x_I

$$x_i(x_j) = 50 - 0.5 \cdot x_j - 0.5 \cdot c_i(r_i, r_j),$$

Due to the symmetry of the reaction functions of i and j, we find the profit-maximizing output quantity in the Nash equilibrium:

$$\begin{aligned} x_i &= 50 - 0.5 \cdot [50 - 0.5 \cdot x_i - 0.5 \cdot c_j(r_i, r_j)] - 0.5 \cdot c_i(r_i, r_j) \\ x_i &= 25 + 0.25 \cdot x_i + 0.25 \cdot c_j(r_i, r_j) - 0.5 \cdot c_i(r_i, r_j) \\ &\Rightarrow x_i^* = \frac{1}{3} \cdot [100 - 2 \cdot c_i(r_i, r_j) + c_j(r_i, r_j)] \end{aligned}$$

Substituting x_i^* into (8.2) gives the profits depending on the costs:

$$\pi_i(r_i, r_j) = \frac{1}{9} \cdot [100 - 2 \cdot c_i(r_i, r_j) + c_j(r_i, r_j)]^2 \quad \text{for } i \neq j \quad \text{and} \quad i, j = 1, 2 \tag{8.3}$$

In the second step, the optimal investment r_i is now determined, which is set simultaneously by both companies at the first stage. Hence, we insert the cost functions $c_i(r_i, r_j) = 50 - r_i - 0.5 \cdot r_j$ into the profit function (8.3) and also consider the costs for R&D, $0.5 \cdot \lambda_i^2$ to calculate the symmetric Cournot-Nash equilibrium of the first stage:

$$\begin{aligned}\pi_i(r_i, r_j) &= \frac{1}{9} \cdot \left[100 - 2 \cdot (50 - r_i - 0.5 \cdot r_j) + (50 - r_j - 0.5 \cdot r_i)\right]^2 - 0.5 \cdot r_i^2 \\ &= \frac{1}{9} \cdot (50 - 1.5 \cdot r_i)^2 - 0.5 \cdot r_i^2 \\ &= \frac{1}{9} \cdot \left(2500 - 150 \cdot r_i + 2.25 \cdot r_i^2\right) - 0.5 \cdot r_i^2\end{aligned}$$

The first derivative with respect to r_i is

$$\frac{\partial \pi_i(r_i, r_j)}{\partial r_i} = \frac{50}{3} - 0.5 \cdot r_i$$

If we set this equal to zero, we find $r_i = 100/3$, i.e., the research decision is independent of the competitor's decision.

Part b)

To find out more about the influence of the external effect, we determine the solution in general as a function of s.

$$\begin{aligned}\pi_i(r_i, r_j) &= \frac{1}{9} \cdot \left[100 - 2 \cdot (50 - r_i - s \cdot r_j) + (50 - r_j - s \cdot r_i)\right]^2 - 0.5 \cdot r_i^2 \\ &= \frac{1}{9} \cdot \left[50 - (2 - s) \cdot r_i + (2 \cdot s - 1) \cdot r_j\right]^2 - 0.5 \cdot r_i^2\end{aligned}$$

The first-order condition, i.e., the first derivate of the profit with respect to r_i, results in:

$$\frac{\partial \pi_i(r_i, r_j)}{\partial r_i} = \frac{2}{9} \cdot \left[50 + (2 - s) \cdot r_i + (2 \cdot s - 1) \cdot r_j\right] \cdot (2 - s) - r_i \overset{!}{=} 0$$

From the resulting reaction functions

$$r_i(r_j) = \frac{(2 - s) \cdot \left[100 - 2 \cdot (1 - 2 \cdot s) \cdot r_j\right]}{1 + 8 \cdot s - 2 \cdot s^2}$$

we find the Cournot-Nash equilibrium of the first stage,

$$r_i^* = \frac{100 \cdot (2 - s)}{5 + 2 \cdot (s - 1) \cdot s}.$$

If we take a closer look at the expression, we can see that in the relevant range (s between 0 and 1) it is decreasing in s, i.e., the more research spillover there is, the less companies will invest. This means that too little research is being carried out.

8.2.6 Company and Competition Authority

Problem Text in Sect. 8.1.6.

Part a)
The game trees for the different move orders of the players are as follows (please note that to avoid confusion the company is referred to as F for firm):

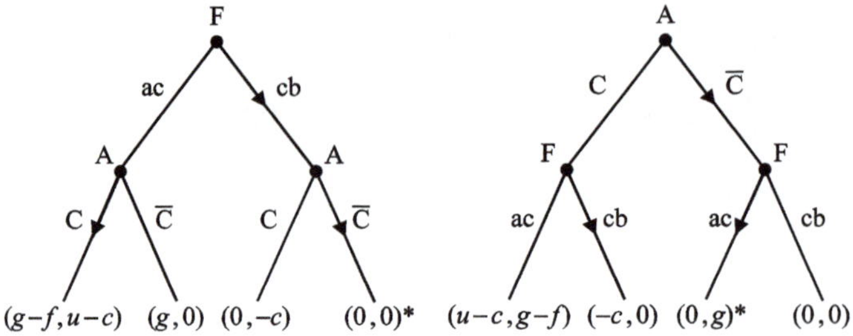

i. Company decides first ii. Authority decides first

The company's action *ac* denotes anti-competitive behavior and *cb* denotes compliant behavior. The authority can choose between C (control) and $\overline{C}$ (no control).

If the company moves first (situation i), the subgame perfect Nash equilibrium is $\{cb, \{C(ac), \overline{C}(cb)\}\}$. On the equilibrium path, the company will therefore behave compliantly, and the authority will not control.

If, on the other hand, the authority moves first, the result is $\{\overline{C}, \{cb(C), ac(\overline{C})\}\}$, meaning, in the result, the authority will not control, and the company will behave in an anti-competitive manner.

In this game, there is therefore a second-mover advantage, i.e., the last player to move can realize the more advantageous result.

Part b)
Since this is a decision with imperfect information, we choose the matrix form:

F, A	Control	No control
Anti-competitive	$(g-f, u-c)$	$(g, 0)$
Compliant	$(0, -c)$	$(0, 0)$

There is no Nash equilibrium in pure strategies here, as for each strategy combination one player has an incentive to deviate. Let us start, e.g., with the strategy choice {no control, compliant} that occurs on the equilibrium path with perfect information. Here, the authority has no incentive to change its behavior, but the company can improve its position by behaving anti-competitive—this strategy combination can therefore not be an equilibrium. If the company chooses anti-competitive behavior, the authority has an incentive to control, which also means that {no control, anti-competitive} cannot be an equilibrium. If the authority controls, it is better for the company to behave compliant—thus, we can also rule out the constellation {control, anti-competitive}. If the company behaves compliantly, there is no incentive for the authority to control it. Therefore, {control, compliant} can also not be an equilibrium. Finally, we end up again with the initial combination, which we know is not an equilibrium. Therefore, there is no Nash equilibrium in pure strategies.

We now determine the Nash equilibrium in mixed strategies both via the optimization approach and via the expected utility comparison. Both paths must lead to the same result. Here, p_F stands for the probability with which the company plays the strategy "anti-competitive", and p_A stands for the probability with which the authority plays the strategy "control".

The optimization approach yields for the company

$$\begin{aligned}
E[u_F(p_F, p_A)] &= p_F \cdot p_A \cdot (g - f) + p_F \cdot (1 - p_A) \cdot g \\
&= p_F \cdot p_A \cdot g - p_F \cdot p_A \cdot f + p_F \cdot g - p_F \cdot p_A \cdot g \\
\frac{\partial E[u_F(p_F, p_A)]}{\partial p_F} &= p_A \cdot g - p_A \cdot f + g - p_A \cdot g \overset{!}{=} 0 \\
&\Rightarrow p_A^* = g/f
\end{aligned}$$

and for the competition authority

$$\begin{aligned}
E[u_A(p_F, p_A)] &= p_F \cdot p_A \cdot (u - c) + (1 - p_F) \cdot p_A \cdot (-c) \\
&= p_F \cdot p_A \cdot u - p_F \cdot p_A \cdot c - p_A \cdot c + p_F \cdot p_A \cdot c \\
\frac{\partial [u_A(p_F, p_A)]}{\partial p_A} &= p_F \cdot u - p_F \cdot c - c + p_F \cdot c \overset{!}{=} 0 \\
&\Rightarrow p_F^* = c/u
\end{aligned}$$

In the expected utility comparison, we figure out the action with which the players are better off depending on the probability choice of their opponent. Thus, it is better for the company to play "anti-competitive", i.e., choose $p_F = 1$, if the following inequality holds:

$$\begin{aligned}
E[u_F(p_F = 1, p_A)] &> E[u_F(p_F = 0, p_A)] \\
(g - f) \cdot p_A + g \cdot (1 - p_A) &> 0 \\
p_A &< g/f.
\end{aligned}$$

The authority will decide for "control", i.e., choose $p_A = 1$, if

$$\begin{aligned}
E[u_A(p_F, p_A = 1)] &> E[u_A(p_F, p_A = 0)] \\
(u - c) \cdot p_F + (-c) \cdot (1 - p_F) &> 0 \\
p_F &> c/u.
\end{aligned}$$

We thus obtain the Nash equilibrium in mixed strategies with the strategy combination ($p_F^* = c/u$, $p_A^* = g/f$). As we see, the optimal probability with which player i plays a strategy depends on the preference parameters, i.e., the elements of the utility function, of player j and vice versa. The company will play the strategy "anti-competitive" with a higher probability the higher c and the smaller u is. So, if for the authority the control is particularly costly (high c) or if a successful conviction of the company only brings little utility (low u), the company will take advantage of these favorable conditions from its point of view. Thus, the probability of anti-competitive behavior increases. The authority, in turn, will control more often (more precisely: control will be more likely to be expected), the larger g and the smaller f are. As we have seen from the matrix, a high difference $g - f$ means that the company has a high incentive to behave in an anti-competitive way. But then control is particularly promising, so the authority should control particularly often.

The following figure shows the reaction correspondences of this game and the Nash equilibrium in mixed strategies that occurs at the intersection of the curves.

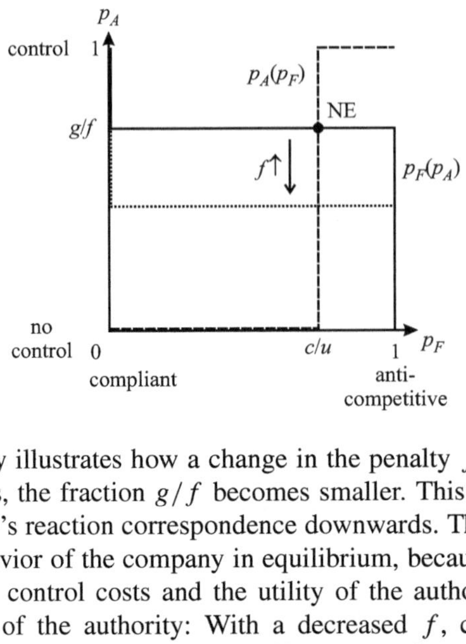

The graph clearly illustrates how a change in the penalty f affects the equilibrium: If f increases, the fraction g/f becomes smaller. This shifts the horizontal part of the company's reaction correspondence downwards. Thus, the penalty does not change the behavior of the company in equilibrium, because this only depends on the (unchanged) control costs and the utility of the authority but reduces the control probability of the authority: With a decreased f, control must be less frequent to keep the company indifferent. The expected penalty finally depends on the amount of the penalty and its probability, so with a higher penalty and simultaneously reduced probability, the expected penalty remains unchanged.

8.2.7 Collusion Among Many Companies

Problem Text in Sect. 8.1.7.

Part a)
In the case of a homogeneous good, price competition has an interesting effect: If a company undercuts its competitor(s), it gets the entire market, since all companies offer the identical good, consumers will always buy from the cheapest provider—however, this assumes that the company also has the capacity to supply the market. The companies will therefore lower their prices until (at identical marginal costs of all companies) $p = c$ and thus also $\pi = 0$ applies—regardless of how many consumers buy from them (at identical price, a uniform distribution of consumers to all companies is usually assumed). Starting from $p = c$, no improvement is possible for the companies as the following considerations show:

- If a company increases its price to $p > c$, no customer will buy from it anymore, i.e., its profit remains unchanged at $\pi = 0$. There is no improvement.
- If a company lowers its price to $p < c$, all consumers will buy from it, but it makes a loss with each sold unit, so that $\pi < 0$. The company would harm itself by doing this!

Thus, $p = c$ is the Nash equilibrium. We can also imagine that we start at a price $p > c$ and the companies finally undercut each other to $p = c$. This situation is also known as the *Bertrand paradox*: Two companies which compete on price for homogeneous goods effectively have no market power, so that the result is the same (price equals marginal cost, zero profit) as in the case of perfect competition with an infinite number of companies.

For further analysis, we first formulate the grim trigger strategy in concrete terms: "Starting with the first period, choose the collusion price in each period, unless a player chose a different price in the previous period; then choose the competition price ($p = c$) for all eternity". Precise information about the level of the collusion price is not necessary (it would correspond to the monopoly price), as the test of whether the strategy can be supported only depends on the payoff for the company. In the case of adherence to the strategy, the collusion profit π^C must be shared with all other n companies, i.e., the single company gets $\pi^C \cdot 1/n$, whereas a marginal deviation from the collusion price downwards practically transfers the entire collusion profit to the deviating player. However, the company will no longer make a profit from the following period onwards for all eternity. We apply the one-shot deviation principle, i.e., we check whether a deviation is profitable. The further the deviation takes place in the future, the lower the increase in benefit from a deviation due to discounting, which is why we only need to check whether it is profitable to deviate in the first round. If this is not the case, it will never be profitable again:

$$\sum_{t=0}^{\infty} \delta^t \cdot \pi^C \cdot \frac{1}{n} \geq \pi^C + \sum_{t=1}^{\infty} \delta^t \cdot 0$$

$$\frac{\pi^C}{1-\delta} \cdot \frac{1}{n} \geq \pi^C$$

$$\Rightarrow \delta \geq \tilde{\delta} \equiv \frac{n-1}{n}$$

Note that $\sum_{t=n}^{\infty} \delta^t = \frac{\delta^n}{1-\delta}$. We obtain the not very surprising result that with an increasing number of companies in the market, collusive behavior becomes increasingly difficult—the higher n, the higher the discount factor of the companies must be to support collusive behavior with the grim trigger strategy. For example, for $n = 2$ the threshold is $\tilde{\delta} = 1/2$, it increased for $n = 3$ to $\tilde{\delta} = 2/3$, and for $n = 4$ already to $\tilde{\delta} = 3/4$. The result is also independent of the level of the collusion profit.

Part b)
As an extension to part a), we consider the growth rate of the market g[2]:

$$\sum_{t=0}^{\infty} \delta^t \cdot g^t \cdot \pi^C \cdot \frac{1}{n} \geq \pi^C + \sum_{t=1}^{\infty} \delta^t \cdot g^t \cdot 0$$

$$\frac{\pi^C}{1-\delta \cdot g} \cdot \frac{1}{n} \geq \pi^C$$

$$\Rightarrow \delta \geq \tilde{\delta} \equiv \frac{n-1}{n} \cdot \frac{1}{g}$$

We see that the right side of the inequality is smaller than in part (a) if $g > 1$. This means that the company's discount factor does not have to be as high to maintain collusion. Collusion is therefore more stable in growing markets ($g > 1$) as the future is associated with higher profits. A deviation in favor of a comparatively small profit in the future is only profitable for very short-sighted companies. Conversely, in shrinking markets ($g < 1$) it becomes very difficult to maintain collusion as the companies prefer to (completely) skim off today's higher profits rather than share the future smaller profits.

[2] The technical assumption that $g \cdot \delta < 1$ was made so that the formula $\sum_{t=1}^{\infty} q^t = q/(1-q)$ remains applicable. This is only valid for $0 \leq q < 1$, which assures convergence of the geometric series.

8.2.8 Cartels and Coordinated Behavior Among Many Companies

Problem Text in Sect. 8.1.8.

Part a)

The procedure for Cournot competition among many companies was described in Sect. 3.4. Since more than two companies are considered in (c), we present the approach generally depending on n. First, we determine profit maximization calculation, using the (inverse) market demand, i.e., the demand solved for the price, $p = 24 - X$:

$$\pi_i = \left[24 - \left(x_i + \sum_{j \neq i} x_j\right)\right] \cdot x_i = 24 \cdot x_i - x_i^2 - x_i \cdot \sum_{j \neq i} x_j$$

$$\frac{\partial \pi_i}{\partial x_i} = 24 - 2 \cdot x_i - \sum_{j \neq i} x_j \stackrel{!}{=} 0$$

Substitution in the symmetric equilibrium yields

$$\frac{\partial \pi_i}{\partial x_i} = 24 - 2 \cdot x_i - (n-1) \cdot x_1 \stackrel{!}{=} 0$$
$$\Rightarrow x_{i(n)} = \frac{24}{n+1}$$

With this, we determine price and profit:

$$p(n) = 24 - n \cdot x_i(n) = \frac{24}{n+1}$$
$$\pi_i(n) = p(n) \cdot x_i(n) = \frac{24}{n+1} \cdot \frac{24}{n+1} = \frac{576}{(n+1)^2}$$

For the payoffs at (collusion, collusion) we need the collusion solution. Collusion is characterized by the companies maximizing the joint profit, i.e., they behave together like a monopolist. Thus, we can easily determine the payoff by looking at the solution for $n = 1$:

$$x_M(1) = 12, \; p(1) = 12, \; \pi_M(1) = 144$$

The two companies will then divide production and profit evenly ($x_A = x_B = 12/2 = 6$ and $\pi_A = \pi_B = 144/2 = 72$). The solution in competition arises for $n = 2$:

$$x_i(2) = 8, \; p(2) = 8, \; \pi_i(2) = 64$$

Thus, we have determined the symmetric payoffs. For the asymmetric payoffs (collusion and Cournot), we need to calculate the profits of the two companies that arise when one deviates. If company i deviates, that is, it chooses the Cournot quantity $x_i^C = 8$ instead of half the monopoly quantity $\frac{x_i^M}{2} = 6$, then it receives as profit

$$\pi_i = (24 - 8 - 6) \cdot 8 = 80.$$

The other company, which adheres to the agreement and thus chooses $\frac{x_i^M}{2} = 6$, has a profit of

$$\pi_j = (24 - 6 - 8) \cdot 6 = 60.$$

Thus, the matrix can be completed as follows:

A, B	Collusion	Cournot
Collusion	(72, 72)	(60, 80)
Cournot	(80, 60)	(64, 64)

If a binding contract is possible (i), the two companies will choose the collusion solution, as they receive a higher payoff (72, 72) than in competition (64, 64). Since the contract is assumed to be binding, none of the companies can deviate from the strategy choice "collusion".

If a binding contract is not possible due to a cartel prohibition (ii), both will choose the competitive price and the classic prisoner's dilemma occurs. A ban on coordinated behavior is not necessary, as "Cournot" represents a dominant strategy, meaning that even if the two companies should coordinate in advance and both decide for "collusion", none will adhere to it and deviate.

Part b)

The Grim-Trigger strategy can be formulated as follows: "Starting with the first period, choose the collusion quantity in each period, unless a player has chosen a different quantity in the previous period, then choose the Cournot quantity for all eternity".

To determine the critical discount factor, up to which collusion can be supported with this strategy, we apply the one-shot deviation principle (where $\sum_{t=1}^{\infty} \delta^t = \delta/(1-\delta)$):

$$\sum_{t=0}^{\infty} \delta^t \cdot 72 \geq 80 + \sum_{t=0}^{\infty} \delta^t \cdot 64$$

$$\frac{72}{1-\delta} \geq 80 + \frac{64 \cdot \delta}{1-\delta}$$

$$\Rightarrow \delta \geq \tilde{\delta} \equiv 0.5$$

The discount factor must therefore be at least $\tilde{\delta} = 0.5$ to support collusion through the Grim-Trigger strategy.

If the companies have a discount factor of $\delta^l = 0.25$ as in (i), they attribute only minor importance to future payoffs. Since also $\delta^l < \tilde{\delta}$ applies, collusion cannot be supported in this case. This also means that a ban on coordinated behavior is not necessary, as the companies have no willingness for collusive behavior.

The situation is different in (ii) with the significantly higher discount factor $\delta^h = 0.75$. This is greater than $\tilde{\delta}$, which means that collusion can be supported. Since there is now an incentive for collusive behavior, a ban on coordinated behavior is necessary.

In a finite repetition of the game, backward induction can be applied to check whether the companies are acting collusively. In this case, we see that in the last period no punishment is possible if a company does not adhere to the agreement. This means that both will choose "Cournot". Both know this already in the penultimate period and therefore also choose "Cournot"—there is no possibility to punish in the last period where both play "Cournot". We can continue these considerations up to the first period, whereby we can show that collusion is not possible in finite repetition. This also means that a ban on coordinated behavior is not necessary with finite repetition.

Part c)

Collusion requires that the monopoly quantity of 12 is produced evenly by the n companies. Thus, the deviating company knows that the remaining quantity in the market is at $(n-1) \cdot 12/n$. Taking this into account, the profit maximization calculation of the deviating company (d) is

$$\pi_d(x_d) = \left[24 - x_d - (n-1) \cdot \frac{12}{n}\right] \cdot x_d$$

$$\frac{d\pi_d}{dx_d} = 24 - 2 \cdot x_d - \frac{12 \cdot (n-1)}{n} \overset{!}{=} 0$$

$$\Rightarrow x_d = \frac{6 \cdot (n+1)}{n}$$

At this quantity, the deviating company makes a profit of:

$$\pi_d = \left[24 - \frac{6 \cdot (n+1)}{n} - (n-1) \cdot \frac{12}{n}\right] \cdot \frac{6 \cdot (n+1)}{n} = \frac{36 \cdot (n+1)^2}{n^2}$$

This profit is higher than with collusive behavior, $144/n$ (see (a)).

We now know how high the profit is in the case of deviation and can consider this accordingly when determining the critical discount factor using the grim trigger strategy:

$$\sum_{t=0}^{\infty} \delta^t \cdot \frac{144}{n} \geq \frac{36 \cdot (n+1)^2}{n^2} + \sum_{t=1}^{\infty} \delta^t \cdot \frac{576}{(n+1)^2}$$

$$\frac{1}{1-\delta}\cdot\frac{144}{n}\geq\frac{36\cdot(n+1)^2}{n^2}+\frac{\delta}{1-\delta}\cdot\frac{576}{(n+1)^2}$$

$$\frac{144}{n}\geq\frac{36\cdot(n+1)^2}{n^2}-\frac{576\cdot n^2-36\cdot(n+1)^4}{(n+1)^2\cdot n^2}\cdot\delta$$

$$\Rightarrow\delta\geq\tilde{\delta}\equiv\frac{(n+1)^2}{1+n\cdot(6+n)}$$

To determine how the critical discount factor depends on n, we can derive this with respect to n and obtain

$$\frac{d\tilde{\delta}}{dn}=\frac{4\cdot\left(n^2-1\right)}{[1+n\cdot(6+n)]^2}>0$$

For $n>1$ the enumerator is positive and since the expression in the denominator is squared, it is always positive. Thus, the critical discount factor depends positively on n (an increase in n increases the critical discount factor). This means, the more companies there are in the market, the more difficult it is to support collusion.

In the next step, we extend the approach to include the growth rate of the market

$$\sum_{t=0}^{\infty}\delta^t\cdot g^t\cdot\frac{144}{n}\geq\frac{36\cdot(n+1)^2}{n^2}+\sum_{t=0}^{\infty}\delta^t\cdot g^t\cdot\frac{576}{(n+1)^2}$$

$$\frac{1}{1-\delta\cdot g}\cdot\frac{144}{n}\geq\frac{36\cdot(n+1)^2}{n^2}+\frac{\delta\cdot g}{1-\delta\cdot g}\cdot\frac{576}{(n+1)^2}$$

$$\Rightarrow\delta\geq\tilde{\delta}\equiv\frac{(n+1)^2}{1+n\cdot(6+n)}\cdot\frac{1}{g}$$

We see that collusion is more feasible in growing markets ($g>1$) than in shrinking markets ($g<1$)– the critical discount factor decreases with increasing g. Therefore, collusion is much more difficult to enforce in highly competitive and shrinking markets than in weakly competitive and growing markets.

8.2.9 Advertising

Problem Text in Sect. 8.1.9.

Part a)

In this game we can identify the two companies Sumsang and Pear as players. Sumsang decides first on its advertising investments and then on its output quantity in competition, $\{a, x_S(a)\}$. Pear can only decide on its production, $\{x_P(a)\}$. The payoff functions of the two players correspond to their profit functions, which can be formulated depending on the strategic variables a, x_S, and x_P:

$$\pi_S(a,x_S,x_P)=(100+a-x_S-x_P)\cdot x_S-400-10\cdot x_S-0.5\cdot a^2$$

$$\pi_P(a, x_S, x_P) = (230 - 2 \cdot x_P - x_S) \cdot x_P - 400 - 10 \cdot x_P$$

The game is sequential and consists of two stages, each with a different information structure. The first stage has perfect and complete information, while the second stage has imperfect and complete information.

Part b)
To determine the reaction functions, we must partially derive the profit functions with respect to the strategic variables of the players. At the time of competition in the sales market, the advertising investments have already been made, which is why this is no longer a strategic variable for Sumsang in the second stage. Sumsang can only decide in this stage about quantity produced x_S:

$$\pi_S(a, x_S, x_P) = (100 + a - x_S - x_P) \cdot x_S - 400 - 10 \cdot x_S - 0.5 \cdot a^2$$
$$\frac{\partial \pi_S(a, x_S, x_P)}{\partial x_S} = 100 + a - 2 \cdot x_S - x_P - 10 \overset{!}{=} 0$$
$$\Rightarrow x_S(x_P) = 45 + 0.5 \cdot a - 0.5 \cdot x_P$$

Similarly, x_P is the strategic variable of Pear:

$$\pi_P(a, x_S, x_P) = (230 - 2 \cdot x_P - x_S) \cdot x_P - 400 - 10 \cdot x_P$$
$$\frac{\partial \pi_P(a, x_S, x_P)}{\partial x_P} = 230 - 4 \cdot x_P - x_S - 10 \overset{!}{=} 0$$
$$\Rightarrow x_P(x_S) = 55 - 0.25 \cdot x_S$$

Part c)
To determine the equilibrium quantities, we substitute the two reaction functions $x_S(x_P)$ and $x_P(x_S)$ into each other. This gives the sales quantity for Sumsang:

$$x_S = 45 + 0.5 \cdot a - 0.5 \cdot (55 - 0.25 \cdot x_S)$$
$$x_S = \frac{8}{7} \cdot (45 + 0.5 \cdot a - 27.5) = 20 + \frac{4}{7} \cdot a$$

This amount can be substituted into $x_P(x_S)$ to determine the amount of Pear,

$$x_P = 55 - 0.25 \cdot \left(20 + \frac{4}{7} \cdot a\right) = 50 - \frac{1}{7} \cdot a.$$

Both equilibrium quantities depend on Sumsang's advertising measures. They have a positive effect on the equilibrium quantity of Sumsang and a negative effect on the quantity of Pear.

The following figure shows the reaction functions as solid lines:

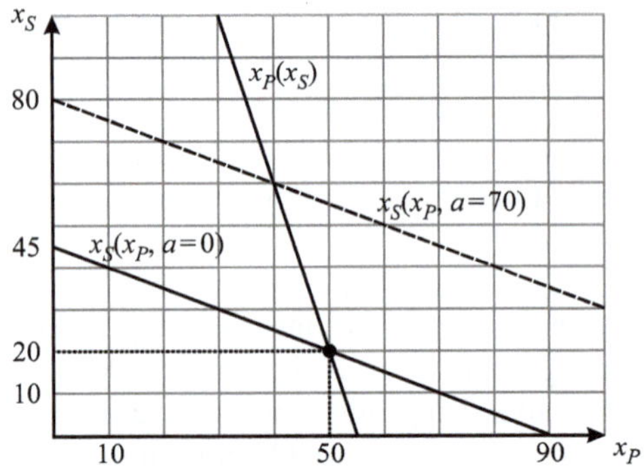

At the intersection of the reaction curves, the chosen quantities are mutually best responses: According to its reaction function, Sumsang reaches its profit maximum, given Pear chooses $x_P = 50$, at $x_S = 45 - 0.5 \cdot 50 = 20$, while Pear maximizes its profit for $x_S = 20$ with $x_S = 55 - 0.25 \cdot 20 = 50$. A deviation at this point will only lead to a reduction in profit, which is why no company will deviate from this quantity choice. For $a = 0$, the Nash equilibrium is therefore $(x_S = 20, x_P = 50)$.

Part d)
For the three available advertising packages, the following values result:

a	x_S	x_P	π_S	π_P
0	20	50	0	4,600
70	60	40	750	2,800
105	80	35	487.5	2,050

Sumsang has the highest profit at $a = 70$ and should therefore opt for this advertising package. If it invests more in advertising, then the exponentially increasing costs of the advertising measures, $0.5 \cdot a^2$, outweigh their only linearly increasing advantage, $100 + a$.

The dashed line in the figure c shows the reaction function of Sumsang for $a = 70$. Compared to the initial function it is shifted outwards: For each given quantity of Pear, Sumsang will now produce more—this is also clearly visible in the reaction function, in which the advertising measures positively contribute: $x_S(x_P) = 45 + 0.5 \cdot a - 0.5 \cdot x_P$. As this decreases the price, Pear will restrict its quantity to reduce its loss of profit.

If Pear could not observe the investment, it would not restrict its quantity and continue to produce 50. Sumsang would indeed expand its quantity according to its reaction function, $x_S(x_P = 50) = 45 + 0.5 \cdot 70 - 0.5 \cdot 50 = 55$, but not to the extent as it would be in the case of observation ($x_S = 60$). This is because in this case the central strategic effect of the investment (= investment leads to a quantity reaction of the other company) is missing.

Part e)
The information structure is now incomplete: Sumsang can have two types, either its advertising measures are successful ("successful type"), or they are not ("unsuccessful type").

Please note that we can no longer apply the solution concept of subgame perfection, as there are no independent subgames. Thus, we must apply the sequential equilibrium as a combination of strategy and probability for each type. The solution of this modified game is the Bayes-Nash equilibrium, in which each player chooses the best response to a given strategy of the other players for a given probability distribution over all player types and under the condition that each player knows their own type.

Pear must consider whether it can infer Sumsang's type from its behavior. However, it must consider the following: The unsuccessful type of Sumsang could—despite the costs of the advertising measure—have an incentive to advertise just to convince Pear in a deceptive way that it is the successful type. As a result, Pear would reduce its quantity—to the advantage of Sumsang. Therefore, Pear cannot unambiguously infer the successful type from the observation that an advertising measure has been taken. However, conversely, a successful type will always advertise. Thus, from the observation that no advertising was chosen, one can conclude that it is an unsuccessful type, as the successful type has no incentive to present itself as unsuccessful.

8.2.10 Lobbying

Problem Text in Sect. 8.1.10.

Part a)
The two corporations A and B can be identified as players, who can each decide on the number of their advisory sessions, h_A and h_B, which yields the following payoff matrix:

A, B	$h_B = 30$	$h_B = 90$
$h_A = 30$	(<u>2250</u>, <u>2250</u>)	(<u>5850</u>, -1350)
$h_A = 90$	(−1350, <u>5850</u>)	(4050, 4050)

Both players have the strictly dominant strategy $h_i = 30$, so the unique Nash equilibrium ($h_A = 30, h_B = 30$) results. However, this equilibrium is pareto-dominated by ($h_A = 90, h_B = 90$) where both players would receive a higher payoff of 4050 instead of just 2250.

Part b)
To determine the reaction functions, we must set up the profit function first. This is dependent on one's own and the foreign advisory hours. In general, the profit of

cooperation i is

$$\pi_i(h_i, h_j) = \frac{1}{2} \cdot [90 \cdot (h_i + h_j) + h_i \cdot h_j] - h_i^2.$$

The strategic variable of player i are the number of advisory hours h_i. Thus, it is necessary to partially derive with respect to these and set this derivative equal to zero,

$$\frac{\partial \pi_i(h_i, h_j)}{\partial h_i} = \frac{90}{2} + \frac{1}{2} \cdot h_j - 2 \cdot h_i \stackrel{!}{=} 0.$$

Next, we solve this condition for h_i to obtain the reaction function

$$h_i(h_j) = \frac{90}{4} + \frac{1}{4} \cdot h_j.$$

To get the equilibrium advisory hours, we insert the reaction functions into each other and get

$$h_i = h_j = 30.$$

Therefore, the Nash equilibrium is $(h_1 = 30, h_2 = 30)$ and both players each have a payoff of 2250.

Part c)
If the corporations abandon their competitive attitude and cooperate in providing advice, the optimization problem changes to maximizing the joint profit $\pi_{\text{total}}(h_A, h_B) = \pi_A + \pi_B$. This means that the corporations advise together, thereby maximizing the joint profit and then sharing the resulting total payoff. The derivation of the reaction functions is done in the same way as in part (b). For the calculations, we obtain

$$\pi_{\text{total}}(h_A, h_B) = \pi_A + \pi_B = 90 \cdot (h_A + h_B) + h_A \cdot h_B - h_A^2 - h_B^2$$
$$\frac{\partial \pi_{\text{total}}(h_A, h_B)}{\partial h_A} = 90 + h_B - 2 \cdot h_A \stackrel{!}{=} 0$$
$$\frac{\partial \pi_{\text{total}}(h_A, h_B)}{\partial h_B} = 90 + h_A - 2 \cdot h_B \stackrel{!}{=} 0$$

Since the situation is symmetric (due to identical cost function, both firms will offer the same number of consulting hours $h_A = h_B$), we can see the solution immediately: The number of advisory hours that each firm contributes to the cooperation is $h_A^* = h_B^* = 90$. By substituting into the profit function, we obtain a total profit of $\pi_{\text{total}}^* = 8100$; thus, the profit for the two firms is $\pi_A^* = \pi_B^* = 4050$.

An external effect is generally understood to be the impact of one's own actions on uninvolved third parties. In the specific case, an external effect (positive or negative) would exist if each corporation, when maximizing profit individually (part b), ignores the (positive or negative) impact of its own chosen number of advisory hours on the competitor. From the result of joint profit maximization, we see that the companies offer significantly more advising in cooperation. This means that there must be a positive external effect here: The government derives greater value (and accordingly promotes this with subsidies) from the joint advisory services of the corporations than from the individual services. We also recognize from the first-order conditions that the incentive for each company to provide advisory is higher in cooperation than when it only looks at "its half of the cake": The first-order condition for company i is $0.5 \cdot (90 + h_i)$, which is higher compared to individual profit maximization. This amount corresponds to the "second half of the cake" that it only considers when maximizing profit jointly. Whenever an external effect exists, the Nash equilibrium is not Pareto-optimal. Thus, the Nash equilibrium determined in part (b) is inefficient according to the Pareto criterion.

8.2.11 Holdup Problem

Problem Text in Sect. 8.1.11.

Part a)
In this game, the two players Blubberama and Nerdsoft interact. In the first stage, Nerdsoft can decide to create the software or to reject, and in the third stage to accept or decline the offer. Blubberama can either make a high or a low offer in the second stage. The information structure is complete and perfect, so we choose the extensive form (game tree):

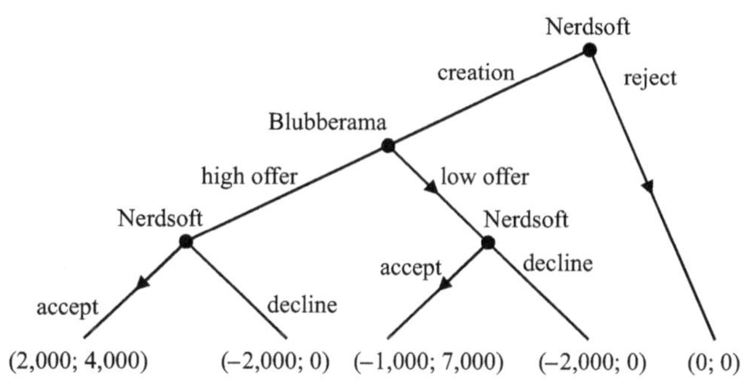

The subgame perfect Nash equilibrium is obtained by backward induction (see small arrows in the figure) and is

{{reject, accept (creation, high offer), accept (creation, low offer)}, {low offer}}.

The rather unsatisfactory result that the software is not created is an example of the holdup problem: Blubberama's dependence after the creation of the software, which cannot be sold to any other company (the software is a "factor-specific investment" for the transaction), leads to Blubberama fully exploiting its bargaining power to Nerdsoft's disadvantage. However, since Nerdsoft anticipates this, it will not engage in business with Blubberama, thus preventing an economically reasonable activity (the overall economic advantage would be 6,000 euros—the utility of Blubberama minus the development costs of Nerdsoft).

The penalty changes the payoff structure of the game as follows:

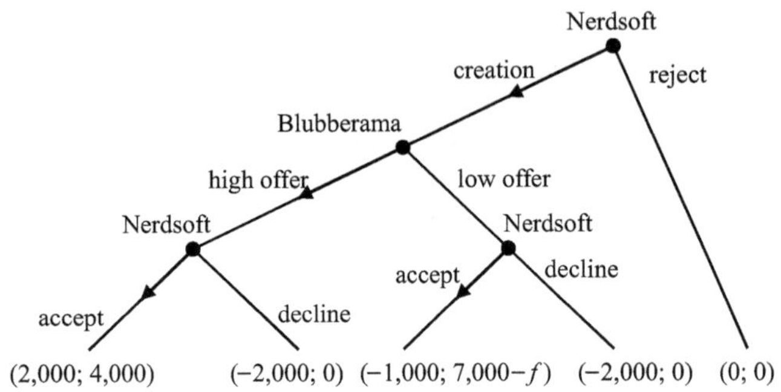

Once Nerdsoft has created the software, it will accept Blubberama's offer. We must therefore compare the payoff for a high, accepted offer, 4000, with that for a low, accepted offer, $7000 - f$. In doing so, the penalty must prevent Blubberama from deciding to submit a low offer. It will therefore opt for a high offer if the penalty is $f > 3000$ euros. Because Blubberama will in this case submit a high offer, Nerdsoft will create the software, from which both ultimately benefit: Through Blubberama's self-commitment, both receive a higher payoff. The holdup problem is overcome as Blubberama can now credibly commit to submitting a high offer.

Part c)
The changed payoff structure is as follows:

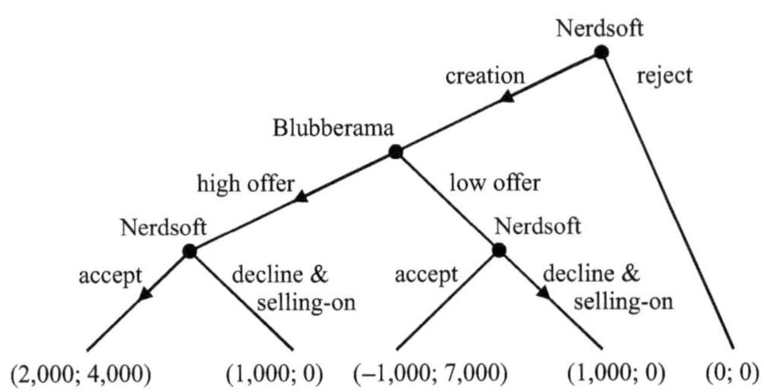

This modification also allows for the equilibrium that is more beneficial for both: If Blubberama would submit the low offer, Nerdsoft would reject it, which is not in Blubberama's interest. Due to the additional competition, Blubberama's bargaining power has decreased, which, however, as in (b), allows it a credible self-commitment to the high offer. Anything that strengthens the bargaining power of the dependent party reduces the holdup problem.

With a bid of 6,000 euros, Nerdsoft will always sell to the other company and Blubberama will not get the software. In this case, at least for Blubberama, the competition is harmful. Blubberama can only obtain the software by making an even higher bid—the lucrative third-party option means that Nerdsoft now has considerable negotiating power.

8.2.12 Competition with Incomplete Information

Problem Text in Sect. 8.1.12.

Part a)
Since the companies only differ in their marginal costs, but face the same price-response function, the profit maximization calculation can initially be set up in general terms (with $i, j = 1, 2$ and $i \neq j$):

$$\pi_i(x_i, x_j) = (12 - x_i - x_j) \cdot x_i - c_i \cdot x_i$$
$$\frac{\partial \pi_i(x_i, x_j)}{\partial x_i} = 12 - 2 \cdot x_i - x_j - c_i \stackrel{!}{=} 0$$
$$\Rightarrow x_i^*(x_j) = 6 - 0.5 \cdot c_i - 0.5 \cdot x_j$$

The second partial derivative, $\partial^2 \pi_i(x_i, x_j)/\partial x_i^2 = -2$, is less than zero, thus the condition for a maximum is fulfilled.

Company A only has one type with unit costs of $c_A = 3$. However, it is uncertain which type company B has and therefore, what reaction to expect. Company A will anticipate the expected quantity decision $E(x_B)$ of company B. It assumes that B has high costs with probability ρ and therefore produces the quantity x_B^h, and has low costs with probability $1 - \rho$, which would result in the quantity x_B^l:

$$x_A^*[E(x_B)] = 6 - 0.5 \cdot 3 - 0.5 \cdot E(x_B) = 0.5 \cdot [9 - E(x_B)]$$
$$= 0.5 \cdot \left[9 - \rho \cdot x_2^h - (1 - \rho) \cdot x_2^l\right]$$

Company B can have two possible types. Since it knows its type, it can determine the optimal reaction for each of the two types. If its marginal costs are high, $c_B^h = 6$, its optimal reaction is

$$x_B^h(x_A) = 3 - 0.5 \cdot x_A.$$

and with low marginal costs $c_B^l = 0$ accordingly

$$x_B^l(x_A) = 6 - 0.5 \cdot x_A.$$

Part b)

Company A will base its quantity choice on the expected reaction of company B. This can be determined by weighing the reactions of the two types of company B with their respective probabilities. This results in an expected reaction function of company B, which depends both on the quantity chosen by company A and on the probability of the two types of company B:

$$E[x_B(x_A)] = \rho \cdot x_B^h(x_A) + (1-\rho) \cdot x_B^l(x_A)$$
$$= \rho \cdot [3 - 0.5 \cdot x_A] + (1-\rho) \cdot [6 - 0.5 \cdot x_A] = 6 - 3 \cdot \rho - 0.5 \cdot x_A$$

In the specific example, the probability of high costs is $\rho = 0.5$, so the expected reaction is $E[x_B(x_A)] = 4.5 - 0.5 \cdot x_A$.

The optimal quantity choice of company A results from substituting the expected reaction $E[x_B(x_A)]$ into the reaction function of company A:

$$x_A[E(x_B)] = 4.5 - 0.5 \cdot E[x_B(x_A)] = 4.5 - 3 + 1.5 \cdot \rho + 0.25 \cdot x_A$$
$$\Rightarrow x_A^*(\rho) = 2 + 2 \cdot \rho$$

The higher ρ, i.e., the more likely company B has high costs and thus produces less efficiently, the greater is x_A. For $\rho = 0.5$, specifically, $x_A^* = 3$.

Company B considers the expectations of company A in its reaction. Company B will therefore put itself in A's position, i.e., it anticipates A's expectation regarding its (uncertain) decision. This means that company B knows that company A will choose $x_A^* = 3$ and therefore—depending on its own type—will orient itself accordingly. Specifically, if company B has high costs, it will respond with $x_B^h(x_A^*) = 3 - 0.5 \cdot (2 + 2 \cdot \rho) = 2 - \rho$ or for $\rho = 0.5$ with $x_B^{h*} = 1.5$ and at low costs accordingly with $x_B^l(x_A^*) = 6 - 0.5 \cdot (2 + 2 \cdot \rho) = 5 - \rho$ or $x_B^{l*} = 4.5$, respectively. The more efficient type of company B with the low costs will therefore choose a higher output quantity. The Bayesian equilibrium is

$$\left(x_A^* = 3, \left\{x_B^{h*} = 1.5, x_B^{l*} = 4.5\right\}\right).$$

In the following figure, we graphically represent the considerations. The reaction function of company A, the reaction functions of the two types of company B, and the reaction function of company B expected by company A are plotted. The expected reaction function lies between the two reaction functions of the types—in this specific case exactly in the middle, as both types are equally likely. The more likely a type is, the closer the expected reaction function will be to the reaction function of this type.

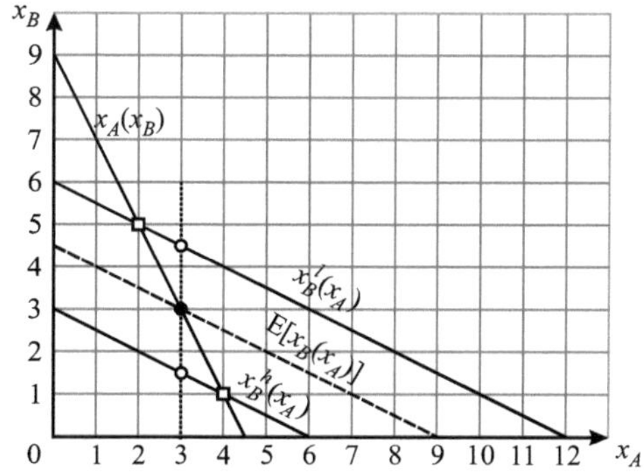

We find the optimal quantity chosen by company A, $x_A^* = 3$, From the intersection of company B's expected reaction function $E[x_B(x_A)]$ with the reaction function of company A, $x_A(x_B)$. The two types of company B react to this quantity according to their reaction functions, $x_B^l(x_A)$ and $x_B^h(x_A)$, with $x_B^{l*} = 4.5$ or $x_V^{h*} = 1.5$ (indicated in the figures by the white circles on the reaction functions).

Part c)

With complete information, company A knows which type of company B it is competing with and will choose its quantity accordingly. If company B has high costs, company A will react to this,

$$x_A\left(x_B^h\right) = 4.5 \quad 0.5 \cdot (3 - 0.5 \cdot x_A) \ \Rightarrow x_A = 4, \ \ x_B^h = 1.$$

Compared to the quantity choice under incomplete information, company A will increase its quantity, and company B will decrease its. This is since both companies are aware of the high costs and thus the less efficient production of company B, which will lead the more efficient supplier (in this case company A) to expand its quantity. The situation is different when company B has low costs:

$$x_A\left(x_B^l\right) = 4.5 - 0.5 \cdot (6 - 0.5 \cdot x_A) \ \Rightarrow x_A = 2, \ \ x_B^h = 5$$

Here, the more efficient company B expands its quantity, while company A offers less. We can thus see that incomplete information fundamentally leads to an inefficient result—the more productive type (with low costs of B, company B, and with high costs of B, company A) will produce too little and the inefficient type will produce too much. Incomplete information therefore mainly benefits the inefficient producer. The efficient producer, on the other hand, would have an incentive to reveal its type.

In the figure, the two equilibria are marked by the white squares on the reaction functions.

Public Finance Application Examples 9

9.1 Problems

9.1.1 Public Good I

The two neighboring municipalities Worrytown (W) and Gamblercity (G) want to jointly establish a casino to promote local tourism. Both municipalities would benefit equally from the revenues, regardless of how much they contribute to the construction costs. A publicly available report shows the following net monetary payoffs for both municipalities:

W, G	No participation	Participation
No participation	(9, 2)	(25, 1)
Participation	(4, 4)	(16, 3)

The expected utility functions of both municipalities are also known with $U_W(x) = \sqrt{x}$ and $U_G(x) = x^2$. Both municipalities make their decision unobserved by each other.

a. Determine the payoff matrix based on the respective utility functions! What type of game is this (order of moves, information structure)? Briefly describe the solution concepts you know for this type of game and explain which one can be applied here!
b. Is the solution Pareto-optimal? Discuss what the problem is in this game! Would anything change if one municipality could observe the other municipality's strategy choice (reasoning)?
c. The casino license runs for 10 years, with specified amounts being paid each year. Does this allow the project to be realized? If not, how long should the license be valid?

Solution in Sect. 9.2.1.

F. Bartholomae and M. Wiens, *Game Theory and Applications*, Classroom Companion: Economics, https://doi.org/10.1007/978-3-658-44608-6_9

9.1.2 Public Good II

A fireworks display is to take place at a party with n people. Each party guest derives a utility of $u > 0$ from the fireworks. However, the fireworks must be ignited by one person who thereby incurs opportunity costs of $c > 0$ as this one can only enjoy the fireworks to a limited extent. However, one person is sufficient to start and operate the fireworks.

a. Formulate the situation as a game: Who are the players? What actions do they have available? What payoffs result?
b. Assume $n = 2$ people. Represent the game in a suitable form! In terms of incentives, is it possible to provide the fireworks?
c. How does the probability that the fireworks will take place change when the number of party guests increases and $u > c$ applies?

Solution in Sect. 9.2.2.

9.1.3 Median Voter Theorem

In the Republic of Democratia, elections are being held in which the two candidates Bernie Bait and Peggy Catcher are running for president. The goal of both candidates is not to stick to election programs but rather to win the election. The candidate who gets the most votes wins the election (simple majority). They can choose between three political positions: Left, center, or right. The 20,000 voters are divided among the individual positions as follows:

Left	Center	Right
5,000	10,000	5,000

Each voter gives their vote to the candidate who is closest to their position, i.e., left-wingers would rather vote for a centrist candidate than a right-wing one. If both candidates are equally far away, the voter decides "spontaneously" (e.g., by tossing a coin); thus, in this case, they "vote" for both candidates with a probability of 50%.

a. Represent the game (players, strategies, and payoffs) in strategic form (matrix)! The trade unionist Bernie Bait feels more on the left, while the ultra-capitalist Peggy Catcher sees herself as a right-winger. Who wins in this case if both position themselves ideologically clearly according to their preferences for the voter?
b. If both are eager to win and put ideological issues aside, how will they behave? Compare your result with a) and explain in this context what the concept of Nash equilibrium is!
c. The initial situation is now changed in two points:

– The political spectrum—and thus also the strategy set of the candidates—is now continuous between 0 and 1, that is, each candidate can take any position (0, 1) in addition to the edge positions (left = 0, right = 1).
– All voters are evenly distributed over this spectrum.

Let s^* be the median of the voter distribution, that is, 50% of the voters are on both sides of this position. Show that the strategy combination (s^*, s^*) represents the unique Nash equilibrium of the game (Median Voter Theorem)! Also illustrate your result graphically using the reaction correspondences!

d. Continue to assume a continuous political spectrum. Show that no Nash equilibrium in pure strategies exists if three candidates were to run!

Solution in Sect. 9.2.3.

9.1.4 Election Campaign

Bowing to the pressure of the international community, the Caribbean dictator Che Castro is forced to organize "free" elections in the Republic of Coconut Paradise. However, Che is now of advanced age and does not want to actively intervene in the election campaign. He hopes that the citizens will honor his previous achievements. If he wins the election, he can continue to govern, which will allow him to divert another 5 million coconut dollars from tax revenue to a Swiss numbered account for his retirement provision. If he loses, however, his successor will send him into exile empty-handed. According to independent election observers, he will win the election with $\rho = 0.6$. Che's utility function is given by $u(y) = 10 \cdot y/(y + 3)$, where y indicates the income in millions.

a. Draw Che's utility function and determine numerically and graphically the expected payoff, the expected utility, and the utility of the expected payoff! What is the dictator's attitude towards risk?
b. General José Matador offers to conduct an "election campaign" for Che. This would ensure that he wins the election. How much would the dictator be willing to pay for the help at most? Describe your approach! Explain the specific meaning of "risk premium" and "certainty equivalent" in this context and indicate both in your graph from a)!

Solution in Sect. 9.2.4.

9.1.5 Tax Evasion

To sustainably reduce their national debt, the government has decided to take tougher action against tax evasion. As public sector jobs are to be cut at the same time, a tax auditor from the private sector has been appointed. Reckless Tech, the largest company in the country, becomes the focus of his investigations. Thanks to "creative accounting", Reckless Tech can increase its profit after tax by 200 million euros compared to a properly prepared financial statement. However, if the company is audited by the tax auditor, this tax evasion will be uncovered. Reckless Tech then must make a payment (back taxes and penalty) of t, of which 5% goes to the tax auditor as a success bonus. The audit costs the tax auditor 10 million euros. If he refrains from the audit, he incurs no costs. The decisions of the tax auditor and the company are made immediately before the end of the tax year.

a. Formulate the situation presented as a game: Who are the players? What are their strategies? Characterize the information structure and present the payoffs depending on the chosen strategy combinations in a suitable form!
b. Characterize the players' strategies for $t < 200$ million euros and $t > 200$ million euros and determine all Nash equilibria!
c. Present the reaction correspondences for $t = 500$ million euros numerically and graphically! Which game that you know is present in this specific case?
d. A government consultant suggests that the decision on whether to audit a company should be made binding and publicly announced at the beginning of a tax period. Using a suitable graphical representation for $t = 500$ million euros, show the game structure and determine the resulting equilibrium! Is it actually advantageous in this case to commit in advance?
e. Prime Minister Sam Smart is annoyed that the success bonus for the private tax auditor causes only 95% of the additional tax revenue to reach the government and that costs are incurred for the audit even if it is carried out by government bodies. Instead, Sam proposes to announce to every company that it will be audited at the end of the year to give companies an incentive to be tax honest. The actual audit could then be omitted, as no company would evade taxes because of the announced audit.

   What problem do you see with this "ideal" solution, assuming rational companies? In what other contexts involving government policy do you encounter a comparable problem structure? Why does the problem not occur when the solution with the private tax auditors is chosen? How can this problem be avoided with pure government control in the case of annually incurred taxes?

Solution in Sect. 9.2.5.

9.1.6 Political Reforms

Once again, elections are being held in the Republic of Democratia. As the country is suffering from a severe economic crisis, voters place particular emphasis on the economic positioning of the parties. The current governing party "Progressive Union" is considering how to conduct the upcoming election campaign.

As possible actions, it can announce that it will set up a reform program in the coming legislative period or refrain from reforms. Subsequently, the population can decide whether to vote for or against the party. If the party has announced a reform program and is elected, it can finally decide whether it wants to implement the program.

An election victory always gives the party a gross utility of 5, while a vote against leads to a utility of 0. If there are no reforms in the country, the economic situation continues to deteriorate, which is rated by the population with –10. The alternative to Progressive Union is extremist parties whose politics will lead to political paralysis and international isolation of the country, which is rated by the population with –10. If the reforms are tackled by the ruling party, the whole country benefits, which the population rates with 10. However, the reform design is very complex and requires long night sessions of the party officials. This inconvenience leads to a loss of utility of –3 for the party. However, if the government breaks its election promises, i.e., it does not implement announced reforms, political disillusionment sets in among the population, which reduces the utility of the population by an additional –15.

a. Formulate the situation as a game: Who are the players? What actions do they have? How is the information structure described? Present the game in a suitable form!
b. Justify whether the path "reform program—election—implementation" can be realized! Which equilibrium will be established? Specify this equilibrium completely!
c. Democratia is a member of a larger union of countries, which has a great interest in reforms in Democratia. To initiate the necessary reforms, various measures are being considered: (i) a penalty payment by the government if it breaks election promises, (ii) a penalty payment by the government if it does not carry out reforms, and (iii) financial aid for the government if it implements the announced reforms. Since the payments directly affect the budget and thus also the probability of re-election of the party, a penalty reduces the utility of the government, while aid increases it.

   How high must the change in utility resulting from the penalty/aid be in each of the three measures for them to be successful? Justify whether all measures are equally suitable for ensuring that reforms are carried out in the country!
d. Now assume that there are two factions within the party: a reform-minded wing and a radical wing. If the radical wing prevails, the party's utility changes to 7 if an announced reform is not implemented; if the reform-minded wing prevails, the utility in the event of implementation of the reform is 7. Who prevails cannot be observed by the population. Justify whether it makes sense from the population's point of view to assume that the reform-minded wing has prevailed when a reform is announced!

Solution in Sect. 9.2.6.

9.1.7 Public Goods and Clarke-Grove Mechanism

In the soccer-enthusiastic small town of Tikitaka, there are two soccer clubs: The big established successful club 1. FC Tikitaka, as well as the smaller—regionally at least as popular—amateur club TSV 1816 Tikitaka. Since the traditional soccer field is very old, both clubs are considering jointly building a small stadium. Such a stadium costs $C = 120$, but the two clubs would derive different utilities from it: The 1. FC has a willingness to pay of $v_A = 100$ and the TSV has a willingness to pay of $v_B = 50$. Each club knows its own willingness to pay, but also that of the other club due to good contacts in the soccer scene. The mayor of Tikitaka has the final say in this project and will only approve the stadium construction if both clubs can raise the required 120 together. For this purpose, she asks both clubs independently about their willingness to pay v_i, which she unfortunately does not know. The willingness to pay stated by the clubs is binding, i.e., they must also contribute to the construction costs in this amount if the money is sufficient. Each club now has two options: Either it honestly tells the mayor the actual amount of its own willingness to pay ($s_i = v_i$), or it speculates that the other club does this, whereby it only must provide for the difference to the required 120. In the latter case, the concerned club is therefore dishonest, as it states a lower willingness to pay than it is ($s_i < v_i$). In the case that the sum of the stated willingness to pay exceeds the amount of 120, then the clubs will participate in the costs in proportion to their stated willingness to pay, for example, if both make honest statements in the amount of v_A and v_B in the ratio 2:1.

a. Determine the payoffs for all strategy combinations and present the game in matrix form!
b. Determine a plausible solution to the simultaneous game using dominance considerations! Name the methods used for this! Is the solution a Nash equilibrium? If yes, does the method ultimately leading to the solution ensure that the Nash equilibrium is unique? Determine all Nash equilibria in pure strategies! Is there a conflict of interest and/or a coordination problem in this case? Explain your answer!
c. The mayor is not a soccer expert, but she is a trained economist with profound knowledge in finance. In her search for a solution to her problem, she comes across the Clarke tax, which works as follows: Each of the two clubs states its willingness to pay bindingly. If the stadium can be built (i.e., if the sum of the willingness to pay reaches the costs of 120), then each of the two clubs must pay a tax with $t_i = C - s_j$. The tax that the club i must pay, therefore, depends only on the stated willingness to pay of the club j. The payoff of each player then results as the difference from his actual willingness to pay and the tax: $u_i = v_i - t_i$.

Determine the payoffs for all strategy combinations under the conditions of the Clarke tax and present the game in matrix form! Find all Nash equilibria in pure strategies and provide a plausible solution to the game! Compare your result with that from part (b) in terms of Pareto efficiency!

Solution in Sect. 3.

9.2 Solutions

9.2.1 Public Good I

Problem Text in Sect. 9.1.1.

Part a)
If the monetary payoffs of the matrix given in the problem are evaluated with the respective utility functions of the two municipalities, the following matrix results:

W, G	No participation	Participation
No participation	(3, 4)	(5, 1)
Participation	(2, 16)	(4, 9)

Since the municipalities decide unobserved from each other, there is imperfect information. The report itself is public, so we can assume complete information. It is therefore a simultaneous game.

Dominance considerations or the direct search for the Nash equilibrium are possible solution concepts. Dominance considerations either look for a dominant strategy that can be played or for dominated strategies that are not chosen (iterated elimination of dominated strategies). When searching for the Nash equilibrium, we look for a pair of mutually best responses. We see in the present game that both Gamblercity and Worrytown have the dominant strategy "no participation".

Part b)
The solution (no participation, no participation) is not Pareto-optimal, as both could be better off with the strategy combination (participation, participation): Worrytown would have a utility of 4 instead of 3 and Gamblercity would have 9 instead of 4. The problem is that the Pareto optimum cannot be reached because both have the strictly dominant strategy "no participation" and will therefore always deviate from this optimum. The incentive structure of the game is therefore that of a prisoner's dilemma.

The order of moves has no effect on the result when strictly dominant strategies are present: (Participation = p, no participation = np):

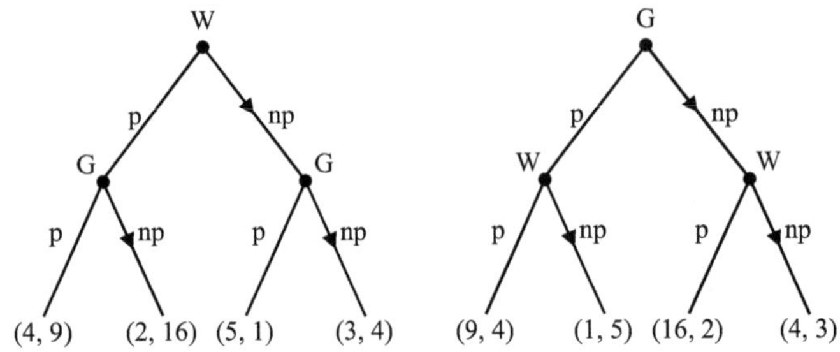

We see that in both cases it happens that no municipality participates in the construction of the casino.

Part c)
No, from the theorem on finitely repeated games, which has to be applied in the case of 10 years, we know that only the repetition of the Nash equilibrium of the base game is subgame perfect. Thus, even in this case, the casino would not be built. Only if the game is repeated an infinite number of times, the Folk theorems state that any feasible individually rational payoff vector can be realized. Since the casino project is Pareto-optimal, it fulfills this condition, so that we could support it (with sufficiently high discounting) as an equilibrium. The license must therefore be valid forever.

9.2.2 Public Good II

Problem Text in Sect. 9.1.2.

Part a)
We can identify the n party guests as players who each have the two actions of "contributing" or "not contributing" to the fireworks. The resulting payoffs of each player are:

$$u_i = \begin{cases} 0 & \text{if no one contributes} \\ u - c & \text{if player } i \text{ contributes} \\ u & \text{if someone else contributes} \end{cases} \tag{9.1}$$

Part b)
Since it is a simultaneous game, we represent the game in matrix form.

A, B	Contribute	Not contribute
Contribute	$(u-c, u-c)$	$(u-c, u)$
Not contribute	$(u, u-c)$	$(0, 0)$

Since no precise information about u and c was given, three cases must be distinguished here: $u < c$, $u = c$, and $u > c$:

- With $u < c$, "not contribute" is a strictly dominant strategy for both players: If the other player contributes, it is better not to contribute, the same applies if the other player does not contribute. Thus, (not contribute, not contribute) represents the only Nash equilibrium of the game.
- With $u = c$, "not contribute" is a weakly dominant strategy for both players and there are a total of three Nash equilibria: (not contribute, contribute), (contribute, not contribute), and (not contribute, not contribute).
- With $u > c$, there are two Nash equilibria in pure strategies: (not contribute, contribute) and (contribute, not contribute). Since two Nash equilibria in pure strategies exist, there must be another one in mixed strategies exist. This can be easily found using the expected utility comparison. Due to the symmetry, it is also sufficient to only consider one player. Let p_B be the probability that player B chooses "contribute", then player A prefers "contribute" to "not contribute" if $p_B \cdot (u-c) + (1-p_B) \cdot (u-c) > p_B \cdot u + (1-p_B) \cdot 0$. Thus, player A contributes if $p_B < (u-c)/u$. The Nash equilibrium in mixed strategies in this case is $\left(p_A - \frac{u-c}{u}, p_B - \frac{u-c}{u}\right)$, with p_A denoting player A's probability to choose "contribute".

For none of these cases is it guaranteed that the fireworks would take place: If $u < c$ there will be no fireworks for sure, for $u = c$ it is also not guaranteed, as only in two of the three Nash equilibria will the fireworks take place. For $u > c$ the fireworks will take place, provided the two players manage to coordinate on one of the equilibria in pure strategies. If they fail to do this and the strategic uncertainty persists, the fireworks will take place with a probability of $1-c/u$—and is therefore more likely the larger u is compared to c.

Part c)

We know that it is generally only better to contribute if none of the other players contribute (when the fireworks take place, you have a utility, but if you are the one contributing, you also have a cost). The question is, if the number of players increases, is it more likely that one of the other players will contribute? After all, there are more people who could contribute.

To determine the equilibrium, we again use the expected utility comparison and consider the situation from the perspective of player A:

$$\begin{aligned}
\text{Contribute :} \quad & E[u_A(p_A = 1, p_B, \dots, p_n)] = u - c \\
\text{Not contribute :} \quad & E[u_A(p_A = 0, p_B, \dots, p_n)] \\
& = u \cdot \text{prob("at least one of the } n-1 \text{ players contributes")} \\
& = u \cdot \left[1 - \text{prob("none of the } n-1 \text{ players contributes")}\right] \\
& = u \cdot [1 - (1 - p_B) \cdot (1 - p_C) \cdot \ldots \cdot (1 - p_n)].
\end{aligned}$$

Player A will therefore contribute if $u - c \geq u \cdot [1 - (1 - p_B) \cdot (1 - p_C) \cdot \ldots \cdot (1 - p_n)]$ or $(1 - p_B) \cdot (1 - p_C) \cdot \ldots \cdot (1 - p_n) \geq c/u$. We can take advantage of the symmetry of the game, that is, assume that $p_A = p_B = \ldots = p_n = p$, i.e., $(1 - p_B) \cdot (1 - p_C) \cdot \ldots \cdot (1 - p_n) = (1 - p) \cdot (1 - p) \cdot \ldots \cdot (1 - p)$. This is a total of $n - 1$ expressions, as there are n players, but one of them is player A, who is not included here. This simplifies the condition to $(1 - p)^{n-1} \geq c/u$. To solve for the probability p played in Nash equilibrium to keep the players indifferent, we need to take the $(n - 1)$-th root: $1 - p \geq \sqrt[n-1]{c/u}$ or

$$p \leq 1 - \left(\frac{c}{u}\right)^{\frac{1}{n-1}}.$$

Note that the exponent $\frac{1}{n-1}$ is only another way of writing the $(n - 1)$-th root. Thus, the probability that any party guest contributes to the fireworks in the symmetric equilibrium is

$$\begin{aligned}
1 - \text{prob("no one contributes")} = 1 - (1 - p)^n = 1 - \left(\frac{c}{u}\right)^{\frac{n}{n-1}} &= 1 - \left(\frac{c}{u}\right)^{1+\frac{1}{n-1}} \\
= 1 - \frac{c}{u} \cdot \left(\frac{c}{u}\right)^{\frac{1}{n-1}}&.
\end{aligned}$$

As we can see, the more party guests, the larger the root, which means that the expression gets smaller and smaller. This leads to the interesting but somehow disappointing result that the probability of fireworks decreases(!), the more party guests participate. The reason for this is that each guest—with an increasing number of guests—assumes with increasing certainty that someone else will take care of it. Psychologists refer to this problem in such contexts as the problem of responsibility diffusion or "volunteer's dilemma".

9.2.3 Median Voter Theorem

Problem Text in Sect. 9.1.3.

Part a)
The number of expected votes can be considered as the payoff of the players. Thus, we get the following matrix:

Bernie, Peggy	Left	Center	Right
Left	(10,000, 10,000)	(5,000, 15,000)	(10,000, 10,000)
Center	(15,000, 5,000)	(10,000, 10,000)	(15,000, 5,000)
Right	(10,000, 10,000)	(5,000, 15,000)	(10,000, 10,000)

If both candidates position themselves according to their preferences, they will both get 10,000 votes and thus neither of them will emerge as the winner from the election.

Part b)
For both players "center" represents a strictly dominant strategy: Given the competitor chooses an extreme position, "center" is preferable to any other position, as then both the voters from the "center" camp and from the other extreme position can be won. If the competitor also chooses "center", "center" is also preferable, as otherwise the election is certainly lost. This results in ("center", "center") as unique Nash equilibrium of the game and both have (as in a) a payoff of 10,000 votes each. Although neither of the candidates has improved their situation, in the Nash equilibrium neither of the players has an incentive to deviate. For example, if Peggy assumes that Bernie follows his ideology, she will have an incentive not to follow her ideology, but to choose "center" instead, as this would allow her to improve her situation.

Part c)
The two players now have a continuous strategy choice, i.e., Bernie and Peggy can choose any position between 0 (left) and 1 (right), $s_B, s_P \in [0, 1]$. Since the voters are uniformly distributed, the median of the distribution is at 0.5, i.e., we must show that the strategy combination ($s_B = 0.5, s_P = 0.5$) represents the Nash equilibrium of the game. We therefore check whether a unilateral deviation from this strategy combination is worthwhile. Due to symmetry, it is sufficient to consider only one player.

Specifically, the distribution can be described graphically as follows (with a density function, the total area adds up exactly to 1, i.e., if the width is 1 the height/density is 1 as well):

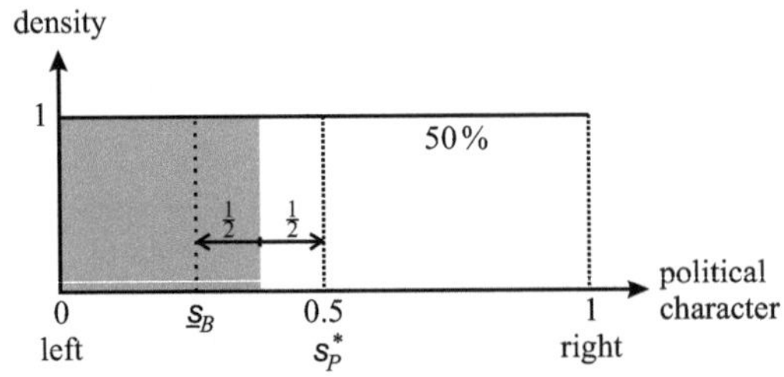

Let us assume that Peggy has positioned herself on the median, $s_P^* = 0.5$.

- In the figure we see the case when Bernie chooses a position $\underline{s}_B < 0.5$ and has therefore taken a more left-wing position than Peggy. His payoff is represented by the gray area. On the one hand he receives the votes of all voters who are more left-wing than he is (in the range $[0, \underline{s}_B]$) and on the other hand the votes of all voters who are more right-wing than he is but closer to his position than to that of Peggy. This applies to all who are to the left of the midpoint of the distance between him and Peggy, $\left[\underline{s}_B, \frac{1}{2} \cdot (s_P^* - \underline{s}_B)\right]$. His payoff is therefore:

$$\left[\underline{s}_B + \frac{1}{2} \cdot (s_P^* - \underline{s}_B)\right] \cdot 20,000 = \left[\underline{s}_B + \frac{1}{2} \cdot (0.5 - \underline{s}_B)\right] \cdot 20{,}000 = 5{,}000 + 10{,}000 \cdot \underline{s}_B.$$

We see that the payoff increases in $\underline{s}_B$. Furthermore, in the range $\underline{s}_B \in [0, 0.5)$ it is always less than 10,000. With $s_B^* = 10{,}000$, however, his votes would be 10,000 and thus higher than with $s_B < 0.5$.

- If Bernie chooses a position $\overline{s}_B > 0$ and has therefore taken a more right-wing position than Peggy, he receives as payoff all votes of the voters who are more right-wing than he is (in the range $[\overline{s}_B, 1]$) and all votes of the voters who are more left-wing than he is but closer to his position than to that of Peggy, $\left[\frac{1}{2} \cdot (\overline{s}_B - s_P^*), \overline{s}_B\right]$. His payoff is therefore:

$$\left[(1 - \overline{s}_B) + \frac{1}{2} \cdot (\overline{s}_B - s_P^*)\right] \cdot 20{,}000 = \left[1 - \overline{s}_B + \frac{1}{2} \cdot (\overline{s}_B - 0.5)\right] \cdot 20{,}000$$
$$= 15{,}000 + 10{,}000 \cdot \overline{s}_B.$$

We see here that the payoff decreases in $\overline{s}_B$. Furthermore, in the range $\overline{s}_B \in (0.5, 1]$ it is always less than 10,000. Thus, $s_B^* = 0.5$ also always yields more votes than $s_B > 0.5$.

In summary, we see that if Peggy chooses $s_P^* = 0.5$ a deviation from $s_B^* = 0.5$ neither to $s_B < 0.5$ nor to $s_B > 0.5$ is worthwhile for Bernie: In both cases, Bernie will certainly lose the election, as he can gather fewer votes than Peggy. The strategy combination $(s_B = 0.5, s_P = 0.5)$ therefore represents a Nash equilibrium.

To graphically represent the equilibrium, we need reaction correspondences, i.e., we need to determine how the optimal reaction looks like given a certain positioning of the opposing candidate. Basically, there are always three options: a more left-wing positioning, a more right-wing positioning, or a positioning at the position of the opposing candidate. As we have seen, different payoffs result depending on which position Peggy takes. If Peggy is more left-wing than the median, it makes no sense for Bernie to position himself even more left-wing than Peggy, as this would only secure him a few votes to the left of Peggy. On the other hand, Peggy would get all the votes from voters who are more right-wing than her and thus win the election. Bernie's best response here is to present himself as more right-wing than Peggy but to stay as close as possible to her position. This way, Bernie gets the votes of all voters who are more right-wing than he is and loses very few votes from voters who are more left-wing than he is, but more right-wing than Peggy, with a very narrow positioning towards Peggy. He would therefore win the election. The same consideration applies if Peggy has positioned herself further to the right than the median. In this case, Bernie should position himself more to the left, but as close as possible to Peggy's position. The rule is therefore always to position oneself closer to the median than the other, but still very close to the position of the competitor. Only if the opposing candidate has chosen a position exactly in the median, the same position should be chosen, as otherwise the election would be lost.

Formally, the reaction correspondences of the candidates are

$$s_B(s_P) = \begin{cases} s_P + \varepsilon & s_P \in [0, 0.5) \\ 0.5 & s_P = 0.5 \\ s_P - \varepsilon & s_P \in (0.5, 1] \end{cases}, \quad s_P(s_B) = \begin{cases} s_B + \varepsilon & s_B \in [0, 0.5) \\ 0.5 & s_B = 0.5 \\ s_B - \varepsilon & s_B \in (0.5, 1] \end{cases}$$

where ε represents a very small value and means that the positioning should be very close to that of the other candidate (but not directly at their position). The following figure represents the reaction correspondence graphically. These are very close to each other and only intersect at the Nash equilibrium, $(s_B = 0.5, s_P = 0.5)$.

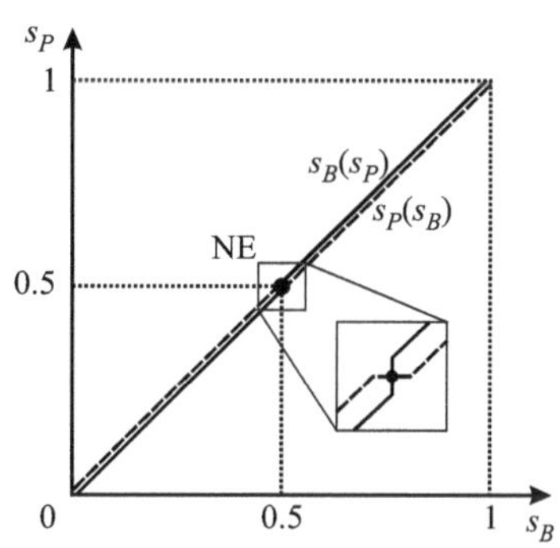

The *median voter theorem* states that in majority elections, there is an incentive for candidates/parties to position themselves in the center to receive as many votes as possible and thus win the election. However, as it turns out, this makes the parties hardly distinguishable for the voters and the election success is not guaranteed, as the voters are evenly distributed between the two candidates/parties.

Part d)
Based on the following figure, we want to better illustrate a situation with three candidates:

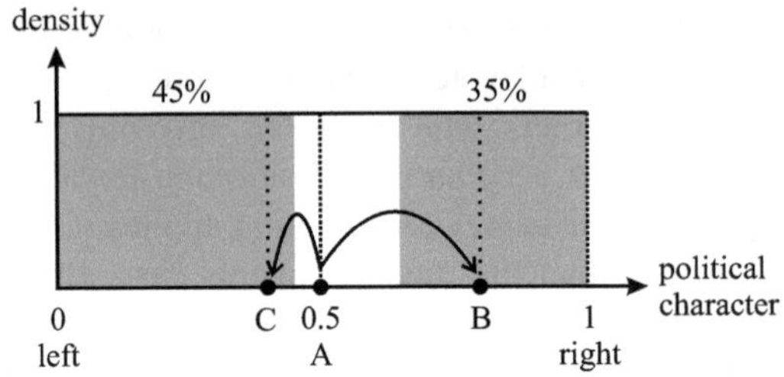

Let us assume that, in the initial situation, the three candidates position themselves at the median (position A in the figure). In this case, the voters are evenly distributed between the three candidates and each therefore receives 33% of the votes. However, a candidate could choose position B, i.e., adopt a more right-wing position than the median. At B, the candidate would get all the votes of the voters who are more right-wing than his position, as well as all the votes of the voters who are more left-wing than he is, but more right-wing than the median. In this specific case, candidate B would receive 35% of the votes and thus win the election (the right gray area; the remaining 65% would be evenly distributed among the other two candidates, so each would only have 32.5% of the votes). Now, however, one of the other two candidates could switch to position C, i.e., adopt a left-wing counter-position to B. At this position, he would even receive 45% of the votes (left gray area) and win the election (B would get 35% and A the remaining 20%). Now A could in turn adopt a more left-wing position than C and thereby for example achieve 44% of the votes (then C would have 21% and B still 35%). Each candidate can thus always improve given the position choice of the other candidates, so there is always an incentive to deviate from his initial position. Therefore, there is no Nash equilibrium in pure strategies here.

9.2.4 Election Campaign

Problem Text in Sect. 9.1.4.

Part a)
We calculate the expected value with

$$E(a) = 0.6 \cdot 5 + 0.4 \cdot 0 = 3,$$

the expected utility with

$$E[u(a)] = 0.6 \cdot 10 \cdot \frac{5}{5+3} + 0.4 \cdot 10 \cdot \frac{0}{0+3} = 3.75,$$

and the utility of the expected payoff with

$$u[E(a)] = 10 \cdot \frac{3}{3+3} = 5.$$

Since the expected utility is less than the utility of the expected value, Che is risk-averse. Accordingly, his utility function is concave:

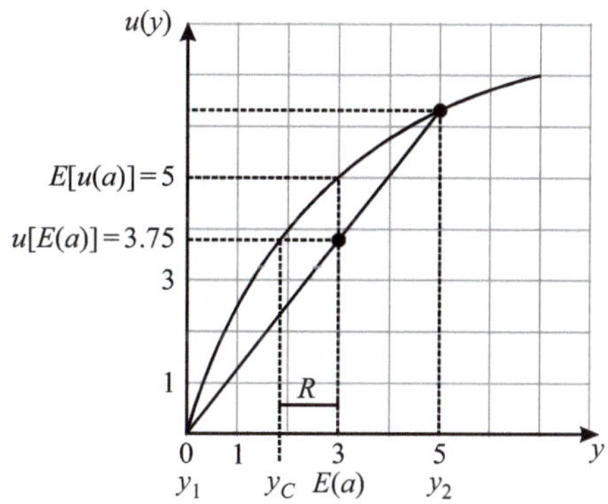

Part b)
Che would be willing to pay at most as much for the electoral assistance so that he would receive the same utility as with the lottery. That is, he would just be willing to pay the risk premium of the lottery. To calculate the risk premium, the certainty equivalent is needed, i.e., the certain payoff that generates the same utility for him as the lottery. This is calculated as

$$u(y_C) = E[u(a)] \quad \Leftrightarrow \quad 10 \cdot \frac{y_C}{y_C + 3} = 3.75 \quad \Rightarrow \quad y_C = 1.8.$$

As we can see, the risk premium is the certain payoff that provides the same utility as the uncertain lottery, i.e., the individual is indifferent between the lottery and the certainty equivalent.

The difference between the expected value and the certainty equivalent finally provides the risk premium,

$$R = E(a) - y_C = 3 - 1.8 = 1.2.$$

The risk premium indicates how much the individual is willing to give up receiving the safe payoff of the safety equivalent instead of the uncertain lottery. However, it can also be interpreted the other way round: How much must the decision-maker be paid to be willing to take risks (i.e., an uncertain lottery instead of a certain payoff)? The risk-taking Che is therefore willing to hand over up to 1.2 million coconut dollars, i.e., to pay to the General, to avoid the lottery—the uncertain outcome of the election.

In the figure in (c), both the certainty equivalent and the risk premium are already plotted.

9.2.5 Tax Evasion

Problem Text in Sect. 9.1.5.

Part a)
We can identify Reckless Tech and the tax auditor as players. Reckless Tech can choose between the strategies "creative accounting" and "correct financial statement" and the tax auditor between "control" and "no control". Since the decisions are made unobserved from each other, there is imperfect information but, at the same time, there is common knowledge about the payoffs and strategies of the players. Due to the imperfect information structure, we represent the payoffs in matrix form:

Reckless Tech, Tax Auditor	Control	No control
Creative accounting	$(200 - t, 0.05 \cdot t - 10)$	(200, 0)
Correct financial statement	(0, – 10)	(0, 0)

Part b)
With $t < 200$ million euros, "creative accounting" represents a strictly dominant strategy for Reckless Tech: In the case of "control", the profit is higher than the penalty, so a "correct financial statement" is not worth it. If there is "no control", "creative accounting" is preferred anyway. Conversely, "no control" is a strictly dominant strategy for the tax auditor: His control costs exceed the premium in case of control success—for control to be profitable it must apply: $0.05 \cdot t - 10 > 0$ or $t > 10/0.05 = 200$. Thus, there is a clear Nash equilibrium in dominant strategies at (creative accounting, no control).

With $t > 200$ million euros, there is no Nash equilibrium in pure strategies: If the tax auditor "controls", Reckless Tech will prepare a "correct financial statement". But, with correct accounting, the tax auditor does not want to control. If there is no control, Reckless Tech, however, has an incentive for creative accounting. Here, the tax auditor would like to control, which brings us back to the beginning of the cycle. Therefore, an equilibrium must exist in mixed strategies.

Let p_R be the probability with which Reckless Tech chooses "creative Accounting" and p_T be the probability with which the tax auditor controls. Reckless Tech decides for "creative accounting", i.e., $p_R = 1$, when

$$\begin{aligned} E[u_R(p_R = 1, p_T)] &> E[u_R(p_R = 0, p_T)] \\ (200 - t) \cdot p_T + 200 \cdot (1 - p_T) &> 0 \cdot p_T + 0 \cdot (1 - p_T) \\ p_T &< 200/t. \end{aligned}$$

The tax auditor will decide for "control", i.e., $p_T = 1$, when

$$\begin{aligned} E[u_T(p_R, p_T = 1)] &> E[u_T(p_R, p_T = 0)] \\ (0.05 \cdot t - 10) \cdot p_R - 10 \cdot (1 - p_R) &> 0 \cdot p_R + 0 \cdot (1 - p_R) \\ p_R &> 200/t. \end{aligned}$$

The Nash equilibrium in mixed strategies is thus $(p_R = 200/t, p_T = 200/t)$. A higher penalty thus equally reduces the probability with which the authority controls as well as the probability for creative accounting. This is because the penalty t is included in both utility functions and at the same time makes fraud and control less attractive.

Part c)

For $t = 500$ million euros (> 200 million euros), there is a Nash equilibrium in mixed strategies at $(p_R = 0.4, p_T = 0.4)$ with the reaction correspondences:

$$p_R(p_T) = \begin{cases} 1 & p_T \in [0, 0.4) \\ [0, 1] & p_T = 0.4 \\ 0 & p_T \in (0.4, 1] \end{cases}, \quad p_T(p_R) = \begin{cases} 0 & p_R \in [0, 0.4) \\ [0, 1] & p_R = 0.4 \\ 1 & p_R \in (0.4, 1] \end{cases}$$

Graphically, this results in the reaction correspondences with only one intersection point at $(p_R = 0.4, p_T = 0.4)$:

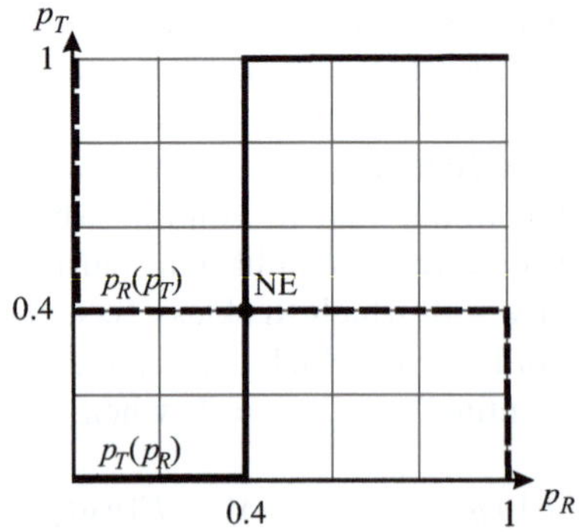

This game is an inspection game, which has the characteristics of an anti-coordination game (such as penalty kicks, matching pennies, etc.).

Part d)
With this change we now have perfect information for which we use the extensive form or game tree for representation:

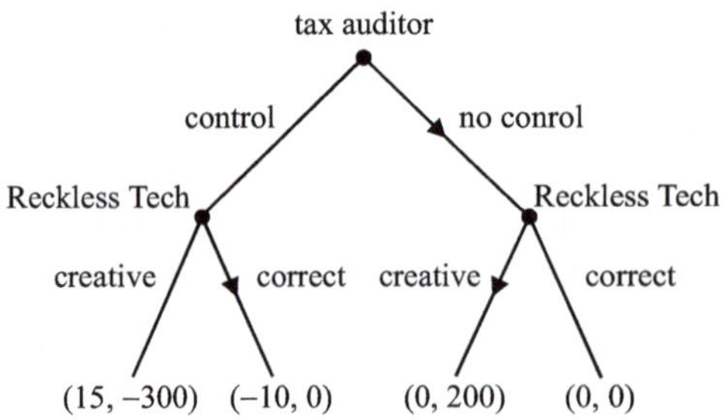

If we apply backward induction, we find the subgame perfect Nash equilibrium ("no control", ("correct financial statement" ("control"), "creative accounting" ("no control")). In this game, there is a second-mover advantage, meaning the second player (Reckless Tech) can determine the outcome that is more favorable for him. Therefore, it is not advantageous for the tax auditor to commit first.

Part e)
The proposed "ideal solution" is based on the problem of *time inconsistency*: The government has the incentive to announce the control of companies but finally will refrain from control to avoid costs. The government's announcement is therefore not credible. A rational company will anticipate this inconsistency and therefore opt for creative accounting, which means the solution originally sought by the government is not achieved.

A comparable problem structure exists, for example, in monetary policy: The central bank has the incentive to announce monetary measures to influence the expectations of economic subjects; however, once it has achieved its goal, it will not want to implement them. Another example is patent protection, which is intended to encourage companies to innovate. Once the innovation has occurred, it is desirable from society's perspective that it is available to everyone, and the innovative company does not claim the monopoly rent for itself. There is therefore an *ex post* incentive to abolish patent protection.

This problem does not occur with private tax auditors: They always have the incentive to inspect, as they only receive payment if they can provide their client with evidence of the inspections.

The government can circumvent the problem by building a reputation for always controlling. This is possible with repeated audits, as companies know from the past that the government's announcement was not an empty threat. To reduce costs, the audit can also only take place with a certain probability, which is high enough to ensure that the companies account correctly.

9.2.6 Political Reforms

Problem Text in Sect. 9.1.6.

Part a)

We can identify the governing party "Progressive Union" and the population as players. Before the election, the party can choose between the actions "announce reform program" and "no announcement", and after the election between "implement reforms" and "not implement reforms". The population, in turn, can choose between "vote" and "vote out". All decisions are made observably from each other, meaning the population can observe which program the party announces and the party can observe whether it was elected. Thus, perfect information is present. At the same time, there is *Common Knowledge* about the payoffs and strategies of the players. Due to the perfect information structure, it is recommended to represent the payoffs in extensive form (game tree):

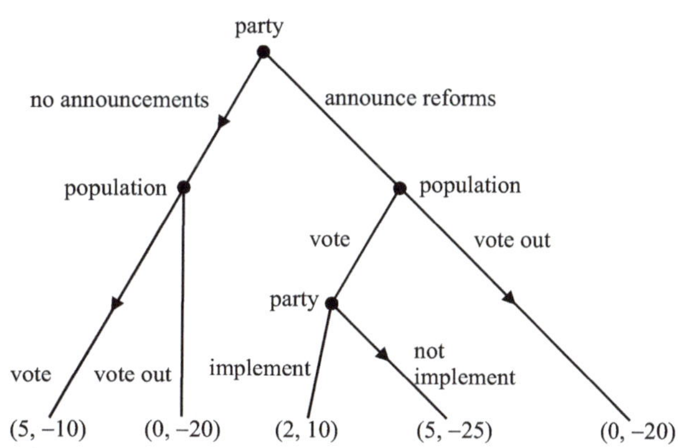

Part b)

The indicated path does not represent an equilibrium of the game. It is based on the incredible promise to implement reforms. However, once the party has been elected, it has no incentive to do so, which is anticipated by the population and results in the party being voted out. Only when the party does not announce a reform program is this perceived as credible by the population and leads to electoral success (and thus to a higher payoff) for the party.

The subgame perfect Nash equilibrium can be determined using backward induction (as already shown in the figure in a). This is here: ("not announcement", "not implement"), ("vote" ("no announcement"), "vote out" ("announce reforms"))).

Part c)

To encourage the party to implement the announced reforms under measure (i), its utility change in the case of non-implementation of the announced reforms must be at least -3 (this is the only way to ensure a lower utility in the case of non-implementation compared to implementation). The party will then always implement its announced reforms and will also be elected in this case, but it will still not announce any reforms (the party's payoff in this case is 5 and is therefore higher than the payoff of 2, which it would receive with the reform program).

For measure (ii), the party's utility change must be at least -3. This penalty is due if it does not announce reforms as well as if it breaks announced reforms (as in the case of measure (i)). If it carries out announced reforms, it receives a payoff of 2; thus, the payoff when it does not announce reforms and is elected must be less than 2, which is the case with a penalty of at least 3. In equilibrium, the party will then announce and implement reforms.

Measure (iii) is the counter variant to measure (ii) and increases the government's utility when it announces and implements reforms. Just as a penalty of -3 is required for measure (iii), the utility of implementing the announced reform must be increased by 3, as the party then receives a higher payoff than when not announcing the reforms as well as when not implementing announced reforms. In this case, too, the desired result can be achieved.

In summary, we see that measures (ii) and (iii) are suitable for achieving the goal of the union of states, because only then will the government be willing to announce and implement reforms.

Part d)

Now there is incomplete information, as the party can be of two types—reform-minded or radical. While the reform-minded type will always implement announced reforms, the radical type will not consider this.

To check whether it makes sense from the population's point of view to assume that the reform-minded wing has prevailed when a reform program is announced, it must be examined whether the radical type has an incentive to present itself as a reform-minded type by announcing reforms. If the population assumes that the party is reform-minded, it will always vote for the party, as it will also implement the reforms. A radical party can increase its utility if it announces reforms that it does not implement (it then receives a payoff of 7 instead of 5). Hence, it has an incentive to present itself as reform-minded and announce reforms. Therefore, the population can never be sure which of the two wings has prevailed, as both have an incentive to announce reforms. Therefore, it is not possible to infer the reform-minded type of the party from the observed action "announcing reforms".

9.2.7 Public Goods and Clarke-Grove Mechanism

Problem Text in Sect. 9.1.7.

Part a)
The two players (1. FC Tikitaka as player A and TSV 1816 Tikitaka as player B) each have two strategies: To state their own willingness to pay "honestly" ($s_i = v_i$) or to make "dishonest" statements in the hope that the other club j contributes its actual willingness to pay ($s_i = C - v_j$) and club i only has to add the rest to enable the financing of the stadium.

If both make dishonest statements, the stadium cannot be built as they only contribute 90 together and thus do not reach the required 120. If both make honest statements, then they would contribute more than necessary. If the shares of the contributions are distributed according to the costs, then player A contributes 80 and player B contributes 40. In the two asymmetric constellations, where always one club is "honest" and the other is "dishonest", the contributions always add up to the required sum $C = 120$. The payoff of player i is then calculated as the difference between their own willingness to pay and their own contribution: $u_i = v_i - s_i$. We thus obtain the following matrix:

A, B	Honest $s_{B1} = 50\ (= v_B)$	Dishonest $s_{B2} = 20\ (= C - v_A)$
Honest $s_{A1} = 100\ (= v_A)$	(20, 10)	(0, 30)
Dishonest $s_{A2} = 70\ (= C - v_B)$	(30, 0)	(0, 0)

Part b)
As we see in the matrix, both clubs have the weakly dominant strategy to be "dishonest": If player j makes "dishonest" statements, player i is indifferent, as he receives a payoff of zero in both honesty and dishonesty. If player i behaves "honestly" while player j is "dishonest", the latter can reap the entire utility from the stadium construction. If player i also behaves "dishonestly", the stadium cannot be built, resulting again in a payoff of zero.

Dominance considerations thus lead to the determination of an equilibrium in dominant strategies, which is (dishonest, dishonest). An equilibrium found with this method always represents a Nash equilibrium, but this method does not ensure the uniqueness of the equilibrium (i.e., there may be other Nash equilibria). In contrast, an equilibrium in *strictly* dominant strategies would always be unique. However, the underline method allows us to find out whether there are other Nash equilibria and where they are. As the matrix shows, there are two more Nash equilibria, one at (honest, dishonest) and the reverse constellation (dishonest, honest). Since the number of Nash equilibria in pure strategies is already odd, there is no further one in mixed strategies.

Since the two asymmetric Nash equilibria are Pareto-optimal, it is not a cooperation problem, but a coordination problem. The coordination problem involves a conflict of interest, as there is no compromise possible between the equilibria. Overcoming the problem would only be conceivable through an exogenous focal point, for example, by knowing that one of the two clubs has higher financial reserves.

Part c)

If we apply the described procedure to the problem, player i receives a payoff of $u_i = v_i - t_i$ and for $t_i = C - s_j$ we get accordingly $u_i = v_i - C + s_j = v_i + s_j - 120$. Each player thus pays the *Clarke tax* t_i, which is exactly measured by the amount still needed to finance the stadium. If, for example, both clubs are "honest", then player A pays $u_A = v_A + s_B - 120 = 100 + 50 - 120 = 30$. In the same way, all payoffs can be transformed. With the tax, the game then changes as follows:

A, B	Honest $s_{B1} = 50$	Dishonest $s_{B2} = 20$
Honest $s_{A1} = 100$	$(\underline{30}, \underline{30})$	$(\underline{0}, \underline{30})$
Dishonest $s_{A2} = 70$	$(\underline{30}, \underline{0})$	$(\underline{0}, \underline{0})$

As we see, all four strategy combinations now represent a Nash equilibrium in pure strategies. Since this number is even, there also exists an equilibrium in mixed strategies, which, however, will not be determined here. Both players are now indifferent between their strategies, which on the one hand has exacerbated our coordination problem by another equilibrium (in pure strategies). On the other hand, however, the new equilibrium at (honest, honest) is the only one that is Pareto-optimal and thus represents a very convincing endogenous focal point for both clubs to overcome the dilemma.

The mechanism of the Clarke tax, better known in the literature as the *Clarke-Groves mechanism*, is a *truthtelling mechanism*, i.e., a mechanism that removes the incentive for players to lie. As we see from the game, it now makes no sense for either of the two clubs to make a dishonest statement, as "honest" and "dishonest" lead to the same payoff. This is because the tax that club i must pay depends only on the declared willingness to pay of club j. The Clarke tax thus has the property of exactly offsetting the (positive) external effect that club i causes to club j: By contributing too little and the costs C not being covered, the stadium project would be brought down, for which club j has a positive willingness to pay.

Military Application Examples 10

10.1 Problems

10.1.1 Battle in the Bismarck Sea

In the Second World War, a Japanese admiral was ordered in 1943 to move troops and material from the Bismarck Archipelago to the mainland of Papua New Guinea.[1] The Japanese had the choice between a rainy northern route and a sunnier southern route. The U.S. Air Force was informed about the planned relocation and wanted to bomb the convoy. However, they did not know which sea route the Japanese would choose. To find this out, reconnaissance flights had to be made. However, there were not enough aircraft available for simultaneous checking of both routes. The convoy would be on the move for three days and if it was on the course that the Americans first investigated, they could immediately start the attacks. On the rainy route, the Air Force would have to expect that air strikes would have to be omitted due to poor visibility on one day. If reconnaissance flights were first made on the northern route, two days of bombing would result, regardless of the decision of the Japanese; if the southern route was investigated first, in contrast, three days would be available for air strikes if the Japanese did choose this route, and only one day if they instead opted for the northern course.

a. Formulate this situation as a game: Who are the players? What strategies do they each have? What payoffs result depending on the chosen strategy combinations? Can the situation be adequately analyzed within the framework of a matrix game, or does it have to be represented in extensive form (game tree)?

[1] See also Haywood Jr, O. G. (1954): Military decision and game theory, Journal of the Operations Research Society of America, 2(4), 365–385.

F. Bartholomae and M. Wiens, *Game Theory and Applications*, Classroom Companion: Economics, https://doi.org/10.1007/978-3-658-44608-6_10

b. How can a plausible solution to the game be determined using dominance? Is this solution unique?
c. Does the solution from b) also represent a Nash equilibrium?
d. Would the application of the maximin rule result in a different solution? Does the maximin criterion represent a reasonable solution concept in this specific case? Is it plausible in economic applications?

Solution in Sect. 10.2.1.

10.1.2 Counterterrorism

Due to recent threats, the government is forced to reconsider surveillance standards in object protection. The protection of a certain object has a value of V for the government, whereas the terrorists assign a value of T to the destruction of this good. The government can decide to carry out weak controls (w), in which the attack can be prevented with a probability of ρ_w or to control strongly (s), whereby an attack can be prevented with ρ_s but additional costs c are incurred. If the control is successful, the attack is prevented, and the terrorists receive a penalty F. The decisions of the government and the terrorists are made simultaneously. The payoffs are summarized in the following matrix:

Government, Terrorists	Attack	No attack
Weak control	$(\rho_w \cdot V, (1-\rho_w) \cdot T - \rho_w \cdot F)$	$(V, 0)$
Strong control	$(\rho_s \cdot V - c, (1-\rho_s) \cdot T - \rho_s \cdot F)$	$(V - c, 0)$

a. Briefly define the two characteristics of the information structure and describe the current situation! Do you consider the given information structure plausible in the context of the analysis?
b. Show that for $\rho_w = \rho_s$, the strategy “weak control” is a dominant strategy for the government! Briefly explain why this result seems plausible! Explain whether a sequential game structure would affect your result!

The government commissions a private security service provider to explicitly analyze the situation. For this purpose, the following values are determined: $V = 100$, $\rho_w = 0.8$, $\rho_s = 0.95$, $c = 10$, $T = 50$, and $F = 10$.

c. Present the resulting payoff matrix and show that there is no Nash equilibrium in pure strategies!
d. Define the term “mixed strategy”! Explain why there must be a Nash equilibrium in mixed strategies in this game! Determine numerically and graphically the Nash equilibrium in mixed strategies!

e. Discuss whether the concept of mixed strategy in the given situation seems plausible from a game theoretical perspective and is desirable from the perspective of the population (as an uninvolved but affected party)!
f. The security service provider also proposes two options to the government to further reduce the risk of an attack: Option 1 involves doubling the penalty, whereas option 2 targets the cost side and halves it. Discuss which of the two options you consider to be more target-oriented!

Solution in Sect. 10.2.2.

10.1.3 Battle at Dawn

In the Second Punic War, the Carthaginians led by Hannibal and the Romans fought for the strategically important Capua. A battle at dawn was to bring the decision. Both combatants have the choice between "attack" and "hesitate". Since the battle takes place very early in the morning, the soldiers are either "rested" or "tired" with equal probability—the realization of the type of soldiers of both parties is independent of each other. Each commander only knows the type of his own soldiers. The capture of Capua is valued at 100 by both sides. However, a capture is only possible if the enemy either hesitates or the own soldiers are rested while the enemy is still tired. If the soldiers of both armies are in the same state, an attack is not successful. If a battle occurs, losses occur with rested soldiers, which are valued at -40. If the soldiers are tired, a disaster occurs, which is valued at -200. If there is no battle, no losses occur.

Determine all strategies of the players! Determine all Bayes-Nash equilibria in pure strategies!

Solution in Sect. 10.2.3.

10.2 Solutions

10.2.1 Battle in the Bismarck Sea

Problem Text in Sect. 10.1.1.

Part a)

In this case, we can specifically identify the Japanese and the U.S. Airforce as players. The Japanese can choose between two routes, which is equivalent to their strategy: the northern route or the southern route. The same applies to the U.S. Airforce.

The information structure reveals that information is imperfect, i.e., the action of the opposing player is not observable, but complete, i.e., all payoffs and strategies are *common knowledge*. Hence, we have a simultaneous game. This allows us to represent the payoffs in a matrix. A representation in extensive form would be also possible, but this does not provide any additional information and unnecessarily complicates the evaluation of the game.

The payoffs result depending on the strategy combinations, i.e., there are $2 \cdot 2 = 4$ combination possibilities. Since it was not specified how the given information is to be transferred into payoffs, there are two equivalent possibilities of how the game can be represented:

Option 1

JAP, USA	North route	South route
North route	(−2, 2)	(−1, 1)
South route	(−2, 2)	(−3, 3)

Option 2

JAP, USA	North route	South route
North route	(1, 2)	(2, 1)
South route	(1, 2)	(0, 3)

Option 1 directly compares the gains and losses: For example, in the combination (North route, North route), the US has 2 days in which they can bombard the Japanese, whereas the Japanese are exposed to the bombardment for exactly these 2 days. Option 2 presents the situation somewhat "more positively": The Japanese survive the bombardment on one of the three possible days. However, the basic statements or the results we will obtain are independent of which of the two variants we choose.

A closer look at the matrices reveals that they represent constant-sum games. Since the payoffs of the respective strategy combination add up to zero in matrix 1, we have the special case of a zero-sum game: the gain of one player is the immediate loss of the other player. In matrix 2, the sum is 3, which is exactly the number of maximum possible days. The constant-sum property of the game will be of relevance in the further course of the analysis, as some of our results can be traced back to this peculiarity.

Part b)
Let us use matrix 1 in the following analysis (the considerations for matrix 2 are analogous). In the case that the USA chooses the North route, Japan has a payoff of -2 when choosing the North route and -2 when choosing the South route. If the USA chooses the South route, the Japanese receive -1 on the North route and -3 on the South route. Thus, choosing the North route represents a weakly dominant

strategy, as it is strictly better ($-1 > -3$) when the USA chooses the South route and equally good ($-2 = -2$) when the USA chooses the North route. For the USA, however, we cannot find a dominant strategy. This is also plausible in the context of the described situation: Since the USA wants to bombard the Japanese, they always want to choose the route that the Japanese choose. The Japanese, of course, want to avoid this. However, on the North route, they have the chance of lower losses due to the bad weather conditions than on the South route.

If the Japanese choose their weakly dominant strategy "north route", the USA will also respond with "north route", i.e., (North route, North route) will be the solution to the game. Since this is only an equilibrium with a weakly dominant strategy, this solution is not necessarily unique.

Part c)

To determine the Nash equilibrium of the game, we consider the best responses of each player to their opponent's strategy (again, the considerations are identical for both matrices). We can represent these using reaction correspondences. The results are

$$r_{\text{JAP}}(\text{North}) = \{\text{North, South}\} \quad r_{\text{JAP}}(\text{South}) = \{\text{North}\}$$
$$r_{\text{USA}}(\text{North}) = \{\text{North}\} \qquad r_{\text{USA}}(\text{South}) = \{\text{South}\}$$

Similarly, we can also mark our considerations in the table by underlining the best responses (the highest payoff given the opponent's strategy):

JAP, USA	North route	South route
North route	($\underline{-2}$, $\underline{2}$)	($\underline{-1}$, 1)
South route	($\underline{-2}$, 2)	(−3, $\underline{3}$)

As we can see from the reaction correspondences as well as in the matrix, (North route, North route) represents the unique Nash equilibrium of the game: The best response of Japan to the US choice of the North route is to choose the North route and the best response of the U.S. Airforce to the Japanese choice of the North route is likewise to choose the North route. We thus have a pair of mutually best responses, i.e., this represents a Nash equilibrium.

Part d)

When applying the maxim rule, we first look for the lowest value (minimum) for each player's strategy and then determine the highest value (maximum) from all these values. The player will then choose the corresponding strategy according to this procedure.

In the following table, the lowest values of each strategy are marked in bold:

JAP, USA	North route	South route
North route	(**−2**, **2**)	(−1, **1**)
South route	(−2, **2**)	(**−3**, 3)

If the Japanese choose the North route, they must expect a payoff of −2 in the worst case (the USA chooses the North route), whereas if they choose the South route, they must expect a payoff of −3 (the USA chooses the South route). If Japan wants to minimize the worst case, it will choose the North route, as the potential damage is smallest there (its worst payoff is minimal). We can make the same considerations for the USA: If they choose the North route, they must expect a payoff of 2 in the worst case and 1 if they choose the South route. They are therefore best off choosing the North route.

We thus obtain the maximin solution (North route, North route) and find that this exactly corresponds to the Nash equilibrium.

The reason why the Nash equilibrium and the maximin solution are identical is that this is a constant-sum game. If we consider what characterizes the maximin strategy and the Nash equilibrium, we find that in constant-sum games these two concepts always coincide: The basic structure of these strictly competitive games is that one player's gain is always the other player's loss. The maximin strategy now states that the best strategy should always be chosen based on all the worst results. But the worst results of one player are the best possible results of the other player in a constant-sum game. Thus, the maximin strategy is nothing more than the best response to the other player's best response, which—by definition—results in a Nash equilibrium! Thus, in this specific case, the maximin solution is a reasonable solution concept as it corresponds to the Nash equilibrium. In addition, the maximin solution is always Pareto-efficient here: No player can be made better off without making another player worse off. This property is also generalizable for constant-sum games (if you want to increase the payoff of one player and keep the co-player's payoff as it is, the sum of both payoffs increases and hence cannot remain constant).

In general, however, the maximin strategy is not a recommendable solution concept for economic issues, as it is based on the very pessimistic attitude that all other players will try to minimize the player's utility, even at the expense of their own advantage. Another reason is that the player applying this strategy shows extreme risk aversion and wants to avoid losses at all costs. As we have seen, this is indeed promising in constant-sum games, but this type of game is very rarely encountered, especially in economic contexts. An economic example could be that companies compete directly for market shares in a stagnant market, so the gain of additional market shares by one company is only possible at the expense of other companies. However, from a company's point of view, it makes more sense to formulate (medium and long-term) profit rather than (short-term) market share as the target variable, and here an additional euro earned by one company does not necessarily reduce the profit of the other companies by exactly this euro.

10.2.2 Counterterrorism

Problem Text in Sect. 10.1.2.

Part a)
The information structure can be characterized on the one hand as to whether it is perfect (versus imperfect) and on the other hand whether it is complete (versus incomplete). Information is perfect when the players can observe the strategy choice of the other player(s), and it is complete when the payoffs and strategies of all players are *common knowledge*. In this case, the information is thus imperfect but complete. In the context of the analysis, this seems plausible, as both sides know what goals the other side is pursuing but cannot observe what actions they are taking.

Part b)
For $\rho_w = \rho_s = \rho$ the payoff matrix changes to

Government, Terrorists	Attack	No attack
Weak control	$(\rho \cdot V, (1-\rho) \cdot T - \rho \cdot F)$	$(V, 0)$
Strong control	$(\rho \cdot V - c, (1-\rho) \cdot T - \rho \cdot F)$	$(V - c, 0)$

Thus, "weak control" becomes the dominant strategy of the government: If the terrorists decide to attack, "weak control" provides a higher payoff ($\rho \cdot V > \rho \cdot V - c$) as well as in the case when the terrorists do not carry out an attack ($V > V - c$). The result is plausible, as strong control comes with additional costs, but does not provide any additional benefit in the form of better protection (higher probability of preventing an attack).

In the case of a dominant strategy, the sequential structure also does not change the strategy choice, as a dominant strategy is always played regardless of what the opponent chooses.

Part c)
The payoff matrix is as follows:

Government, Terrorists	Attack	No attack
Weak control	$(80, \underline{2})$	$(\underline{100}, 0)$
Strong control	$(\underline{85}, -7)$	$(90, \underline{0})$

As we can see from the underlining in the matrix, there is no Nash equilibrium in pure strategies.

Part d)
A mixed strategy assigns a probability to each pure strategy of the player, with which this is played.

Since the number of Nash equilibria is always odd and there is no Nash equilibrium in pure strategies, there must be a Nash equilibrium in mixed strategies in the present game. Since we must draw the reaction correspondences, the method of expected utility comparison is recommended. Let p_G be the probability with which the government chooses "weak control" and p_T the probability with which the terrorists carry out an "attack".

The government prefers "weak control", i.e., $p_G = 1$, when

$$\begin{aligned} E[u_G(p_G = 1, p_T)] &> E[u_G(p_G = 0, p_T)] \\ 80 \cdot p_T + 100 \cdot (1 - p_T) &> 85 \cdot p_T + 90 \cdot (1 - p_T) \\ p_T &< 2/3. \end{aligned}$$

The reaction correspondence of the government is therefore

$$p_G(p_T) = \begin{cases} 0 & p_T \in \left(\frac{2}{3}, 1\right] \\ [0, 1] & p_T = 2/3 \\ 1 & p_T \in \left[0, \frac{2}{3}\right) \end{cases}$$

Similarly, it is necessary that the terrorists prefer to "attack", i.e., $p_T = 1$, when

$$\begin{aligned} E[u_T(p_G, p_T = 1)] &> E[u_T(p_G, p_T = 0)] \\ 2 \cdot p_G + (-7) \cdot (1 - p_G) &> 0 \cdot p_G + 0 \cdot (1 - p_G) \\ p_G &> 7/9. \end{aligned}$$

The reaction correspondence of the terrorists is therefore

$$p_T(p_G) = \begin{cases} 0 & p_G \in \left[0, \frac{7}{9}\right) \\ [0, 1] & p_G = 7/9 \\ 1 & p_G \in \left(\frac{7}{9}, 1\right] \end{cases}$$

The graphical representation and solution is as follows:

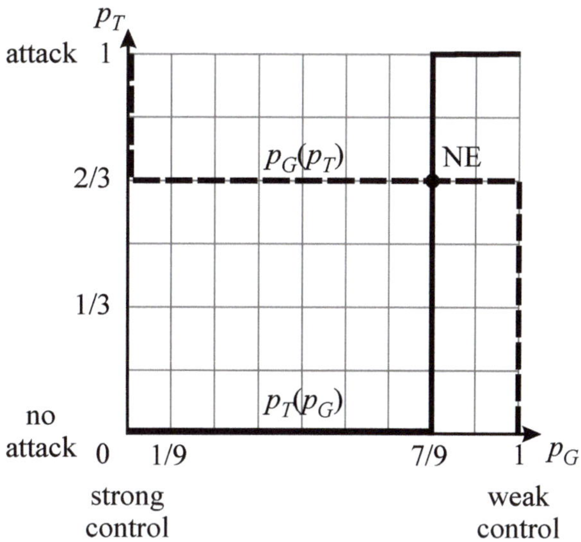

Part e)
As this is an inspection game or an anti-coordination problem, the solution is plausible: While the government wants to guarantee the protection of the object—but not at any cost—the terrorists want to carry out their attack as successfully as possible.

From the population's perspective, however, the possibility that the government does not put all its efforts into protecting against the attack is not desirable. It is not known how the government has considered possible victims in the population due to the attack in its utility calculation. The population would therefore prefer the greatest possible protection—but it must be weighed up to what price they are willing to pay for this (for example in the form of higher taxes or greater restriction of privacy). If you want to take these aspects into account, then you would have to expand the above game to include the population as a third player and explicitly list their opportunity costs resulting from the "strong control" amounting to c.

Part f)
The payoff matrix for option 1 is as follows:

Government, Terrorists	Attack	No attack
Weak control	(80, −6)	(<u>100</u>, <u>0</u>)
Strong control	(<u>85</u>, −16.5)	(90, <u>0</u>)

The payoff matrix for option 2 is as follows:

Government, Terrorists	Attack	No attack
Weak control	(80, 2)	(100, 0)
Strong control	(90, −7)	(95, 0)

Option 1 results in "no attack" becoming the dominant strategy for the terrorists, and thus the government can guarantee the security of the object even with weak control. Option 2, on the other hand, does not solve the fundamental problem, as the terrorists still want to carry out an attack. Only the government now has a greater incentive to control strongly. Therefore, option 1 is significantly better suited to achieving the goal as it directly affects the terrorists' payoff.

10.2.3 Battle at Dawn

Problem Text in Sect. 10.1.3.

Each of the two players, the Carthaginians and the Romans, have four strategies, which depend on their type:

- "Attack" when "rested"; "Attack" when "tired" (strategy AA)
- "Attack" when "rested"; "Hesitate" when "tired" (strategy AH)
- "Hesitate" when "rested"; "Attack" when "tired" (strategy HA)
- "Hesitate" when "rested"; "Hesitate" when "tired" (strategy HH)

Thus, there are a total of 16 payoff combinations to determine to find the equilibrium of the game.

Let us determine the payoff for each type if they decide either for "attack" or for "hesitate" and encounter either a "rested" or "tired" enemy. Since the situation is symmetrical for both combatants, we only consider the Carthaginian side.

First, let us consider "rested" Carthaginians who decide to attack. If the Romans decide to "hesitate", the Carthaginians win regardless of the type of Romans without losses and thus have a payoff of 100, while the Romans receive a payoff of 0. However, if the Romans also decide to attack, it is relevant for the Carthaginians which type the Romans have. If they are "rested", no one wins (utility of 0), but losses of 40 occur, so the payoff is $0 - 40 = -40$ for the Carthaginians and -40 for the Romans as well. If the Romans are "tired", the Carthaginians win and receive a payoff of $100 - 40 = 60$, while the Romans receive a payoff of -200 due to higher losses.

Next, let us consider the "tired" Carthaginians who decide to attack. If the Romans "hesitate", the Carthaginians win and have a payoff of 100 and the Romans of 0. If "rested" Romans decide to attack, the Carthaginians lose and receive a payoff of -200, while the victorious Romans get $100 - 40 = 60$. However, if the attacking Romans are also "tired", no one wins, and both receive a payoff of -200.

If the Carthaginians decide to "hesitate", their type does not matter, and they always receive a payoff of 0. If the Romans also hesitate, their payoff is also 0. However, if the Romans attack, they will always receive a payoff of 100 regardless of their type.

In summary, the following payoffs result:

Carthage, Rome	Attack (rested)	Attack (tired)	Hesitate
Attack (rested)	−40, −40	60, −200	100, 0
Attack (tired)	−200, 60	−200, −200	100,0
Hesitate	0, 100	0, 100	0, 0

Weighted with the probabilities, the following payoff matrix finally results:

C, R	AA	AH	HA	HH
AA	(−95, −95)	(−10, <u>5</u>)	(15, −100)	(<u>100</u>, 0)
AH	(<u>5</u>, −10)	(<u>15</u>, <u>15</u>)	(<u>40</u>, −25)	(50, 0)
HA	(−100, 15)	(−25, <u>40</u>)	(−25, −25)	(50, 0)
HH	(0, <u>100</u>)	(0, 50)	(0, 50)	(0, 0)

In the calculation, the values of the previous considerations regarding the payoffs of the individual types were weighted with their respective probabilities. The following three examples should clarify this:

- Strategy combination (AA, AA): This means that every type of Carthaginian and every type of Roman attacks. If well-rested Carthaginians attack, they encounter well-rested, attacking Romans with a probability of 0.5 and have a payoff of −40 and with a probability of 0.5 tired, attacking Romans with a payoff of 160. Thus, well-rested Carthaginians have an expected payoff of $0.5 \cdot (-40) + 0.5 \cdot 60 = 10$. However, the Carthaginians can also be tired and then have an expected payoff of −200 regardless of the type of attacking Romans. The Carthaginians themselves are 0.5 well-rested and 0.5 tired, that is, the strategy gives them an expected payoff of $0.5 \cdot 10 + 0.5 \cdot (-200) = -95$. The same applies to the Romans.
- Strategy combination (AH, AH): Here only the well-rested Carthaginians attack, while the tired Carthaginians hesitate. The same applies to the Romans. Thus: In their attack, well-rested Carthaginians meet with a probability of 0.5 well-rested, attacking Romans and have a payoff of −40, while with 0.5 they face tired, hesitating Romans, which gives them a payoff of 100. The expected payoff of the well-rested Carthaginians is therefore $0.5 \cdot (-40) + 0.5 \cdot 100 = 30$. If the Carthaginians are tired, they do not attack and have an expected payoff of 0. Thus, the AH strategy provides the expected payoff $0.5 \cdot 30 + 0.5 \cdot 0 = 15$. The same applies to the Romans.

- Strategy combination (AH, HA): In this strategy combination well-rested Carthaginians attack, while tired Carthaginians hesitate. Conversely, well-rested Romans hesitate, while the tired Romans attack. An attack by well-rested Carthaginians thus leads to a payoff of 100 with a 0.5 probability against tired, hesitant Romans, and to a payoff of 60 with a 0.5 probability against tired, attacking Romans. The expected payoff is therefore $0.5 \cdot 100 + 0.5 \cdot 60 = 80$. Tired Carthaginians hesitate and have a payoff of 0, leading to an expected payoff for the Carthaginians of $0.5 \cdot 80 + 0.5 \cdot 0 = 40$. On the Roman side, it looks different: well-rested Romans will hesitate and consequently have a payoff of 0. Tired Romans will attack and meet with a 0.5 probability of well rested, attacking Carthaginians, resulting in a payoff of –200, and with a 0.5 probability of tired, hesitant Carthaginians, generating a payoff of 100 for them. They thus have an expected payoff of $0.5 \cdot (-200) + 0.5 \cdot 100 = -50$. Thus, the expected payoff of the Roman strategy is $0.5 \cdot 0 + 0.5 \cdot (-50) = -25$.

As we see in the table, this game has only one Bayes-Nash equilibrium in pure strategies at (AH, AH), meaning an attack will only occur when well rested.

Miscellaneous Application Examples 11

11.1 Problems

11.1.1 Quiz Show

Githa Greedy reached the final round of the quiz show "Who Wants to Be a Millionaire". Unfortunately, she has no idea how to answer the 1 million euros question. The "Ask-the-Audience" and the "Phone-a-Friend" lifelines are no longer available. She uses the 50:50 lifeline and now has two options: Either she stops and takes 500,000 euros home with certainty, or she tries to answer the question despite her lack of knowledge. If the answer is correct, she gets 1 million euros, if the answer is wrong, she falls back to the last safe prize level of 16,000 euros and is eliminated (classic version of the quiz show). Her utility function is $u(y) = \ln y$.

a. What is Githa's risk attitude? Will she continue to play or stop?
b. A television viewer (who, of course, knows everything better) claims: "Any candidate who is in Githa's situation and continues to play, can only be risk-seeking!" Can you prove him wrong? (*Hint: How would a risk-neutral candidate behave?*) What is your result in the risky version of the quiz show, in which the last safe prize level is 500 euros?

Solution in Sect. 11.2.1.

11.1.2 Insurance and Insurance Demand

A decision-maker with the utility function $u(y) = \sqrt{y}$ has a wealth of $W = 25$ euros. With a probability of $\rho = 0.25$, they suffer a damage of $D = 16$ euros.

F. Bartholomae and M. Wiens, *Game Theory and Applications*, Classroom Companion: Economics, https://doi.org/10.1007/978-3-658-44608-6_11

a. Determine the relevant indifference curve of the decision-maker! Sketch the curve as well as the certainty line in the state preference diagram!
b. The decision-maker can buy insurance policies for $q = 0.25$ euros each, which pay out 1 euro each in the event of a damage claim. Draw the budget line in the state preference diagram! (*Hint: The budget condition describes how wealth can be transferred from one state of the world to another.*) How many policies will the individual buy? Also draw the indifference curve for the situation after the purchase of the insurance!
c. Generally, assume an individual with utility function $u(y)$ and wealth W, which with probability ρ can suffer damage $D < W$. For each q euro, it can buy insurance policies that pay out 1 euro each in the event of damage. Under what condition will a risk-averse decision-maker ($u'(y) > 0$ and $u''(y) < 0$) fully insure themselves, i.e., strive for a situation in which they have the same wealth in both states of the world?

Solution in Sect. 11.2.2.

11.1.3 Negative Value of Information

New developments in genetic engineering have made it possible to carry out a test that can determine with certainty which risk group a person belongs to. It is generally known that 20% of the population have a high risk of suffering from a certain cardiopulmonary disease during their lives. The cost of treatment amounts to 100,000 euros. Correspondingly, 80% of the population is at low risk. Individuals in the high-risk group have a probability of developing the disease of $\rho^h = 0.1$, while in the low-risk group the probability is only $\rho^l = 0.05$. Without the test, no one knows which risk group they belong to.

For the analysis, consider a representative individual with the expected utility function $u(y) = \sqrt{y}$ and a wealth of $W = 16{,}000$ euros.

a. Given that the two types of individuals cannot be distinguished, determine the fair insurance premium, i.e., the premium for which the income just equals the expected payoffs! What would be the expected utility of a representative individual contracting this insurance?
b. A free genetic test is now available and can clearly assign everyone to a risk group. The result is revealed to both the individual and the insurance company. Determine the fair premium for both types of individuals! How does the expected utility of the representative individual change due to the possibility of genetic testing? Explain why the individual has no incentive to be tested!
c. How much would the individual pay to avoid the genetic test?

Solution in Sect. 11.2.3.

11.1.4 Rock Paper Scissors

The players can choose between the three strategies "rock", "paper", and "scissors". The rules are: If both players choose the same strategy, the game ends in a draw. If both players choose different strategies, then "scissors beat paper", i.e., the player who chooses "scissors" wins, while the player who has chosen "paper", loses. Furthermore "paper beats rock", and "rock beats scissors".

Represent this game in matrix form and determine all Nash equilibria!

Solution in Sect. 11.2.4.

11.1.5 Adverse Selection and Warranty

The stores of two competing used car dealers are located next to each other on the main road. The first, Harry's Car Salon, sells high-quality cars that have been subjected to careful inspections by an independent expert. These inspections and the purchase of a used car cost Harry an average of 8,000 euros. Leo's Car Market, on the other hand, sells vehicles of lower quality, which he purchases for an average of 5,000 euros.

With full information about the quality of the cars, the buyers are ready to pay 10,000 euros for high-quality cars and 7,000 euros for cars of lower quality. However, both businesses are still too new for a reputation to have formed. Therefore, the buyers are not informed about the true value of the cars. Thus, the buyers expect a 50% chance of getting a high-quality car, regardless of which dealer they choose, and are willing to pay 8,500 euros.

Harry has an idea: He offers his buyers a warranty against rust on all vehicles purchased from him. As is generally known, the expected costs of such a warranty for high-quality cars are 500 euros per year, while for cars of lower quality they are estimated at 1,000 euros per year. For simplification, assume that the warranty only has a signaling effect and does not increase the value of the vehicles from the consumers' perspective.

a. Suppose Harry grants a one-year warranty on all vehicles. Would this be a credible signal for the quality of his cars? Will Leo also give a one-year warranty, or will he leave it out and accept that buyers will only value his cars at 7000 euros?
b. Assume that Harry grants a two-year warranty. Would this be a credible signal for the quality of his cars? What about a three-year warranty? What period of warranty would you advise Harry to offer? Explain your answer!

Solution in Sect. 11.2.5.

11.1.6 Vickrey Auction (Sealed-Bid Second-Price Auction)

In a sealed-bid second-price auction, the person with the highest bid wins but must only pay the second highest bid as the price (if the bids are identical, it is drawn who wins). Suppose n people participate in the auction. The value that everyone assigns to the object to be auctioned is $0 \leq v_A \leq v_B \leq \cdots \leq v_n$. For simplification, let us assume that every v_i are known to all. The bids b_i are submitted simultaneously (e.g., in a sealed envelope). The player who wins the bid receives a payoff of $u_i = v_i - \max_{j \neq i} b_j$, while the others get $u_i = 0$.

a. Suppose there are only two bidders in the auction, Alice and Bob. Alice's valuation is $v_A = 500$ Euros. Is there a valuation v_B from Bob where Alice is strictly better off bidding 450 euros than bidding her own valuation of 500 euros?
b. Now prove for the general formulation that the weakly dominant strategy for each player is to place a bid equal to their own valuation of the item ($b_i^* = v_i$)! Who wins the auction? What is the payoff of this bidder?
c. Does anything change in the result if each bidder only knows their own valuation of the item?

Solution in Sect. 11.2.6.

11.1.7 All-Pay Auction

Two risk-neutral players, Alice and Bob, participate in an auction, bidding simultaneously for an object with a value of v euros. The higher bid wins but both players must pay their submitted bid. If both bid the same amount, neither of the two gets the object. Bids can only be made in integer euro amounts.

a. Determine the payoff structure of the auction and give the general strategies b_i that a rational bidder will choose!
b. For $v = 1$ euro, $v = 2$ euros, and $v = 3$ euros, give all rationalizable strategies and represent the game in a suitable form! Determine all Nash equilibria in pure strategies!
c. Explain intuitively that there can be no equilibrium in pure strategies for $v > 2$!
d. Determine the equilibrium in mixed strategies for $v > 2$!

Solution in Sect. 11.2.7.

11.1.8 Tullock Contest

Innovation is one of the key driving forces behind economic and social development. Companies competing in R&D often find themselves in a competitive situation that is also referred to as a "tournament". In general, tournaments are a special form of the so-called all-pay auction (see previous Sect. 11.1.7) with the feature that winners and losers alike forfeit the resources invested to win the "prize". In the context of innovations, the prize corresponds to a technological or organizational breakthrough that gives a company a clear competitive advantage.

The classic *Tullock contest* describes the situation as follows: n companies invest effort x at linear cost $c_i \cdot x_i$ to compete simultaneously for a prize of common and fixed value v. The prize corresponds to a rival technological innovation, which the winner can protect by a patent. The losers receive nothing and must write off their investments. The probability p_i for firm i to win the contest is dependent on its relative effort according to $p_i = x_i / \left(x_i + \sum_{j \neq i} x_j \right)$.

a. Consider the case of two companies i and j. Determine the objective function and derive formally the first-order condition (FOC) for company i as well as the reaction functions for both companies! Determine the Nash equilibrium! Is the classic Tullock contest a realistic description of an innovation competition? Is such a tournament efficient? Discuss!
b. Draw the reaction functions of both companies for $c_i = 1$! What happens to the curves and the point of intersection if only the costs of firm j increase to (i) $c_j = 2$ or (ii) $c_j = 3$?
c. Let us now assume that the investment also influences the value of the innovation: a higher effort therefore not only increases the probability of a successful innovation but also increases the value of the innovation (e.g., a higher willingness of customers to pay for the new feature or product). We assume that the success of innovation is random, like a lottery, and that each participating company decides on the innovation effort (or innovation intensity) by the number of lottery tickets purchased. Specifically, the innovation intensity could correspond to the number of hired scientists or innovators: The higher the number, the higher the chance for a valuable innovation. The following figure shows the probability density function of the value of innovative success:

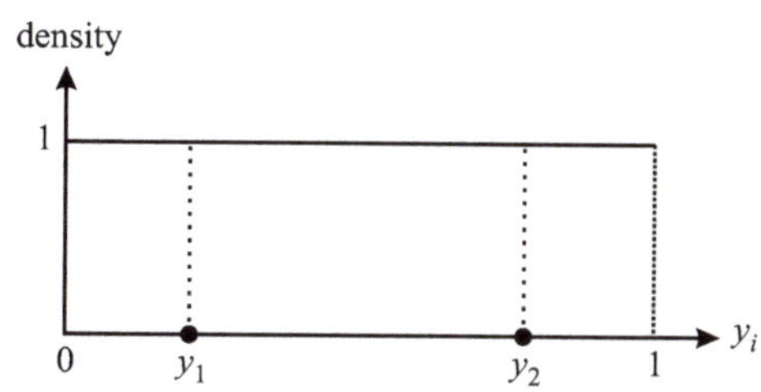

In the "lottery box", the values of the tickets are uniformly distributed (i.e., the density is the same) between 0 and 1. For example, if a company invests in two lottery tickets, the value of these two tickets could be $y_1 = 0.21$ or $y_2 = 0.78$. As the ticket-values are normalized values on the interval [0, 1] and thus "fractions of the highest prize" v, this fraction is multiplied with v to get the expected value of the prize. Hence, if a firm invests in k tickets, the ticket with the highest expected value is given by $E(v^{\max}) = \frac{k}{k+1} \cdot v$. I.e., if the innovation has a value of $v = 90$, e.g., a purchase of two tickets is expected to have a maximum value of $\frac{2}{2+1} \cdot 90 = 60$ and a purchase of nine tickets is expected to have a value of $\frac{9}{9+1} \cdot 90 = 81$. Consider again, now for this modified Tullock contest, the case of two companies i and j. Determine the objective function and derive formally the first-order condition (FOC) for firm i! Interpret the difference to the FOC of part a). Can such a tournament be efficient? Explain!

Solution in Sect. 11.2.8.

11.1.9 Credibility of Monetary Policy

On the island of Monetaria, the central bank (player A) is responsible for monetary policy and thus also for the inflation rate p. It is assumed that p can be perfectly controlled by the central bank. The island's population (player B) aims to precisely anticipate the decision of the central bank in advance to make as few mistakes as possible when concluding labor and rental contracts.

The population forms an expectation p^e about the inflation rate. Whenever the actual inflation rate deviates from the inflation expectation, it incurs costs amounting to $(p - p^e)^2$. Taking these costs into account, the utility function of player B is thus $u_B(p^e) = 16 - (p - p^e)^2$.

On the one hand, the central bank must consider in its decision that high inflation itself causes economic costs (efficiency losses) amounting to $C^E(p) = p^2$. On the other hand, inflation can temporarily alleviate the problem of unemployment, as it reduces real wages and makes labor cheaper for companies. For this to succeed, however, the central bank must "surprise" the population, i.e., choose an inflation rate that exceeds the population's expectation. The central bank knows from experience that the efficiency of inflation on employment is highest when $p = p^e + 4$, i.e., when the actual inflation rate exceeds the expected one by exactly 4 percentage points. If the central bank misses this difference of 4 percentage points, it incurs costs due to a too low efficiency according to the function $C^L(p) = \left[p - (p^e + 4)\right]^2$. The central bank weighs both cost components equally, i.e., $C = 0.5 \cdot C^E(p) + 0.5 \cdot C^L(p)$. Considering the costs, its utility function is therefore $u_A(p) = 32 - \left[0.5 \cdot C^E(p) + 0.5 \cdot C^L(p)\right]$.

a. Consider the simultaneous game with discrete strategies. Assume that the central bank can choose between $p_{A1} = 0$ (no inflation) and $p_{A2} = 4$ (inflation). The population accordingly has the strategies $p^e_{B1} = 0$ and $p^e_{B2} = 4$. Determine the payoffs for all strategy combinations and represent the resulting game in matrix form!
b. Now determine a plausible solution to the game using dominance considerations! Is the solution a Nash equilibrium? If yes, does the procedure ultimately leading to the solution ensure that the Nash equilibrium is unique? Determine all Nash equilibria in pure strategies using the reaction correspondences $r_i(s_j)$!

c. Assume continuous strategies. Determine the reaction functions and then graphically and numerically determine the Nash equilibrium of the simultaneous game including the payoffs achieved!

Now assume the following sequential structure: The Central Bank commits to a strategy in the first stage; the population observes the strategy choice and determines its strategy in the second stage.

a. Determine the equilibrium of this sequential game! What are the payoffs?
b. The subgame perfect Nash equilibrium from d) represents the Pareto optimum in the present game. However, it can only be realized if the Central Bank's commitment is truly binding. First, use your reaction correspondences from b) to show that there is an incentive for the Central Bank to deviate from the Pareto-optimal solution! Briefly explain for each of the following three proposals whether the socially desirable solution can also be realized with simultaneous strategy choice by the population and the Central Bank:

i. The Central Bank publicly announces to achieve an inflation rate of $p = 0$.
ii. A performance-based compensation is set for the Central Bank President, observable to the population: With high inflation his salary decreases, at $p = 0$ it is at its maximum.
iii. The game between the population and the Central Bank is repeated over a finite or infinite time horizon. Consider the following strategy combination in this context: *The Central Bank always sets the inflation rate to* $p = 0$. *In the first period, the population assumes an inflation rate of zero. If it was not deceived in its expectations in a previous period, it also chooses the action* $p^e = 0$ in the following periods. *However, if the Central Bank deviates from the socially optimal solution in a period, the population forms high inflation expectations for all subsequent periods.*

Solution in Sect. 11.2.9.

11.1.10 Overseas Trade

From the eleventh century, overseas trade in the Mediterranean experienced a significant development, which contributed significantly to the economic growth of Southern Europe, North Africa, and the Middle East. Long-distance trade particularly involved goods such as metals, wax, tar, wood, leather, and textiles being transported over long distances. This form of trade was very lucrative for the merchants, but also extremely risky.

Thus, an overseas merchant was generally faced with the following problem: To make the business profitable, he had to employ foreign traders as commissioners or as "agents" with the crossing and handling to commission, i.e., to entrust them with the ship and goods. If these agents were honest and dutiful, they carried out the order, delivered the ship to the client, and received their wages. However, there was also the possibility that the merchant would encounter a fraudster who sailed away with the ship and goods, sold both, and disappeared with the sales.

With this in mind, consider a simultaneous game in discrete strategies between an overseas merchant, who must decide whether to hire an agent, and an (anonymous) agent, who can behave dutifully or fraudulently. In the case of fraud, the agent embezzles the ship and the merchandise. The proceeds from the trade transaction amount to 3, of which the agent receives 1 as wage if he carries out the order dutifully. The value of the ship, amounting to 2, only plays a role if the merchant encounters a fraudulent agent.

a. Formulate this situation as a game: Who are the players? What strategies do they each have? What payoffs result depending on the chosen strategy combinations?
b. Now determine a plausible solution to the game using dominance considerations. Is the solution a Nash equilibrium? If so, does the method used to determine the solution fundamentally ensure that a determined Nash equilibrium is unique? Identify all Nash equilibria in pure strategies in a comprehensible manner. Do you suspect another Nash equilibrium in mixed strategies?
c. Is the solution determined in b) plausible in your opinion? Given the payoff structure, is this more of a cooperation or coordination problem?
d. The overseas merchants alternatively had the option to entrust the order to a "Maghribi" trader. The Maghribi traders formed a kind of tight trading coalition with extraordinarily efficient information flow and strict code of honor. Membership in this coalition was usually for life and benefited from mutual support in the form of loans and aid. However, if it became known that a Maghribi trader had violated the code of honor—whether against members or non-members—such as through fraud or embezzlement, the trader was permanently expelled from the coalition and this news was spread over several thousand kilometers in the Mediterranean area. If the person concerned fled and did not compensate for the damage caused, even his family was made liable for the settlement.

Now consider a supergame between the overseas merchant and the Maghribi trader. In each period, both can conclude a business deal. Assume that for a Maghribi trader, the monetary damage due to loss of reputation per period is -1, that is, he loses the wage payment because he will no longer be able to do business with anyone. The advantage he gains from membership in the coalition is worth 1 per period to him (in addition to the wage). Determine for which discount factors a dutiful behavior of the Maghribi trader can be realized as an equilibrium of the infinitely repeated game!

Solution in Sect. 11.2.10.

11.2 Solutions

11.2.1 Quiz Show

Problem text in Sect. 11.1.1.

Part a)
Githa is risk-averse, as her utility function is right-curved (concave), i.e., the second derivative is negative:

$$\frac{du(y)}{dy} = \frac{1}{x}$$
$$\frac{d^2u(y)}{dy^2} = -\frac{1}{x^2} < 0$$

If Githa is rational, she will make her decision based on which option provides her with the highest utility: If she decides to continue playing, her expected utility is

$$E[u(a)] = 0.5 \cdot \ln 16{,}000 + 0.5 \cdot \ln 1{,}000{,}000 \approx 11.75.$$

If she stops, she has the utility from the certain payoff of $u(500{,}000) \approx 13.12$. Thus, Githa has a higher utility if she quits and accepts the certain payment of 500,000 euros.

Part b)
A risk-neutral candidate focuses on the expected value of the lottery a, i.e., the utility function is, e.g., $u(y) = y$, which is

$$E(a) = 0.5 \cdot 16{,}000 + 0.5 \cdot 1{,}000{,}000 = 508{,}000$$

and is therefore higher than the certain payoff of 500,000 euros. Hence, every risk-neutral candidate will continue to play! The statement of the television viewer is therefore incorrect (even some not very risk-averse candidates will continue to play here)!

In the risky version, the expected value is

$$E(a) = 0.5 \cdot 500 + 0.5 \cdot 1{,}000{,}000 = 500{,}250$$

and is also higher than 500,000 euros. Thus, a risk-neutral candidate will also continue to play here. As we can see, a risk-neutral candidate will continue to play for any (strictly) positive payout of the last safe prize level!

11.2.2 Insurance and Insurance Demand

Problem text in Sect. 11.1.2.

Part a)
The indifference curve is defined as the (geometric) location of all payoff combinations in the different states of the world that provide the same utility. Hence, we need to find all payoff combinations that provide the same expected utility as the lottery. The expected utility of the lottery is

$$\begin{aligned} E[u(a)] &= (1-\rho)\cdot\sqrt{W}+\rho\cdot\sqrt{W-D} \\ &= 0.75\cdot\sqrt{25}+0.25\cdot\sqrt{25-16}=3.75+0.75 \\ &= 4.5 \end{aligned}$$

Now we need all (y_1, y_2) combinations—where y_G is the payoff in the good state without damage and accordingly y_D is the payoff in the state with damage -, that provide an expected utility of 4.5:

$$0.75\cdot\sqrt{y_G}+0.25\cdot\sqrt{y_D}=4.5$$

To be able to draw the curve, we solve it for the payoff in one state (here the damage state y_D):

$$y_D=\left[\left(4.5-0.75\ \cdot\sqrt{y_G}\right)\cdot 4\right]^2=\left(18-3\cdot\sqrt{y_D}\right)^2$$

We can thus assign a payoff in the good state to each payoff in the state where the damage occurs, which provides the individual with the same utility.

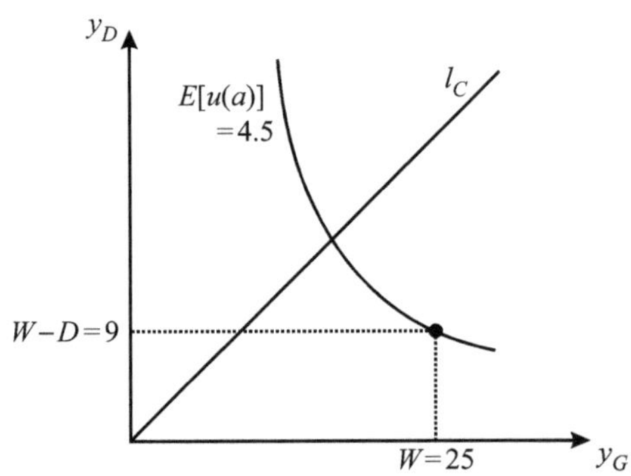

The certainty line l_C is given by a line with a slope of 1 (45°-line) and represents all certain payoff combinations. It represents all safe payoff combinations that are characterized by the same value in both states.

Part b)
The policy allows the decision-maker to give up $q = 0.25$ euros to receive a payment of 1 euro in the bad state, in which the damage occurs. This amount is to be paid regardless of the realized state. This means the insurance premium q must be paid in both the good and the bad state; in return, the damage per policy is reduced by 1 in the bad state (the insurance utility is thus state-dependent). For the budget constraint, prices are needed, which result from the opportunity costs: The decision-maker can either possess q in the good state of the world or $1-q$ in the bad state. Through the insurance, the decision-maker can convert their income (y_G, y_D) into a payoff combination (x_G, x_D). The budget results from the payoffs in the two states weighed with the opportunity costs, $q \cdot W + (1-q) \cdot (W-D) = q \cdot y_G + (1-q) \cdot y_D$. This budget can then be used to buy 1 euro in the bad state (x_D) at a price of $q = 0.25$ euros and 1 Euro in the good state (x_G) at a price of $1-q = 0.75$ euros. Thus, the budget constraint is

$$(1-q) \cdot x_G + q \cdot x_D \leq q \cdot y_G + (1-q) \cdot y_D$$

Solved for the payoff in the bad state:

$$x_D = \frac{1}{q} \cdot [q \cdot y_G + (1-q) \cdot y_D] - \frac{1-q}{q} \cdot x_G$$

In the specific case, $q = 0.25$, so the decision-maker has a budget of $(1 - 0.25) \cdot 25 + 0.25 \cdot (25 - 16) = 21$ and corresponds exactly to the expected income, $0.75 \cdot 25 + 0.25 \cdot (25 - 16) = 21$. This is not surprising, as the price for the good state exactly corresponds to the probability of this state. Thus, the budget line is

$$x_D = \frac{1}{0.25} \cdot 21 - \frac{1-0.25}{0.25} \cdot x_G = 84 - 3 \cdot x_G$$

The decision-maker will maximize the expected utility. This can be achieved by choosing the number of policies n that are purchased at a price of q euros and receiving 1 euro in return in the bad state. The expected utility is

$$\begin{aligned} E[u(a)] &= (1-\rho) \cdot \sqrt{W - q \cdot n} + \rho \cdot \sqrt{W - D - q \cdot n + 1 \cdot n} \\ &= 0.75 \cdot \sqrt{25 - 0.25 \cdot n} + 0.25 \cdot \sqrt{25 - 16 - 0.25 \cdot n + 1 \cdot n} \\ &= 0.75 \cdot \sqrt{25 - 0.25 \cdot n} + 0.25 \cdot \sqrt{9 + 0.75 \cdot n} \end{aligned}$$

Insurance must be paid in both states of the world, but only generates a payoff in the event of damage. As we see, the expected utility depends on the number of policies n, so these must be chosen to maximize the utility. To do this, we derive the utility with respect to n:

$$\frac{dE[u(a)]}{dn} = \frac{0.75 \cdot (-0.25) \cdot 0.5}{\sqrt{25 - 0.25 \cdot n}} + \frac{0.25 \cdot 0.75 \cdot 0.5}{\sqrt{9 + 0.75 \cdot n}}$$

$$= \frac{3}{32} \cdot \left[\frac{1}{\sqrt{9 + 0.75 \cdot n}} - \frac{1}{\sqrt{25 - 0.25 \cdot n}} \right]$$

This condition is set to zero to find the optimal number of policies. As we can see, the term is zero if the difference in the parenthesis is zero, i.e., the first fraction is equal to the second fraction:

$$\frac{1}{\sqrt{9 + 0.75 \cdot n}} = \frac{1}{\sqrt{25 - 0.25 \cdot n}}$$
$$\sqrt{9 + 0.75 \cdot n} = \sqrt{25 - 0.25 \cdot n}$$
$$9 + 0.75 \cdot n = 25 - 0.25 \cdot n$$
$$n^* = 16$$

Thus, the decision-maker will buy 16 policies and pay $16 \cdot 0.25$ euros $= 4$ euros for it. This results in a certain net wealth of $W_C = 21$ in every state, i.e., in the good state, $25 - 4$ and in the bad state, $25 - 4 - 16 + 16$.

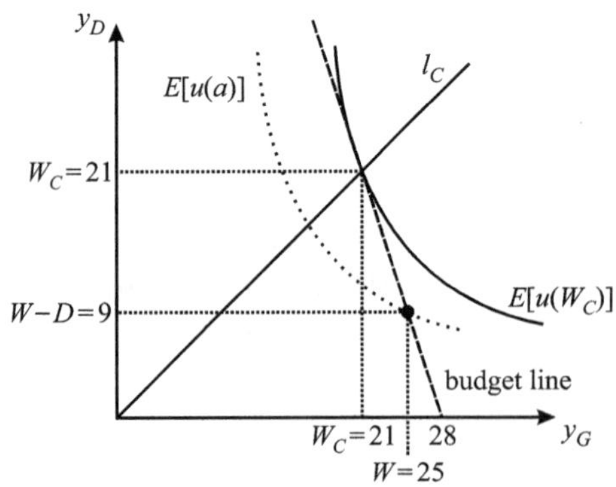

Graphically, the budget line must contain both the endowment point ($y_G = W = 25$, $y_D = W - D = 9$) as well as the certain (net) wealth the decision-maker has after taking out insurance ($y_G = 21$, $y_D = 21$)—or we simply use the calculated budget constraint, $x_D = 84 - 3 \cdot x_G$. The figure shows that the decision-maker with insurance achieves a higher expected utility level $E[u(W_C)]$ than it originally had, $E[u(a)]$.

Part c)
When choosing n insurance policies, the expected utility is as follows:

$$E[u(a)] = (1 - \rho) \cdot u(W - q \cdot n) + \rho \cdot u(W - D - q \cdot n + 1 \cdot n)$$

To determine the optimal number of policies n^*, the expected utility with respect to n must be maximized. Thus, we first determine the derivative

$$\frac{dE[u(a)]}{dn} = (1-\rho)\cdot u'(W-q\cdot n)\cdot(-q)+\rho\cdot u'[W-\mathrm{D}+(1-q)\cdot n]\cdot(1-q)$$

To find the optimum, we set this expression equal to zero.

Full insurance requires that the individual must have the same income in both states, i.e.,$W - q \cdot C = W - D - (1-q)\cdot n$. An equal income also implies equal marginal utility $\overline{u}'$, so we can simplify the first-order condition.

$$(1-\rho)\cdot\overline{u}'\cdot(-q)+\rho\cdot\overline{u}'\cdot(1-q)\overset{!}{=}0 \quad \text{or} \quad \rho\cdot(1-q)\overset{!}{=}(1-\rho)\cdot q$$

This condition is fulfilled when $\rho = q$ applies. The decision-maker will therefore only fully insure themselves if the price of 1 euro matches the probability exactly in the event of damage. If the price is higher than the probability, the decision-maker will demand less insurance, if it is lower, the decision-maker will over-insure.

The decision-maker has therefore fully insured themselves in b) because the price of the insurance policy corresponds exactly to the probability of damage. In such a case, it is said that the insurance is actuarially fair.

11.2.3 Negative Value of Information

Problem text in Sect. 11.1.3.

Part a)

If the insurance company cannot determine the individual risk, it will base its decision on the expected risk. It therefore knows that 20% will fall ill at 10% and 80% at 5%, so the average risk is 6%. The costs that the insurance company would have to cover amount to 100,000 euros, which means that the expected costs are calculated as follows:

$$\left(0.2\cdot\rho^h+0.8\cdot\rho^l\right)\cdot 100{,}000=(0.2\cdot 0.1+0.8\cdot 0.05)\cdot 100{,}000=6{,}000$$

The insurance company will charge exactly these expected costs as an actuarially fair premium. This means that the utility of the individual who must pay this premium is certain, but will also be reimbursed for all costs in the event of illness:

$$E[u(y)]=\sqrt{W-6{,}000}=\sqrt{16{,}000-6{,}000}=100$$

Part b)

If the test is carried out, there are two ways in which the premium can change. If a low risk is determined, the fair premium drops to

$$\rho^l \cdot 100{,}000 = 5{,}000$$

However, if a high risk is determined, the premium increases to

$$\rho^h \cdot 100{,}000 = 10{,}000$$

From the individual's point of view, there is therefore an 80% probability that the premium will fall and a 20% probability that it will rise. The expected utility of taking the test is therefore

$$E[u(a)] = 0.2 \cdot \sqrt{16{,}000 - 10{,}000} + 0.8 \cdot \sqrt{16{,}000 - 5{,}000} \approx 99.4$$

The comparison of utility before and after the genetic test shows that the individual is worse off, $100 > 99.4$.

The chance that the premium will fall cannot compensate the individual for the risk of paying more. For risk-averse individuals, the utility of the expected value (before test) is greater than the expected utility (after test). The test therefore exchanges a certain result (uniform premium) for a "premium lottery", which a risk-averse individual does not want. From the perspective of the (risk-neutral) insurance company, however, nothing would change.

Part c)

To calculate how much the individual would pay to avoid the test, we simply need to calculate the risk premium. Since we already know the expected value of the lottery from a), we now only need the certainty equivalent, i.e., the certain payment that provides the same utility as the lottery from b):

$$\sqrt{y_C} = 0.2 \cdot \sqrt{6{,}000} + 0.8 \cdot \sqrt{11{,}000} \Leftrightarrow y_C \approx 9{,}879.69$$

The risk premium is thus

$$R = 10{,}000 - 9{,}879.69 = 120{,}31$$

This corresponds exactly to the amount that the individual would like to pay to avoid getting the information (and thus the "premium lottery"), i.e., the information has a negative value for the individual!

11.2.4 Rock Paper Scissors

Problem text in Sect. 11.1.4.

If we assign a payoff of 1 to a player's victory, a payoff of 0 to a draw, and a payoff of –1 to a loss, the following payoff matrix results:

Alice, Bob	Rock	Paper	Scissors
Rock	(0, 0)	(−1, 1)	(1, −1)
Paper	(1, −1)	(0, 0)	(−1, 1)
Scissor	(−1, 1)	(1, −1)	(0, 0)

The game does not have a Nash equilibrium in pure strategies, so a Nash equilibrium must exist in mixed strategies. Since only the Nash equilibrium was asked for and not the reaction functions, it is more advantageous in this case with three strategies to choose the maximization approach. Due to the symmetry, it is also sufficient to consider only one player.

The probability that Alice chooses "rock" is p_{A1}, the probability for "paper" is p_{A2}, and for "scissors" it remains $1 - p_{A1} - p_{A2}$. Accordingly, p_{B1}, p_{B2}, and $1 - p_{B1} - p_{B2}$ denote the probabilities for Bob. The maximization approach from Alice's perspective is therefore

$$\begin{aligned}
&E[u_A(p_{A1}, p_{A2}, p_{B1}, p_{B2})] = \\
&\quad p_{A1} \cdot p_{B1} \cdot 0 + p_{A1} \cdot p_{B2} \cdot (-1) + p_{A1} \cdot (1 - p_{B1} - p_{B2}) \cdot 1 + \\
&\quad p_{A2} \cdot p_{B1} \cdot 1 + p_{A2} \cdot p_{B2} \cdot 0 + p_{A2} \cdot (1 - p_{B1} - p_{B2}) \cdot (-1) + \\
&\quad (1 - p_{A1} - p_{A2}) \cdot p_{B1} \cdot (-1) + (1 - p_{A1} - p_{A2}) \cdot p_{B2} \cdot 1 + \\
&\quad (1 - p_{A1} - p_{A2}) \cdot (1 - p_{B1} - p_{B2}) \cdot 0 \\
&= p_{A1} \cdot (1 - 3 \cdot p_{B2}) + p_{A2} \cdot (3 \cdot p_{B1} - 1) - p_{B1} + p_{B2}
\end{aligned}$$

Since Alice has here two strategic variables, p_{A1} and p_{A2}, we have two first-order conditions (derivatives with respect to strategic variables set equal to zero):

$$\begin{aligned}
\frac{\partial E[u_A(p_{A1}, p_{A2}, p_{B1}, p_{B2})]}{\partial p_{A1}} &= 1 - 3 \cdot p_{B2} \overset{!}{=} 0 \quad \Rightarrow p_{B2} = \frac{1}{3} \\
\frac{\partial E[u_A(p_{A1}, p_{A2}, p_{B1}, p_{B2})]}{\partial p_{A2}} &= 3 \cdot p_{B1} - 1 \overset{!}{=} 0 \quad \Rightarrow p_{B1} = \frac{1}{3}.
\end{aligned}$$

It is probably not surprising, but we find that in equilibrium all strategies are equally likely to be played: $p_{A1} = p_{A2} = (1 - p_{A1} - p_{A2}) = 1/3$ and $p_{B1} = p_{B2} = (1 - p_{B1} - p_{B2}) = 1/3$.

11.2.5 Adverse Selection and Warranty

Problem text in Sect. 11.1.5.

Part a)
The warranty is intended to serve as a signal for high quality, meaning it must be designed in such a way that it fulfills this function and is therefore not chosen by both. If both choose "warranty" or both choose "no warranty", they are indistinguishable from the consumers' point of view, meaning there is no signal and pooling occurs. Only if one of the sellers grants the warranty and the other does not, the one with the warranty is credibly perceived as a provider of high-quality cars.

In general, we can determine the payoff matrix that depends on the warranty length n:

Harry, Leo	No warranty	Warranty
No warranty	(500, 3500)	$(-1000, 5000 - n \cdot 1000)$
Warranty	$(2000 - n \cdot 500, 2000)$	$(500 - n \cdot 500, 3500 - n \cdot 1.000)$

With a one-year warranty ($n = 1$) the result is

Harry, Leo	No Warranty	Warranty
No Warranty	(500, 3500)	(−1000, 4000)
Warranty	(1500, 2000)	(0, 2500)

For both, "warranty" represents a strictly dominant strategy. Therefore, the guarantee has no signaling effect here.

Part b)
It only makes sense for both to choose the same length of guarantee, otherwise the one with the shorter warranty period is seen as a provider of cars of lower quality. This means he only incurs the costs of the warranty, without it granting him any advantage. Hence, such a strategy is not rationalizable.

With a two-year guarantee ($n = 2$) we now get

Harry, Leo	No Warranty	Warranty
No Warranty	(500, 3500)	(−1000, 3000)
Warranty	(1000, 2000)	(−500, 1500)

We thus obtain (warranty, no warranty) as the Nash equilibrium. For Harry, "warranty" is still a dominant strategy, whereas for Leo, "no warranty" has now become the dominant strategy. Since Harry now can distinguish himself from Leo as a provider of high-quality cars, a two-year guarantee is therefore a credible signal.

With a three-year guarantee ($n = 3$) the following payoffs result:

Harry, Leo	No Warranty	Warranty
No Warranty	(500, 3500)	(−1000, 2000)
Warranty	(500, 2000)	(−1000, 500)

We now even get two Nash equilibria in pure strategies: (no warranty, no warranty) and (warranty, no warranty). Thus, there even exists a third Nash equilibrium in mixed strategies, which is not very meaningful here. We see that the warranty still has a signaling effect, but this leads to a reduction in Harry's payoff.

Harry should therefore provide a two-year warranty, as this period is sufficient for the warranty to have a signaling effect and an additional year would only reduce Harry's profit.

11.2.6 Vickrey Auction (Sealed-Bid Second-Price Auction)

Problem text in Sect. 11.1.6.

Part a)

We assume that Bob will bid exactly his valuation v_B. Now we check for which $b_B = v_B$ Alice can do better with the bid $b_{A2} = 450$ than with the bid $b^*_{A1} = v_A = 500$.

Let us assume that Bob's valuation is less than 450 euros, e.g., $v_B = 400$ euros. Then Alice would win the bid at both a bid of $b_{A2} = 450$ euros and at $b^*_{A1} = v_A = 500$ euros, as she has made in both cases the highest bid. She would have to pay 400 euros (Bob's valuation) and would thus have a payoff of 500 euros − 400 euros = 100 euros. The lower bid has not improved her position in this case.

The situation is different if Bob's valuation is between Alice's bid (450) and her valuation (500), e.g., at 475 euros. Here she would not win the bid at $b_{A2} = 450$ euros and would thus have a payoff of zero. With a bid of $b^*_{A1} = 500$ euros, however, she does win the bid and must pay 475 euros, resulting in a payoff of 500 euros − 475 euros = 25 euros and thus she is better off than at $b_{A2} = 450$ Euro.

If Bob's valuation is higher than Alice's, it does not matter whether $b_{A2} = 450$ Euro or $b^*_{A1} = 500$ Euro was bid, as in both cases she does not win the auction and therefore has a payoff of zero.

So, there is no valuation v_B from Bob, where Alice could strictly improve her position with a bid of 450 Euro.

Part b)
To prove that bidding your own valuation is a weakly dominant strategy, we need to show that this bid yields at least the same payoff as all other strategies in any situation and is even strictly better (at least once). Although the players have infinitely many strategies (they can make any bid), we can limit the analysis to three relevant constellations as in part a). It is sufficient to show that bidding your own valuation is better than making a higher bid and better than making a lower bid. The resulting payoffs result from the relative position of all other bids. However, only the highest bid of all other bidders, b_j^{max}, is relevant here. It does not matter whether they bid their valuation or not, as we are examining dominance here and a weakly dominant strategy is always chosen regardless of the choice of the other players.

Case 1: Comparison of the strategy to bid your valuation, $b_i^* = v_i$, with the strategies to bid more than your valuation, $b_i > v_i$. For the relative position of $b_j^{\max}$ there are three possibilities:

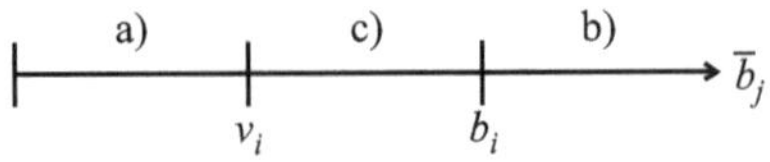

a. The highest bid $b_j^{\max}$ of all other participants is less than your valuation, $b_j^{\max} < v_i$.
   - With bid $b_i > v_i$, you win the auction, and your payoff is $u_i = v_i - b_j^{\max}$.
   - With bid $b_i^* = v_i$, you win the auction, and your payoff is $u_i = v_i - b_j^{\max}$.
b. The highest bid $b_j^{\max}$ of all other participants is greater than your bid, $b_i < b_j^{\max}$.
   - With bid $b_i > v_i$, you do not win the auction, and your payoff is zero.
   - With bid $b_i^* = v_i$, you do not win the auction, and your payoff is zero.
c. The highest bid $b_j^{\max}$ of all other participants is between your valuation and your bid, $v_i < b_j^{\max} < b_i$.
   - With bid $b_i > v_i$, you win the auction, and your payoff is $u_i = v_i - b_j^{\max} < 0$.
   - With bid $b_i^* = v_i$, you do not win the auction, and your payoff is zero.

We see that the strategy $b_i^* = v_i$ is in a) and b) as good as the strategies $b_i > v_i$ but in case c) it is strictly better. Thus, the strategy to bid your own valuation weakly dominates the strategies to bid more than your own valuation.

Case 2: Comparison of the strategy to bid your valuation, $b_i^* = v_i$ with the strategies to bid less than your valuation, $b_i < v_i$. For the relative position of $b_j^{\max}$ there are three possibilities:

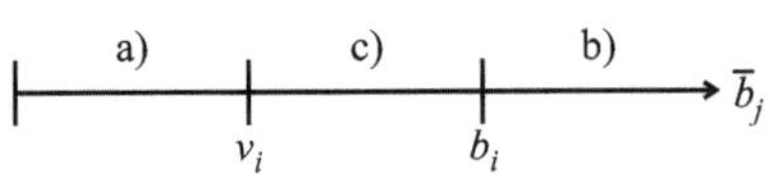

a. The highest bid $b_j^{\max}$ of all other participants is smaller than your valuation, $b_j^{\max} < b_i$.
   - With bid $b_i < v_i$, you win the auction, and your payoff is $u_i = v_i - b_j^{\max}$.
   - With bid $b_i^* = v_i$ you win the auction, and your payoff is $u_i = v_i - b_j^{\max}$.

b. The highest bid $b_j^{\max}$ of all other participants is greater than your valuation, $v_i < b_j^{\max}$.
   - With bid $b_i < v_i$ you do not win the auction, and your payoff is zero.
   - With bid $b_i^* = v_i$ you do not win the auction, and your payoff is zero.

c. The highest bid $b_j^{\max}$ of all other participants is between your bid and your valuation, $b_i < b_j^{\max} < v_i$.
   - With bid $b_i < v_i$, you do not win the auction, and your payoff is zero.
   - With bid $b_i^* = v_i$, you win the auction, and your payoff is $u_i = v_i - b_j^{\max} > 0$.

Here too, we see that the strategy $b_i^* = v_i$ is in a) and b) as good as the strategies $b_i < v_i$, but in case c) it is again strictly better. Thus, the strategy to bid your own valuation weakly dominates the strategies to bid less than your own valuation.

In summary, we were able to show that the strategy $b_i^* = v_i$ is a weakly dominant strategy. Since every player has this weakly dominant strategy, the bidder with the highest valuation will win, that is player n with v_n. The bidder with the second highest valuation v_{n-1} will—in equilibrium—also submit a bid which equals the own valuation, which is why player n receives a payoff of $u_n = v_n - v_{n-1}$.

Part c)

Since each player with $b_i^* = v_i$ has a weakly dominant strategy, which they will choose regardless of the strategy choice of the others, a lack of complete information does not change the outcome. Only with $b_i^* = v_i$ it is ensured that the player, should they have the highest valuation, will also receive the auctioned item (this would not be guaranteed if $b_i < v_i$) and that the player, should they not have the highest valuation, will not receive the auctioned item (this would not be guaranteed if $b_i > v_i$).

11.2.7 All-Pay Auction

Problem text in Sect. 11.1.7.

Part a)

If Alice's bid is greater than Bob's bid, $b_A > b_B$, she receives the object and must pay her bid for it. Her payoff is therefore the difference between the value of the object and her submitted bid, $v - b_A$. The losing Bob, on the other hand, must pay his bid without receiving anything in return and his payoff is therefore $-b_B$. Conversely, for $b_A < b_B$ Alice's payoff is $-b_A$ and Bob's is $v - b_B$. With the same bid, $b_A = b_B$, the payoff of both players is negative, $-b_A$ and $-b_B$.

A rational player will only bid up to the value of the object v. This can be easily shown: If Alice bids $b_A > v$ and wins (for $b_A > b_B$), then she receives the payoff $v - b_A < 0$. If she does not win the auction (for $b_A \leq b_B$), her payoff is $-b_A < 0$. With a bid of $b_A = 0$, it is always ensured that she receives a payoff of 0. Thus, the strategy $b_A = 0$ dominates every strategy $b_A > v$ strictly. The strategy $b_A = v$ is only weakly dominated by $b_A = 0$. All other strategies $b_A < v$ are not dominated, as in the case of winning a positive payoff $v - b_A > 0$ results. The considerations for Bob are analogous.

Part b)
Since it is a simultaneous game, the game is best represented and solved in matrix form.

With $v = 1$ Euro, rational players will either bid 0 or 1 Euro:

Alice, Bob	$b_{B1} = 0$	$b_{B2} = 1$
$b_{A1} = 0$	$(\underline{0}, \underline{0})$	$(\underline{0}, \underline{0})$
$b_{A2} = 1$	$(\underline{0}, \underline{0})$	$(-1, -1)$

Here, both players each have the weakly dominant strategy to bid 0. This results in a total of three Nash equilibria in pure strategies: $(b_{A1} = 0, b_{B1} = 0)$, $(b_{A2} = 1, b_{B1} = 0)$, and $(b_{A1} = 0, b_{B2} = 1)$.

With $v = 2$ euros, rational players will bid either 0, 1, or 2 euros:

Alice, Bob	$b_{B1} = 0$	$b_{B2} = 1$	$b_{B3} = 2$
$b_{A1} = 0$	$(0, 0)$	$(\underline{0}, \underline{1})$	$(\underline{0}, 0)$
$b_{A2} = 1$	$(\underline{1}, \underline{0})$	$(-1, -1)$	$(-1, \underline{0})$
$b_{A3} = 2$	$(0, \underline{0})$	$(\underline{0}, -1)$	$(-2, -2)$

This results in two Nash equilibria in pure strategies: $(b_{A2} = 1, b_{B1} = 0)$ and $(b_{A1} = 0, b_{B2} = 1)$.

At $v = 3$ Euros, rational players will either bid 0, 1, 2, or 3 Euros:

Alice, Bob	$b_{B1} = 0$	$b_{B2} = 1$	$b_{B3} = 2$	$b_{B4} = 3$
$b_{A1} = 0$	$(0, 0)$	$(0, \underline{2})$	$(\underline{0}, 1)$	$(\underline{0}, 0)$
$b_{A2} = 1$	$(\underline{2}, 0)$	$(-1, -1)$	$(-1, \underline{1})$	$(-1, 0)$
$b_{A3} = 2$	$(1, \underline{0})$	$(\underline{1}, -1)$	$(-2, -2)$	$(-2, \underline{0})$
$b_{A4} = 3$	$(0, \underline{0})$	$(0, -1)$	$(\underline{0}, -2)$	$(-3, -3)$

Here, there is no Nash equilibrium in pure strategies.

Part c)
We can already consider this for $v = 3$: Given Bob's bid, Alice will want to outbid it by at most 1 Euro. If she outbids by more than 1 Euro, she only lowers her payoff. The players then continue to outbid each other by 1 Euro. Finally, if Alice makes the highest bid, that is the value of the object v, then Bob has no

incentive to outbid anymore, as he would only make a loss. It is also not an option to make the same bid, as he would then have to pay his bid without receiving anything. Any other positive bid would have the same effect and thus it is best to bid nothing. However, if Bob bids 0, then Alice has no incentive to bid the value of the object, as she would already receive the object with a bid of 1 and her payoff would be significantly higher as a result. Thus, the bidding competition starts all over again. A rest point, i.e., a Nash equilibrium in pure strategies, cannot exist here.

Part d)
With a bid of 0 euros, the player's payoff is always guaranteed to be 0 euros. Thus, all pure strategies that are played with positive probability must have an expected utility of 0.

Let p_{Bj} be the probability with which Bob makes the bid j. Then for Alice, who bids 1 euro, it must hold that

$$0 \stackrel{!}{=} p_{B0} \cdot (v - 1) + (1 - p_{B0}) \cdot (-1).$$

Only when Bob makes a bid of 0—which happens with probability p_{B0}—will Alice receive the object and have the payoff $v - 1$ (value of the object minus her submitted bid of 1). For all of Bob's other bids—which occur with the counter-probability $1 - p_{B0}$—Alice pays her bid without any compensation. If we solve this condition for p_{B0} we see that it is fulfilled for $p_{B0} = 1/v$.

If Alice bids exactly 2 euros, we can analogously conclude that

$$0 = (v - 2) \cdot (p_{B0} + p_{B1}) + (-2) \cdot (p_{B0} + p_{B1}),$$

i.e., she receives the payoff $v - 2$ if Bob bids 0 (with probability p_{B0}) or 1 (with p_{B1}) and -2 in all other cases. Since we know that $p_{B0} = 1/v$, it follows that $p_{B1} = 1/v$.

Due to the symmetry, this can be carried out for all further bids, so it must be the case that

$$p_{B0} = p_{B1} = \ldots = p_{Bv} = 1/v.$$

The procedure can also be applied to Bob in a similar way. Thus, each player will play every single bid in their mixed strategy with the same probability of $1/v$ and expect a payoff of 0.

11.2.8 Tullock Contest

Problem text in Sect. 11.1.8.

Part a)
The objective function (or profit function) of company i is given by

$$\pi_i(x_i, x_j) = v \cdot p_i - c_i \cdot x_i = v \cdot \frac{x_i}{x_i + x_j} - c_i \cdot x_i$$

The first-order condition states that the partial derivative of the profit function with respect to the strategic variable x_i must be zero:

$$\frac{\partial \pi_i(x_i, x_j)}{\partial x_i} = \frac{(x_i + x_j) \cdot v - v \cdot x_i}{(x_i + x_j)^2} - c_i \stackrel{!}{=} 0$$

We find the reaction function by solving the FOC for the strategic variable (x_i). However, this includes tedious calculation—so it is absolutely fine if you cheat at this point:

$$x_i(x_j) = \frac{\sqrt{v} \cdot \sqrt{x_j}}{\sqrt{c_i}} - x_j \text{ and for firm } j: \; x_j(x_i) = \frac{\sqrt{v} \cdot \sqrt{x_i}}{\sqrt{c_j}} - x_i$$

If we insert the two reaction functions into each other, we get the intersection of these two functions and thus the Nash equilibrium of the game:

$$x_i^* = \frac{v \cdot c_j}{(c_i + c_j)^2} \quad \text{and} \quad x_j^* = \frac{v \cdot c_i}{(c_i + c_j)^2}$$

We see that the effort of firm i increases in the fixed value and decreases in the own cost. Whether the effort of firm i increases or decreases for higher cost of firm j depends on the relation of the cost values: For $c_i > c_j$, the equilibrium effort x_i^* increases in c_j but for $c_i < c_j$, the effort goes down if c_j goes up. You can easily verify this by taking the first derivative of x_i^* with respect to c_j and evaluating the sign of the numerator. Hence, if for company i the innovation competition is more costly ($c_i > c_j$), it tries to compensate for this with higher investments.

On the one hand, the Tullock contest illustrates the competitive character of innovations. However, on the other hand, most innovations, in particular incremental innovations, are less characterized by rivalry (= innovation found by firm i cannot be used by firm j) but rather by spillovers between the firms. From an economy-wide perspective, these spillovers have a strong impact on sector productivity. This is why it can be observed that firms which compete in the final market even collaborate in R&D (e.g., Mercedes-Benz and BMW collaborate in the form of a Joint Venture on a large array of themes regarding future smart mobility). Furthermore, most innovative activities directly generate value (either by reduced cost or higher revenue), hence an increase in x has a positive influence on v. This is the focus of part c).

As we assume that effort itself does not increase the value of the prize, the Tullock contest is inefficient: The firms' investments are eventually lost ("wasteful expenditure") while the value of the prize is given and fixed. Hence, the inefficiency could be avoided by randomly assigning the prize to one of the firms right from the start instead of involving them in a tournament that only causes them costly effort.

Part b)
If we want to plot the reaction curves in a x_i-x_j-diagram (with x_j as the vertical axis), we need to convert $x_i(x_j)$ to $x_j(x_i)$, which gives the reaction function of firm i in a different form:

$$x_j(x_i) = \frac{v - 2 \cdot c_i \cdot x_i - \sqrt{v \cdot (v - 4 \cdot c_i \cdot x_i)}}{2 \cdot c_i}$$

The reaction function of firm j was already derived in part a): $x_j(x_i) = \frac{\sqrt{v} \cdot \sqrt{x_i}}{\sqrt{c_j}} - x_i$

The following figure shows the reaction functions for the parameter values $v = 10$, $c_i = 1$, and $c_j = 1$:

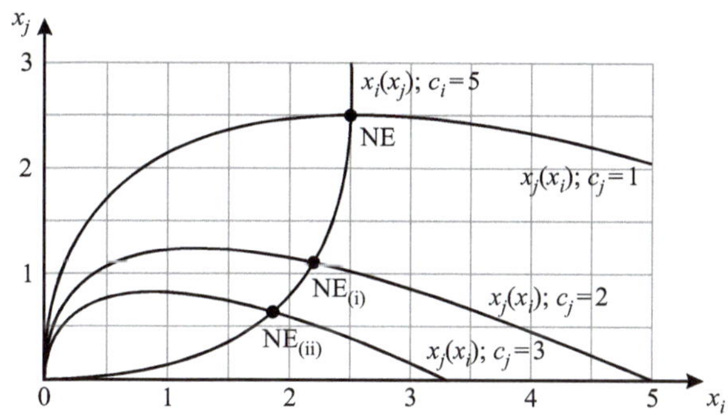

As you can check using the formula expressions from part a), the Nash equilibrium is the intersection of the two curves at the coordinates NE $\left(x_i^* = 2.5, x_j^* = 2.5\right)$. An increase in c_j reduces the investment of both companies as we can also see in the figure: $\text{NE}_{(i)}$ for (i) $c_j = 2$ and $\text{NE}_{(ii)}$ for (ii) $c_j = 3$. It reduces the effort of company i because—as we already found out in part a)—we have the constellation $c_i < c_j$. However, the reduction in the level of effort is much more pronounced at company j, which is directly affected by the cost increase, than at company i.

Part c)
The objective function (or profit function) of company i is given by

$$\pi_i(k_i, k_j) = \text{prob}\left[v_i^{\max} > v_j^{\max}\right] \cdot \text{E}\left[v_i^{\max}\right] - c_i \cdot k_i = \frac{k_i}{k_i + k_j} \cdot \frac{k_i}{k_i + 1} \cdot v - c_i \cdot k_i$$

In this modified model, the number of tickets drawn k_i represents the "innovation effort" x_i of parts a) and b). The first term is—as before—the probability of winning the contest, which is the case if the value of one's own ticket is higher than that of the competitor. Exactly as in the classic Tullock contest, the probability of company i to win the contest against company j is given by $p_i = \text{prob}[v_i^{\max} > v_j^{\max}] = \frac{k_i}{k_i+k_j}$. The second term now reflects the endogenous value of the prize, i.e., the value that is determined by the level of innovation investment.

Deriving the FOC for this modified objective function is significantly more tedious:

$$\frac{\partial \pi\left(k_i, k_j\right)}{\partial k_i} = \frac{2 \cdot k_i \cdot \left(k_i + k_j\right) \cdot (k_i + 1) - k_i^2 \cdot \left(2 \cdot k_i + k_j + 1\right)}{\left(k_i + k_j\right)^2 \cdot (k_i + 1)^2} \cdot v - c_i \stackrel{!}{=} 0$$

You can compare this expression with the solution of part a) and b) by inserting the parameter values $v = 10$, $c_i = 1$, and $c_j = 1$ as well as the former equilibrium values $k_i = 2.5$ and $k_j = 2.5$ which shows that the expression above is larger than zero and becomes zero for a value $k_i^* > 2.5$. This indicates that the modified version with an endogenous prize of the tournament exerts larger innovation incentives. For this reason, such a tournament can be efficient provided that the costs are not too large, and the firms have an incentive to invest ($k_i^* > 0, k_j^* > 0$).

11.2.9 Credibility of Monetary Policy

Problem text in Sect. 11.1.9.

Part a)

The central bank (player A) has two discrete strategies $p_{A1} = 0$ (create no inflation) and $p_{A2} = 4$ (create inflation). The population (player B) has the strategies $p_{B1}^e = 0$ (expect no inflation) and $p_{B2}^e = 4$ (expect inflation). If these strategies are inserted into the utility function of the central bank and that of the population, the payoff matrix for this game is obtained. E.g., for the strategy combination $\left(p_{A1} = 0, p_{B1}^e = 0\right)$ the central bank gets a payoff of $u_A\left(p_{A1} = 0, p_{B1}^e = 0\right) = 32 - [0.5 \cdot C^E(p_{A1} = 0) + 0.5 \cdot C^L(p_{A1} = 0, p_{B1}^e = 0)] = 32 - [0.5 \cdot 0 + 0.5 \cdot 16] = 24$. For the population, the payoff for this strategy combination is correspondingly $u_B(p_{A1} = 0, p_{B1}^e = 0) = 16$. This results in the following payoff matrix:

A, B	$p_{B1}^e = 0$	$p_{B2}^e = 4$
$p_{A1} = 0$	(24, 16)	(0, 0)
$p_{A2} = 4$	(24, 0)	(16, 16)

Part b)
When we search for a solution using dominance considerations, we first check whether there is an equilibrium in dominant strategies. This is obviously not the case here, as only one player, the central bank, has a dominant strategy: Its strategy $p_{A2} = 4$ (creating inflation) is weakly dominant. If the central bank inflates, then the best response of the population is to also expect inflation ($p^e_{B2} = 4$), so that the strategy combination $\left(p_{A2} = 4,\, p^e_{B2} = 4\right)^*$ results as a Nash equilibrium. The payoffs of the players in this equilibrium are each 16. Since only a weakly dominant strategy of one player could be considered in the dominance considerations, the Nash equilibrium determined in this way is not necessarily unique. By checking with reaction correspondences, we can perform a complete search for all Nash equilibria in pure strategies. We get:

$$\begin{aligned}
\text{Central Bank: } & r^*_A\left(p^e_{B1} = 0\right) = \{p_{A1} = 0,\, p_{A2} = 4\} \text{ and} \\
& r^*_A\left(p^e_{B2} = 4\right) = \{p_{A2} = 4\} \\
\text{Population: } & r^*_B(p_{A1} = 0) = \{p_{B1} = 0\} \text{ and} \\
& r^*_B\left(p^e_{A2} = 4\right) = \left\{p^e_{A2} = 4\right\}
\end{aligned}$$

With this, we see that there are two Nash equilibria in pure strategies in this game: one without inflation and with inflation expectation of zero $\left(p_{A1} = 0,\, p^e_{B1} = 0\right)^*$ and one with inflation and positive inflation expectations $\left(p_{A2} = 4,\, p^e_{B2} = 4\right)^*$. We only found the second one through dominance considerations, the first one was only visible with the help of reaction correspondences.

Part c)
The players now have continuous strategies, so they can choose any level of inflation and inflation expectation. We obtain the Nash equilibrium in continuous strategies by deriving the reaction functions of the players and then determining the intersection point of the reaction functions. We obtain the reaction function of a player by determining the first-order condition (partial derivative of the utility function with respect to the respective strategic variables of the player and setting to zero). The utility function of the central bank is:

$$\begin{aligned}
u_A\left(p, p^e\right) &= 32 - \left[0.5 \cdot C^E(p) + 0.5 \cdot C^L\left(p, p^e\right)\right] \\
&= 32 - 0.5 \cdot p^2 - 0.5 \cdot \left[p - \left(p^e + 4\right)\right]^2 \\
&= 30 - p^2 + p \cdot \left(p^e + 4\right) - 0.5 \cdot p^e
\end{aligned}$$

By partially deriving with respect to p we obtain the first-order condition of the central bank; if we then solve this for p we obtain its reaction function:

$$\frac{\partial u_A\left(p, p^e\right)}{\partial p} = p^e + 4 - 2 \cdot p \stackrel{!}{=} 0 \quad \Rightarrow p^*\left(p^e\right) = 2 + 0.5 \cdot p^e$$

For the utility function of the population, $u_B(p, p^e) = 16 - (p - p^e)^2$ we proceed analogously and obtain

$$\frac{\partial u_B(p, p^e)}{\partial p^e} = 2 \cdot p - 2 \cdot p^e \overset{!}{=} 0 \quad \Rightarrow p^{e*}(p) = p$$

The population's reaction function is therefore identical to the 45-degree line: it always expects exactly the inflation rate chosen by the central bank. Whenever the expectation line (as here) corresponds to a 45-degree line, we speak of rational expectation formation or simply "rational expectations".

Graphically, the reaction functions are represented as follows:

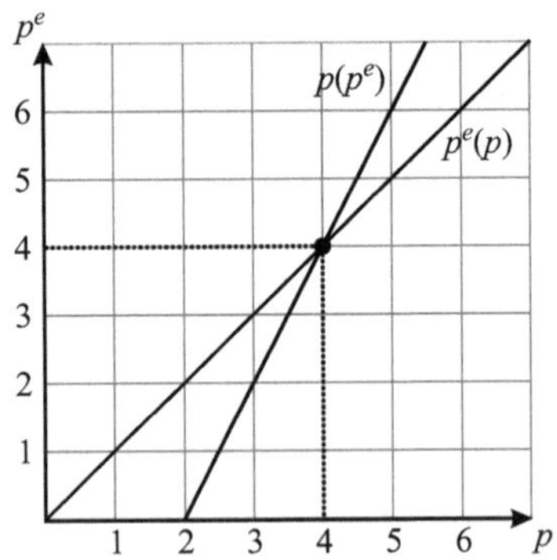

As we see from the figure and confirm by equating the two reaction functions, the Nash equilibrium is ($p = 4$, $p^e = 4$). Both players thus each have a payoff of 16.

Part d)

In contrast to part c), we now have a sequential game, and we use the concept of backward induction to solve the game. To do this, we first determine the reaction function of the second mover (population) and insert it into the utility function of the first mover (central bank). In the second step, the utility function of the first mover is then derived according to the strategic variable. The result is the subgame perfect Nash equilibrium.

We already know the population's reaction in the second stage to the policy choice of the central bank—this corresponds to the reaction function from the simultaneous game from c), $p^{e*}(p) = p$. This reaction can then be inserted into the central bank's utility function in the first stage:

$$u_A(p, p^e) = 24 + 4 \cdot p - p^2 - 4 \cdot [p] + p \cdot [p] - 0.5 \cdot [p]^2 = 24 - 0.5 \cdot p^2$$

To determine the optimal p, the utility of the central bank is to be maximized, $du_A(p, p^e)/dp = p \overset{!}{=} 0$. This results in a subgame perfect Nash equilibrium ($p = 4$, $p^e = 4$) with the resulting payoff combination (24, 16).

Part e)

The reaction function of the central bank shows: If the population chooses $p^e = 0$ then the central bank will respond with $p = 2$, that means deviating from (0, 0).

Thus, (0, 0) cannot establish itself as an equilibrium.

i. The simple, verbal announcement of zero inflation assumed here is not credible ("cheap talk") as it does not solve the incentive problem.
ii. Performance-based compensation increases the incentive of the central banker to act against inflation. A two-stage game, in which the amount of optimal compensation is determined in the first stage, and in the second stage p^e or p are optimally determined, leads directly to the subgame perfect Nash equilibrium. Thus, the Pareto-optimal solution can be realized with performance-based compensation.
iii. In this case, a grim trigger strategy is assumed. In the case of finite repetition, backward induction will show that the incentive problem cannot be overcome, as the players will always deviate in the last period and therefore no punishment is possible. An infinite repetition of the game, however, allows the realization of the socially desirable solution (Folk Theorem).

11.2.10 Overseas Trade

Problem text in Sect. 11.1.10.

Part a)

In this game, there are two players, the merchant and the agent. The merchant has two strategies: He can either "hire" or "not hire" the agent. The agent also has two strategies: he can "fulfill" the order dutifully or "cheat" the merchant. The resulting payoff combinations are listed in the following matrix.

Merchant, Agent	Fulfill	Cheat
Hire	($\underline{2}$, 1)	(-2, $\underline{5}$)
No hire	(0, $\underline{0}$)	($\underline{0}$, $\underline{0}$)

Part b)

As we see, the agent has the weakly dominant strategy "cheat": If he is hired, "cheat" with a payoff of 5 is better than "fulfill" with a lower payoff of 1. If he is not hired, "cheat" is as good as "fulfill"—in both cases he receives 0. The merchant has no (weakly) dominant strategy. Therefore, if he reacts optimally to the weakly dominant strategy of the agent, he will not hire him, so the Nash equilibrium (no hire, cheat) with the payoff combination (0, 0) results. Since the dominant strategy of the agent is only weakly dominant, the Nash equilibrium is not necessarily unique: There could be more. A systematic review, as already done in the matrix by the underlining method, shows us that no other Nash equilibrium exists. Since the number is odd, there can also be no further Nash equilibrium in mixed strategies.

Part c)

The solution determined in b) is plausible. In the considered context, the Nash equilibrium as a solution concept makes sense, especially since it is also the only equilibrium of the game. However, we see that the payoff resulting from the strategy combination (hire, fulfill) (2, 1) would be Pareto-efficient, that is both players would have an interest in realizing this strategy combination instead of the Nash equilibrium with (0, 0). However, this cannot be realized without binding agreements between the two players, as the agent always has the incentive to deviate to "cheat". The underlying cooperation problem here represents an asymmetric prisoner's dilemma, as only the agent has a dominant strategy that leads to a disadvantageous result for both players.

Part d)

The Maghribi traders practice a trigger strategy: To take away the agent's incentive to cheat, which consists of deviating in one period to realize a high profit of 5, a punishment in the form of loss of livelihood and reputation is threatened, which from the subsequent period leads to a payoff of 0 forever. Only if the agent always fulfills the contract, he can expect a consistent payoff of 2.

If we apply the one-stage deviation principle, we thus obtain

$$\sum_{t=0}^{\infty} \delta^t \cdot 2 > 5 + \sum_{t=1}^{\infty} \delta^t \cdot 0$$

$$2 \cdot \frac{1}{1+\delta} > 5$$

$$\Rightarrow \delta \geq \tilde{\delta} \equiv 0.6$$

Thus, the agents must have a moderate preference for the future for the strategy to be successful, and the merchants can trust the Maghribi traders.

Mathematical Basics 12

Game theory has its origins in mathematics, which is very apparent throughout. Although there are many problems that can be solved with simple algebra, optimization is often used, which requires more advanced knowledge. This chapter is intended to introduce readers with little prior knowledge of calculus so that they can solve the problems.

12.1 Introduction to Calculus

The relevant part of calculus for our analysis is the determination of slopes. Why? Let us look at Fig. 12.1. This curve represents the equation $f(x) = 6 \cdot x - x^2$. We are always interested in the optimal, i.e., in this case maximum, solution. Of the three points given, A, B, and C, point C at $x = 3$ is clearly the best. What distinguishes this point from the others and how can we find this point without drawing the curve? A possible, but tedious procedure would be to simply calculate all the points to find out at which x-value the highest function value $f(x)$ results. This would be perfectly feasible in our example. However, if the optimum is not a natural number, this approach makes it much more difficult to find the optimum—e.g., try to find the optimum value for $g(x) = 3 \cdot x^2 - 7 \cdot x$! It may seem a little more absurd at first, but it is much easier to use a different property of this point. While at point C, the slope is zero, point A has a positive slope, and point B a negative slope. The type of slope therefore also determines the direction in which it gets better or worse. The positive slope at point A indicates to increase x to reach higher values for $f(x)$, whereas it is better to reduce x if the slope is negative (point B). At point C, the slope is 0, which means that neither an increase nor a reduction leads to a change. So, the point is optimal. The question now is, how do we determine the slope?

The slope indicates how the dependent variable $f(x)$ changes when there is a change in the independent variable x. This makes it very easy to determine the

F. Bartholomae and M. Wiens, *Game Theory and Applications*, Classroom Companion: Economics, https://doi.org/10.1007/978-3-658-44608-6_12

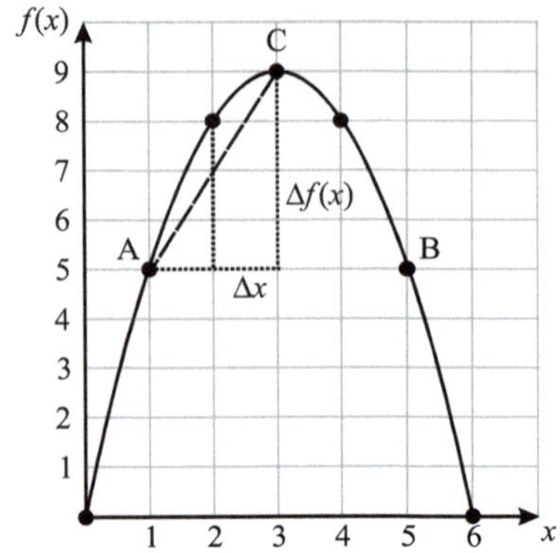

Fig. 12.1 Slope of a non-linear curve

slope of linear lines:

$$s = \frac{f(x_2) - f(x_1)}{x_2 - x_1} = \frac{\Delta f(x)}{\Delta x}$$

Typically, the Greek delta, Δ, denotes the change. For non-linear curves, however, this method is very imprecise. Let us look at our example. Suppose we want to determine the slope at point A. If we increase x there by 2 units, i.e., from 1 to 3, then $f(x)$ increases from 5 to 9, so there is a change of 4 units. The slope would therefore be $(9-5)/(3-1) = 4/2 = 2$. However, if we only increase x by 1 unit, $f(x)$ rises to 8 and the resulting slope would be $(8-5)/(2-1) = 3$. As we can see in the diagram, both calculations underestimate the true slope. To see this, imagine a tangent running through A. Its slope would be greater than all the values calculated here.

And this is where infinitesimal calculus comes into play. This method does not consider such mathematically large changes as before, but just very small ones. We have seen that the smaller the change in x, the more accurate the determination of the slope becomes. As we are now looking at very small changes, we express this with a small d (in contrast to the big Δ). The slope of a function $f(x)$ calculated in calculus is therefore denoted with $df(x)/dx$ or much more convenient, $f'(x)$. The mathematical procedure for determining the slope using infinitesimal calculation is called *derivation*. We will not go into the exact mathematical deduction here. However, for the application, it is important to know some basic derivation rules, which are summarized in the next section.

12.2 Important Derivative Rules

Since many problems require optimization, it is important to be familiar with the most important derivative rules: Note that $f'(x)$ is short for $df(x)/dx$.

Rule	Function	Derivative
Constant	$f(x) = C$	$f'(x) = 0$
Power Rule	$f(x) = a \cdot x^n$	$f'(x) = a \cdot n \cdot x^{n-1}$
Sum Rule	$f(x) = u(x) + v(x)$	$f'(x) = u'(x) + v'(x)$
Product Rule	$f(x) = u(x) \cdot v(x)$	$f'(x) = u'(x) \cdot v(x) + u(x) \cdot v'(x)$
Quotient Rule	$f(x) = \frac{u(x)}{v(x)}$	$f'(x) = \frac{u'(x) \cdot v(x) - u(x) \cdot v'(x)}{[v(x)]^2}$
Chain Rule	$f(x) = u[v(x)]$	$f'(x) = u'[v(x)] \cdot v'(x)$

For all application examples in this book, it is completely sufficient to know and be able to apply the constant, sum, and the product rules. In addition, there are special functions that repeatedly appear in the economic context:

Function	Derivative
$f(x) = \sqrt{x} = x^{0.5}$	$f'(x) = 0.5 \cdot x^{-0.5} = \frac{1}{2 \cdot \sqrt{x}}$
$f(x) = \ln x$	$f'(x) = \frac{1}{x}$

12.3 Critical Points and Partial Derivatives

To determine a critical point, we must therefore derive the function and then set it to zero. This is often referred to as the first-order condition (FOC). This means that we determine the point at which the slope is zero. In our example, for $f(x) = 6 \cdot x - x^2$, the derivative is $f'(x) = 6 - 2 \cdot x$ and, accordingly, the first-order condition $6 - 2 \cdot x \stackrel{!}{=} 0$ (the exclamation mark above the equal sign denotes that this is a condition that must be fulfilled). The solution of the FOC is $x = 3$ which looks very familiar. What about $g(x) = 3 \cdot x^2 - 7 \cdot x$? Here we would get $g'(x) = 6 \cdot x - 7 \stackrel{!}{=} 0$ or $x = 7/6$. Probably we would have searched quite a long time for this unattractive ("crooked") value.

However, we only know that it is a critical point. Without plotting (or lots of inserting), however, we do not know whether it is a maximum or minimum. But this problem can also be solved. Fig. 12.2 shows the two possible curvature behaviors of a function: It can be either concave or convex.[1] Sometimes concave curves are also referred to as right-curved and convex as left-curved: Imagine riding along these curves, then you would always have to turn the handlebars to the right for a concave curve.

We then use the curvature to find out what type of critical point is present: We can see from the graphs on the far right that there is a maximum in the case

[1] A simple mnemonic for distinguishing between concave and the opposite, convex is whether the curve looks like a cave entrance (conCAVE) - at least that is what the third figure in the first line suggests.

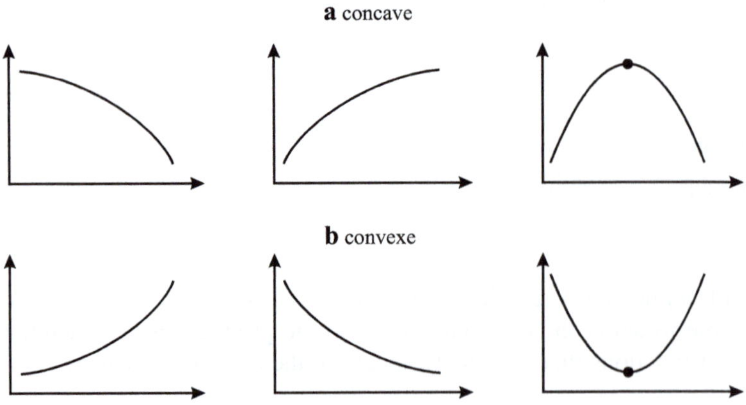

Fig. 12.2 Curvature of non-linear curves

of concavity and a minimum in the case of convexity. The curvature itself can be calculated using the derivative of the derivative, i.e., the second derivative, which therefore corresponds to the "slope of the slope", and indicates how the slope changes: With a concave slope, the slope becomes smaller and smaller or, respectively, the curve becomes flatter and flatter and may even be falling at some point. This means that if we obtain a negative value for the second derivative and know that there is a critical point, then this must be a maximum. The opposite is true for a convex curve, which indicates a minimum: Here the slope keeps increasing, i.e., the second derivative is positive.

Let us summarize:

$$\begin{aligned}\text{Maximum } x_{\max} &: f'(x_{\max}) = 0,\ f''(x_{\max}) < 0 \\ \text{Minimum } x_{\min} &: f'(x_{\min}) = 0,\ f''(x_{\min}) > 0\end{aligned}$$

Let use this to check what critical point we have found in our examples: As the second derivative, we find $f''(x) = -2$. The value is less than 0, which is why we consequently must have a maximum at this point. For the other function, we obtain a positive value, $g''(x) = 6$, so that a minimum is present here.

Since payoff functions depend on more than one variable, $f(x, y, \ldots)$, only a *partial derivative* with respect to the variable of interest can be determined, e.g., x. Partial derivatives are represented by ∂ ("partial"): $\partial f(x, y, \ldots)/\partial x$. Please note that the shorthand notation with the dash would make little sense here, "$f'(x, y, \ldots)$", as it would then be unclear which variable is being derived. However, that is the only difficulty, the derivation itself is analogous to the procedure we know: The variables that are not used for derivation are simply regarded as constant and treated as such. Consider, for example, the function $f(x, y, z) = x^2 + \ln y + x \cdot y^2$. Here, the partial derivatives are $\partial f(x, y, z)/\partial x = 2 \cdot x + y^2$, $\partial f(x, y, z)/\partial y = 1/y + 2 \cdot x \cdot y$, and $\partial f(x, y, z)/\partial z = 0$.

Excursion: Wolfram Alpha. If you are unsure whether you have calculated correctly, we recommend that you check the website of Wolfram Alpha, http://www.wolframalpha.com/. Wolfram Alpha was released in May 2009 and is based on Wolfram Mathematica, a computational platform that encompasses computer algebra, symbolic and numerical computation, visualization, and statistics capabilities. The big advantage is that Wolfram Alpha can handle symbolic computations. You can either ask for the solution or use the syntax you may know from Mathematica. The latter always provides the desired results, which is why we would now like to briefly introduce you to the most relevant ones for the problems in this book.

- D[f,x] gives the (partial) derivative $\partial f / \partial x$. Example: to determine the derivative of $x^2 - x^3$ use D[x^2–x^3, x].
- D[f,{x,n}] gives the n th (partial) derivative $\partial^n f / \partial x^n$. Example: to determine the second derivative of $x^2 - x^3$ use D[x^2–x^3, {x, 2}].
- Solve[expr, vars] attempts to solve the system *expr* of equations or inequalities for the variables *vars*. Example: if you want to solve $f(x) = y_0 + s \cdot x \overset{!}{=} 0$ use Solve[y0 + s*x = = 0, x]. Note "$\overset{!}{=}$" becomes "= =" as a normal "=" is typically used for definitions—you may know this from other programming languages.
- Combination of D and Solve to determine critical points. Example: if you want to determine the critical points of $10 \cdot x^2 - 2 \cdot x$ use Solve[D[10*x^2–2*x, x] = = 0, x].
- Plot[f,{x,xmin,xmax}] generates a plot of *f* as function of *x* from *xmin* to *xmax*. Example: to plot $x^2 - x^3$ in the range between -5 and 5 use Plot[x^2–x^3, {x, -5, 5}].

12.4 Problems

12.4.1 Derivatives

Find the first derivatives of the following functions!

(a)	$f(x) = 7 \cdot x + 15$	(f)	$f(x) = 4 \cdot x^5 - 5 \cdot x^4$
(b)	$f(x) = -7 \cdot x^3$	(g)	$f(x) = (x^2 + 1) \cdot (x^2 + 3 \cdot x + 2)$
(c)	$f(x) = 3 \cdot x^{-\frac{3}{2}}$	(h)	$f(x) = (a - x)^2$
(d)	$f(x) = \frac{1}{2} \cdot \sqrt{x}$	(i)	$f(x) = (x^3 - \sqrt{x})^5$
(e)	$f(x) = 3 \cdot x^2 - 9 \cdot x + 7 \cdot x^{\frac{2}{5}} - 3 \cdot x^{\frac{1}{2}}$	(j)	$f(x) = \frac{x^2+10}{x^3}$

12.4.2 Critical Points

Find all critical points for the following functions!

(a) $f(x) = 20 \cdot x - x^2$
(b) $f(x) = x^3 - 3 \cdot x$
(c) $f(x) = x + \frac{100}{x}$.

12.4.3 Partial Derivatives

Derive all partial derivatives of the given function!

$$f(x_1, x_2, x_3, x_4, x_5) = x_1^2 \cdot x_2 + \frac{x_3}{x_1} - \sqrt{x_2 \cdot x_3} - 5 \cdot x_4^5 - \frac{x_5}{x_1 \cdot x_4}.$$

12.5 Solutions

12.5.1 Derivatives

(a) Application of the power rule: $f'(x) = 7$. Please note that the given function is a linear equation. As we can see, the first derivative gives exactly the slope of the function—i.e., the derivative does exactly what it is supposed to! We can also show this for the general equation, $f(x) = m \cdot x + y_0$, where m denotes the slope and y_0 the intercept. The first derivative is again the slope, $f'(x) = m$.
(b) Application of the power rule: $f'(x) = -7 \cdot 3 \cdot x^{3-1} = -21 \cdot x^2$
(c) Application of the power rule: $f'(x) = 3 \cdot \left(-\frac{3}{2}\right) \cdot x^{-\frac{3}{2}-1} = -\frac{9}{2} \cdot x^{-\frac{5}{2}}$
(d) Application of the power rule: $f'(x) = \frac{1}{2} \cdot 0.5 \cdot x^{0.5-1} = \frac{1}{4 \cdot \sqrt{x}}$
(e) Application of the sum and power rules:

$$\begin{aligned} f'(x) &= 3 \cdot 2 \cdot x^{2-1} - 9 \cdot 1 \cdot x^{1-1} + 7 \cdot \frac{2}{5} \cdot x^{\frac{2}{5}-1} - 3 \cdot \frac{1}{2} \cdot x^{\frac{1}{2}-1} \\ &= 6 \cdot x + 9 + \frac{14}{5} \cdot x^{-\frac{3}{5}} - \frac{3}{2} \cdot x^{-\frac{1}{2}} \end{aligned}$$

(f) Application of the sum and power rules: $f'(x) = 4 \cdot 5 \cdot x^{5-1} - 5 \cdot 4 \cdot x^{4-1} = 20 \cdot x^4 - 20 \cdot x^3$
(g) Application of the product, sum, and power rules:

$$\begin{aligned} f'(x) &= \left(2 \cdot x^{2-1}\right) \cdot \left(x^2 + 3 \cdot x + 2\right) + \left(x^2 + 1\right) \cdot \left(2 \cdot x^{2-1} + 3 \cdot x^{1-1}\right) \\ &= \left(2 \cdot x^3 + 6 \cdot x^2 + 4 \cdot x\right) + \left(2 \cdot x^3 + 3 \cdot x^2 + 2 \cdot x + 3\right) \\ &= 4 \cdot x^3 + 9 \cdot x^2 + 6 \cdot x + 3 \end{aligned}$$

Alternatively, you could have expanded the original function and then derived it.

(h) Application of the chain and power rules: The outer function is $u[v(x)] = [v(x)]^2$ and the inner function is $v(x) = a - x$. Thus, we get $f'(x) = 2 \cdot (a-x)^{2-1} \cdot \left(-x^{1-1}\right) = -2 \cdot a + 2 \cdot x$. Alternatively, you could have also expanded the original function and then derived it.

(i) Application of the chain and power rules: The outer function is $u[v(x)] = [v(x)]^5$ and the inner function is $v(x) = x^3 - \sqrt{x}$. Thus, we get $f'(x) = 5 \cdot \left(x^3 - \sqrt{x}\right)^{4-1} \cdot \left(3 \cdot x^{3-1} - 0.5 \cdot x^{0.5-1}\right) = 5 \cdot \left(x^3 - \sqrt{x}\right)^3 \cdot \left(3 \cdot x^2 - \frac{1}{2 \cdot \sqrt{x}}\right)$.

(j) Application of the quotient and power rules. We have $u(x) = x^2 + 10$ in the numerator and $v(x) = x^3$ in the denominator. Thus, we get

$$\begin{aligned} f'(x) &= \frac{\left(2 \cdot x^{2-1}\right) \cdot \left(x^3\right) - \left(x^2 + 10\right) \cdot \left(3 \cdot x^{3-1}\right)}{\left(x^3\right)^2} \\ &= \frac{2 \cdot x^4 - 3 \cdot x^4 - 30 \cdot x^2}{x^6} = -\frac{30 + x^2}{x^4}. \end{aligned}$$

12.5.2 Critical Points

(a) The first derivative is $f'(x) = 20 - 2 \cdot x$. If we set it equal to zero, $20 - 2 \cdot x \overset{!}{=} 0$, we find $x = 10$. The second derivative is $f''(x) = -2$ which is always negative, thus, there is maximum at $(x = 10, f(10) = 100)$.

(b) The first derivative is $f'(x) = 3 \cdot x^2 - 3$. If we set it equal to zero, $3 \cdot x^2 - 3 \overset{!}{=} 0$, we find two values, $x = 1$ and $x = -1$. The second derivative is $f''(x) = 6 \cdot x$. At $f''(1) = 6$ it is positive, thus, $(x = 1, f(1) = -2)$ is the minimum. At $f''(-1) = -6$ it is negative, thus, $(x = -1, f(1) = 2)$ is the maximum.

(c) The first derivative is $f'(x) = 1 - 100/x^2$. If we set it equal to zero, $1 - 100/x^2 \overset{!}{=} 0$, we find two values, $x = 10$ and $x = -10$. The second derivative is $f''(x) = 200/x^3$. At $f''(10) = 0.2$ it is positive, thus, $(x = 10, f(1) = 20)$ is the minimum. At $f''(-10) = -0.2$ it is negative, thus, $(x = -10, f(1) = -20)$ is the maximum.

12.5.3 Partial Derivatives

We find the following five partial derivatives:

$$\begin{aligned} \frac{\partial f(x_1, x_2, x_3, x_4, x_5)}{\partial x_1} &= 2 \cdot x_1 \cdot x_2 - \frac{x_3}{x_1^2} + \frac{x_5}{x_1^2 \cdot x_4} \\ \frac{\partial f(x_1, x_2, x_3, x_4, x_5)}{\partial x_2} &= x_1^2 - \frac{1}{2} \cdot \sqrt{\frac{x_3}{x_2}} \end{aligned}$$

$$\frac{\partial f(x_1, x_2, x_3, x_4, x_5)}{\partial x_3} = \frac{1}{x_1} - \frac{1}{2} \cdot \sqrt{\frac{x_2}{x_3}}$$

$$\frac{\partial f(x_1, x_2, x_3, x_4, x_5)}{\partial x_4} = -25 \cdot x_4^4 + \frac{x_5}{x_1 \cdot x_4^2}$$

$$\frac{\partial f(x_1, x_2, x_3, x_4, x_5)}{\partial x_5} = -\frac{1}{x_1 \cdot x_4}.$$

Glossary

Action — An action refers to the smallest unit of behavior in decision theory. In game theory, a ***strategy*** consists of a complete plan of all actions of a player. See Sect. 1.1.

Arrow-Pratt Measure — The Arrow-Pratt measure measures the ***risk attitude*** of a decision-maker. See Sect. 1.3.

Backward Induction — A ***Sequential Game*** is solved using the principle of backward induction. Here, one starts at "the end of the game" with the decision of the second (last) player and adjusts the decision of the first (previously moving) player accordingly. With the help of backward induction, the ***Subgame Perfect Nash Equilibrium*** is determined. See Sects. 4.1 and 4.2.

Bayesian Formula — Using the Bayesian formula (also "Bayes' theorem") for two events, A and B, the conditional (posterior) probability $\rho(A|B)$ can be calculated if the prior probabilities $\rho(A)$ and $\rho(B)$ as well as the conditional probability $\rho(B|A)$ are given. Thus, if $\rho(B|A)$ is known, the prior probability $\rho(A)$ can be updated to the posterior probability. See Sect. 6.1.

Bayes-Nash Equilibrium — A Bayes-Nash equilibrium represents a ***Nash equilibrium*** in which, for a given probability distribution over the player types, none of the players or player types has an incentive to deviate. See Sect. 6.1.

F. Bartholomae and M. Wiens, *Game Theory and Applications*, Classroom Companion: Economics, https://doi.org/10.1007/978-3-658-44608-6

Cardinal utility function In a cardinal utility function, quantitative statements about utility differences are economically interpretable. Thus, statements are possible (and economically significant) about whether a decision-maker in situation X has twice or three times as high utility as in a comparison situation Y. See Sect. 1.3.

Complete Information *See **Incomplete Information***

Conflict of Interest A conflict of interest exists, when the favorable ***Payoff*** of one player is unfavorable for the other player, so that they aim for a different game outcome (and vice versa). A very high degree of conflict of interest is present in a ***zero-sum game*** where the gain of one player exactly corresponds to the loss of the other. See Sect. 2.2.

Constant-sum game In a constant-sum game, the ***payoffs*** of the players for all possible strategy combinations always add up to the same value. A ***zero-sum game*** is a special case of the more general category of constant-sum games. See Sect. 2.2.

Cooperation problem A cooperation problem arises in games when a Pareto-efficient and acceptable game outcome for the players (efficient "compromise solution") is not a ***Nash equilibrium***, so the players have an incentive to deviate from this constellation and eventually end up in an inefficient Nash equilibrium. An example of a game with a cooperation problem is the prisoner's dilemma. See Sect. 3.4.

Cooperative game theory In cooperative game theory, all considerations are based on the assumption that the players can make binding agreements. *See **Non-cooperative game theory**.*

Coordination problem In games where multiple ***Nash equilibria*** exist, the players face the problem of equilibrium selection, i.e., they must find a way to explicitly or implicitly agree on an equilibrium. In a coordination problem, there is a risk that the coordination on an efficient ***Nash equilibrium*** fails, either because the possible equilibria are differently attractive for the players (***conflict of interest*** regarding the equilibria) or due to misunderstandings in communication. An example of a game with a coordination problem is the "Battle of the Sexes" game. See Sect. 3.4.

Decision under Risk In a decision under risk, it is not known which state prevails, but the probabilities of the states $\rho(w_j)$ are known. See Sect. 1.3.

Discount Factor A discount factor describes the value of the future for a player or the patience of a player to wait for future ***payoffs***. See Sect. 5.1.

Discount Rate The discount rate corresponds to the interest rate, i.e., the market price used for money and credit transactions. From an economic point of view, the discount rate represents the opportunity costs for a temporary resource waiver. Between the discount rate i and the ***discount factor*** δ there is the following theoretical, fundamental relationship (assuming perfect capital markets): $i = (1 - \delta)/\delta$. See Sect. 5.1.

Dominant Strategy By choosing a dominant ***strategy***, players can never worsen, and by choosing a strictly dominant ***strategy,*** they can always improve, regardless of the ***strategy*** choice of their opponents. A rational player will certainly play a strictly dominant ***strategy***. If all players have a strictly dominant ***strategy***, then the solution of the game is a unique ***Nash equilibrium***. See Sect. 3.1.

Extensive Form The extensive form is the representation of a game as a game tree. See Sects. 2.3 and Chap. 4.

External Effect An external effect refers to the impact of the action of decision-maker i on an uninvolved person (group) j, which decision-maker i does not consider in her decision. See Sect. 3.4.

First-Mover Advantage When the first mover has the opportunity to influence the game result of a ***Sequential Game*** in their favor by being the first to move, this is referred to as a first-mover advantage. See Sect. 4.2.

Focal Point A focal point provides a kind of anchor- or orientation point for the players in the equilibrium selection and can thus help to solve a ***Coordination Problem***. If it arises from the game, for example through particularly attractive ***Payoff*** combinations, or using ***Strategies*** as communication tools, then it is referred to as an *endogenous* focal point. If helpful reference points only exist outside the game (background information, norms, etc.), then it is referred to as an *exogenous* focal point. See Sect. 3.4.

Imperfect Information A player who has imperfect information cannot observe (at least) one ***Strategy*** choice of his opponent. In contrast, if the ***Strategy*** choice of the opponent is observable for the concerned player, then it is referred to as perfect information. In a ***Simultaneous Game***, by definition, all players have imperfect information. See Sect. 2.3 and Chap. 4.

Incomplete information A player with incomplete information cannot identify the type of their opponent. If, on the other hand, the player knows the type of the opponent, this is referred to as complete information. A game with incomplete information is also referred to as a Bayesian game. See Sect. 2.3 and Chap. 6.

Indifference Curve The indifference curve is the geometric location of all ***Payoff*** combinations that provide the individual with the same utility. See Sect. 1.3.

Information Asymmetry An information asymmetry exists when the information status of the players (about ***Actions*** or ***Payoffs***) is different. In contrast, if the information status of the players is identical, it is referred to as symmetric information. See Sect. 2.3; ***Imperfect Information; Incomplete Information***.

Lottery A lottery is determined by every possible ***state of the world***, their probabilities of occurrence, and the associated ***payoffs***. See Sect. 1.3.

Maximin strategy In a maximin strategy, the player assumes that the opponent wants to inflict the greatest possible damage on them, which is why they choose the strategy that results in the least bad outcome for them. With the exception of ***Constant-sum games,*** the application of this strategy is not plausible, as it assumes an extremely pessimistic attitude of the player. See Sect. 1.4.

Mixed Strategy A mixed strategy assigns a probability to each pure strategy. See Sect. 3.3.

Nash equilibrium A Nash equilibrium describes a ***strategy*** profile of mutually best responses. In a Nash equilibrium, no player has an incentive to deviate. A Nash equilibrium is efficient if no ***External effect*** is present. See Sects. 2.4, and 3.4.

Negotiation Game A negotiation game is typically a sequential ***Constant Sum Game***, in which it must be distinguished whether the players can make binding agreements or not. Negotiation games in which no binding agreements are possible or intended are referred to as non-cooperative negotiation games. An example of this would be the Rubinstein bargaining game. The subfield of game theory that deals with negotiations based on binding agreements is the ***Cooperative Game Theory.*** See Sect. 7.1.11.

Non-cooperative game theory In non-cooperative game theory, all considerations are based on the assumption that the players cannot make binding agreements. *See* ***Cooperative game theory***.

One-shot deviation principle A strategy profile s of an infinitely repeated game satisfies the one-shot deviation principle if a player has no incentive after any history h^t of the game to deviate from s—regardless of whether the players are on or off the equilibrium path in the considered period. See Sect. 5.3.

Ordinal utility function For an ordinal utility function, quantitative information about utility differences is not economically interpretable, i.e., the preference order only provides information about whether a bundle of goods is better or worse, but not "by how much" better or worse. In the case of an ordinal utility function, therefore, only the order or ranking of the options (such as bundles of goods or ***payoff***s) matters. See Sect. 1.3.

Pareto optimality A situation is Pareto-optimal when it is no longer possible to improve the situation of one player without simultaneously making another player worse off. See Sect. 3.4.

Payoff The payoffs describe the motivations of the players and thus determine the incentive structure of the game. Payoffs are usually expressed in monetary units, but in general, payoffs can also include non-monetary quantities (such as leisure time or the pleasant feeling of satisfaction). See Sect. 2.2.

Perfect information *See **Imperfect information**.*

Rationality The assumption of (strict) rationality dictates that each player makes a consistent, goal-oriented decision and does not make systematic errors. The assumption of rationality, along with the ***Self-interest*** assumption, is the central characteristic of Homo Oeconomicus. See Sect. 2.3.

Rationalizable strategies Rationalizable ***strategies*** survive the procedure of iterative elimination of strictly dominated ***strategies***. Only rationalizable strategies are considered by a rational player. See Sect. 3.1.

Risk Risk in the context of the expected utility theory (see Sect. 1.3) refers to the variability or volatility of parameters (stochastic uncertainty), which have an impact on the decision-maker's payoff (e.g., volatility of sales, prices). Risk according to this concept can be measured by dispersion metrics like the standard deviation, variance, or, e.g., a mean preserving spread of the respective parameter. A prominent application of this risk concept is portfolio theory.

Risk attitude Risk attitude (or also "risk preference") refers to the degree of an individual's risk propensity. Depending on the risk attitude, risk is evaluated differently: For a risk-averse individual, higher risk leads to a loss of utility, for a risk-neutral individual, it leads to no change in utility, and for a risk-seeking individual, it leads to an increase in utility. See Sect. 1.3.

Second-Mover Advantage If the second mover can influence the game result of a ***Sequential Game*** in their favor by being the second to move, this is referred to as a second-mover advantage. See Sect. 4.2.

Security Line The security line represents all ***Payoff*** combinations, that are safe, i.e., that in every ***State of the World*** deliver the same ***Payoff***. See Sect. 1.3.

Self-interest
The assumption of self-interest assumes that a decision-maker, when evaluating decision options, focuses exclusively on their own benefit, and thus ignores the benefit of other individuals. A strictly self-interested decision-maker therefore has no social motivation. The assumption of self-interest is, along with the ***rationality*** assumption, the central characteristic of Homo Oeconomicus. See Sect. 2.2.

Sequential Game
In a sequential (or multi-stage) game, the players make their ***Actions*** one after the other, making the move of the first mover observable for the second mover. In terms of the observability of the moves, sequential games are always games with ***Information Asymmetry***. See Chap. 4.

Signaling Game
A signaling game is a ***Sequential Game*** with ***Incomplete Information*** of the second mover, who (at least partially) can derive information about their opponent's type from the preceding *Action*—the "signal". See Sect. 6.3.

Simultaneous Game
A simultaneous game is a game with symmetrical and ***Imperfect Information***, in which no players can observe the moves of their opponents. The term "simultaneous" is somewhat inaccurate, as in a simultaneous game it is not the simultaneity, but the observability of the moves that matters. See Chap. 3.

Social Motivation
A player with social motivation not only evaluates their own game result, but also pays attention to the ***payoffs*** of the other players. See Sect. 2.2.

Stackelberg Equilibrium
A Stackelberg equilibrium (or Stackelberg solution) is the ***subgame perfect Nash equilibrium*** of a ***sequential game*** between two companies, where the market leader—the so-called Stackelberg leader—has a ***first-mover advantage***. See Sect. 4.3.

State of the World
The state of the world (or also "contingency") refers to the realizations of individual outcomes. See Sect. 1.1.

Strategic form
The strategic form (also normal form) is the representation of a game as a matrix. See Sect. 2.1 and Chap. 3.

MIX
Papier aus verantwortungsvollen Quellen
Paper from responsible sources
FSC® C105338

If you have any concerns about our products,
you can contact us on
ProductSafety@springernature.com

In case Publisher is established outside the EU,
the EU authorized representative is:
Springer Nature Customer Service Center GmbH
Europaplatz 3, 69115 Heidelberg, Germany

Printed by Libri Plureos GmbH
in Hamburg, Germany